D0850818

Advances in

VIRUS RESEARCH

VOLUME 21

CONTRIBUTORS TO THIS VOLUME

Louis Andral

J. R. Christensen

Jyotirmoy Das

C. G. Hayes

Thomas J. Kelly, Jr.

Osvaldo Lovisolo

Jack Maniloff

Giovanni P. Martelli

Robert G. Milne

Daniel Nathans

Marcello Russo

Bernard Toma

R. C. Wallis

Advances in
VIRUS RESEARCH

Edited by

MAX A. LAUFFER

Andrew Mellon Professor of
Biophysics
University of Pittsburgh
Pittsburgh, Pennsylvania

FREDERIK B. BANG

Department of Pathobiology
The Johns Hopkins University
Baltimore, Maryland

KARL MARAMOROSCH

Waksman Institute
of Microbiology
Rutgers University
New Brunswick, New Jersey

KENNETH M. SMITH

Cambridge, England

VOLUME 21

1977

ACADEMIC PRESS New York San Francisco London
A Subsidiary of Harcourt Brace Jovanovich, Publishers

ACADEMIC PRESS, INC.
111 Fifth Avenue, New York, New York 10003

United Kingdom Edition published by
ACADEMIC PRESS, INC. (LONDON) LTD.
24/28 Oval Road, London NW1

LIBRARY OF CONGRESS CATALOG CARD NUMBER: 53–11559

ISBN 0–12–039821–4

PRINTED IN THE UNITED STATES OF AMERICA

CONTENTS

Epidemiology of Fox Rabies

BERNARD TOMA AND LOUIS ANDRAL

Ecology of Western Equine Encephalomyelitis in the Eastern United States

C. G. HAYES AND R. C. WALLIS

The Genome of Simian Virus 40

THOMAS J. KELLY, JR., AND DANIEL NATHANS

Plant Virus Inclusion Bodies

GIOVANNI P. MARTELLI AND MARCELLO RUSSO

Maize Rough Dwarf and Related Viruses

ROBERT G. MILNE AND OSVALDO LOVISOLO

Viruses of Mycoplasmas and Spiroplasmas

JACK MANILOFF, JYOTIRMOY DAS, AND J. R. CHRISTENSEN

CONTRIBUTORS TO VOLUME 21

Numbers in parentheses indicate the pages on which the authors' contributions begin.

LOUIS ANDRAL, *Centre d'Etudes sur la Rage, Malzéville, France (1)*

J. R. CHRISTENSEN, *Department of Microbiology, University of Rochester School of Medicine and Dentistry, Rochester, New York (343)*

JYOTIRMOY DAS, *Department of Microbiology, University of Rochester School of Medicine and Dentistry, Rochester, New York (343)*

C. G. HAYES,* *Department of Epidemiology and Public Health, Yale University, New Haven, Connecticut (37)*

THOMAS J. KELLY, JR., *Department of Microbiology, The Johns Hopkins University School of Medicine, Baltimore, Maryland (85)*

OSVALDO LOVISOLO, *Plant Virus Laboratory, National Research Council, Turin, Italy (267)*

JACK MANILOFF, *Department of Microbiology and Department of Radiation Biology and Biophysics, University of Rochester School of Medicine and Dentistry, Rochester, New York (343)*

GIOVANNI P. MARTELLI, *Istituto di Patologia Vegetale, University of Bari, Bari, Italy (175)*

ROBERT G. MILNE, *Plant Virus Laboratory, National Research Council, Turin, Italy (267)*

DANIEL NATHANS, *Department of Microbiology, The Johns Hopkins University School of Medicine, Baltimore, Maryland (85)*

MARCELLO RUSSO, *Istituto di Patologia Vegetale, University of Bari, Bari, Italy (175)*

BERNARD TOMA, *Ecole Nationale Vétérinaire d'Alfort, Alfort, France (1)*

R. C. WALLIS, *Department of Epidemiology and Public Health, Yale University, New Haven, Connecticut (37)*

* Present address: Pakistan Medical Research Center, 6 Birdwood Road, Lahore, Pakistan.

Advances in

VIRUS RESEARCH

VOLUME 21

EPIDEMIOLOGY OF FOX RABIES

Bernard Toma

Ecole Nationale Vétérinaire d'Alfort, Alfort, France

and

Louis Andral

Centre d'Etudes sur la Rage, Malzéville, France

I. Introduction

Despite the availability of more effective and less harmful vaccines for protection of humans and animals, rabies remains a very widespread disease in the world, transmitted by both domestic and wild animal vectors. Dogs, especially stray dogs, are responsible for maintaining the enzootic in most of the countries of Africa and Asia where veterinary and sanitary structures are inadequate. Certain wild animals, particularly vampire bats, are the reservoirs of perennial rabies in vast zones of Central America and South America.

In other areas, notably of Europe, where animal rabies control can be effectively applied and where there is practically no rabies infection of bats, another species of wild animal has taken the place of the dog as preferential vector of rabies during the last decades: the fox. In most European countries, there has been a progressive inversion in the frequencies of canine rabies and vulpine rabies, to the advantage of the latter. This change seems to be taking place within the framework of a general increase in the density of the fox population, which, if one accepts the hunting statistics of certain countries, has at present reached a level two to four times higher than at the beginning of the century. This has been due to diverse factors such as the elimination of natural enemies of the fox, changes in hunting practices or of agricultural techniques, and the like.

The methods used to fight vulpine rabies are much more difficult to apply than are available methods for eradicating canine rabies, and results are far more hazardous. Therefore, countries in which standards of livestock maintenance are excellent and which have successfully eliminated all of the other great contagious animal diseases are still powerless to limit expansion of rabies. This problem is directly related to the density of the fox population. There has been a great deal of research during the past 15 years on the epidemiology of fox rabies and possible measures of prophylaxis against it. The results have begun to accumulate, and one can now attempt to present a synthesis of the essential ideas.

II. Descriptive Epidemiology

Different species of fox are involved in fox rabies, depending on the region: in Greenland, northern Canada, and Russia the vector responsible is *Alopex lagopus,* the white or arctic fox; in central and western Europe and in southern parts of Canada it is *Vulpes vulpes,* the red fox; in the United States transmission is assured by *V. vulpes* and also by *Urocyon cinereoargenteus,* the gray fox. Figure 1 represents the general distribution of vulpine rabies in the world (Kaplan, 1969; Chalmers and Scott, 1969; Irvin, 1970).

This study will be concerned most particularly with the European vulpine rabies enzootic which has developed in the course of the last three decades and for which the present geographic distribution is shown in Fig. 2.

The origin of the current enzootic of European vulpine rabies was a focus of rabies in foxes and badgers found in Poland in 1935. During World War II, the enzootic was displaced from the east toward the west, so that most of the countries of the northern portion of Europe are now affected. After East Germany and West Germany, various other countries were successively infected: Denmark in 1964, Belgium, Luxembourg, and Austria in 1966, Switzerland in 1967, France in 1968, and Holland in 1974.

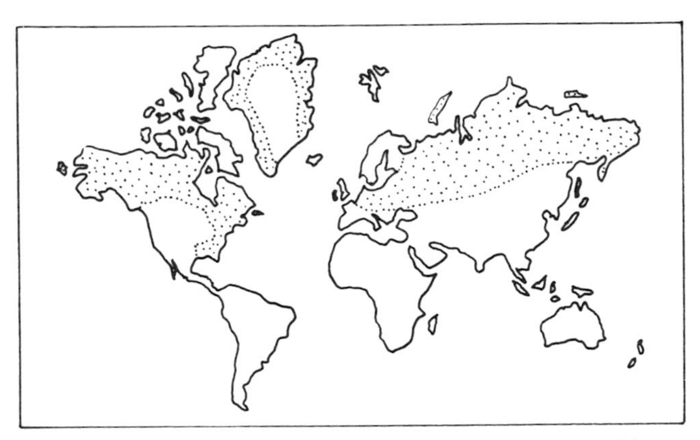

FIG. 1. Map of the world distribution of vulpine rabies.

Toward the east, vulpine rabies spread and intensified in Poland, Hungary, Czechoslovakia, and the Soviet Union (Sansyzbaev, 1975; Shigailo *et al.*, 1975). The statistics from these countries reflect progressive replacement of urban rabies, with a canine vector, by vulpine rabies. For example, the percentage of rabies infection in foxes in relation to rabies infection in other animal species in Poland rose successively: 29.3% in 1966; 52.3% in 1968; 70.6% in 1971 (Mol, 1971).

A. *Distribution in Space*

We will use France as a study model of the spatial progression of enzootic vulpine rabies, since most of the data compiled in this country

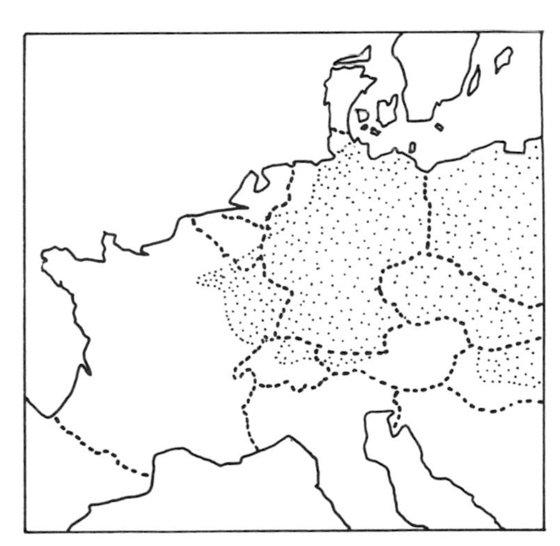

FIG. 2. Map of the areas of western Europe affected by the rabies enzootic in 1975.

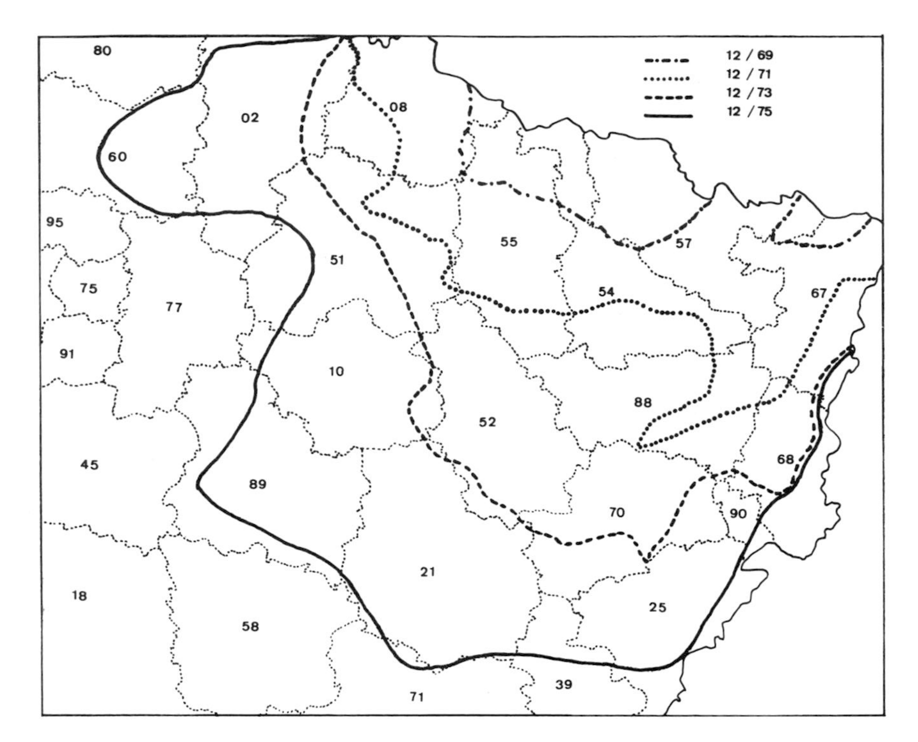

FIG. 3. Progression of the front of the vulpine rabies enzootic by 2-year periods from 1968 to 1975 in France. The numbers refer to provinces.

are obtained equally in the other infected areas. Figure 3 indicates the steps of progression of the rabies front for 2-year periods from 1968 to 1975. There are differences in the shape of the front in the invaded areas during identical periods, though it proceeds in the same direction. These variations are tied to many factors to which we will return after analysis of the details of transmission of the virus.

In general, the major *directions* of the progression are toward the west and south, as confirmed by the course of its inroads into different countries of western Europe (Steck, 1968; Kohn and Maar, 1968; Kauker and Zettl, 1969; Grégoire, 1969; Toma and Andral, 1970; Wandeler *et al.*, 1974a). The *shape* of the front is quite regular, the new cases appearing regularly at short distances from previously recorded cases. Yet, at times, cases appeared many dozens of kilometers from all other foci of rabies.

Finally, the mean rate of *spread* of rabies is on the order of 30–60 km per annum. Thus, the extreme limits of the rabies front in France at the end of 1975 covered a circle of about 300 km, having as a focal point the area where the virus had penetrated into France 7 years earlier. However, there were important variations, since the front progressed very slowly, even retreated at times in certain areas (i.e., the German–Dutch frontier), whereas in others it moved very rapidly, up to 100 km in a given direction in a single year (Andral and Toma, 1973).

B. *Distribution in Time*

The annual incidence and the monthly incidence will be considered successively in the following discussion.

1. *Annual Incidence*

The course of the annual incidence of vulpine rabies varies according to the surface area taken into account in recording rabies cases. Consequently, one should distinguish between its progress in a region or a country of small dimensions and in a region having a greater land area.

a. Areas of Limited Size. Such relatively small areas are exemplified by Belgium and Luxembourg. Figure 4 shows the change in the annual incidence of rabies in Belgium. There was an increase in incidence during the first years, followed by a regression and even disappearance of the disease. However, in 1974 it reappeared at a relatively high level (Costy-Berger and Marchal, 1975). Figure 5 shows a similar sequence for Luxembourg.

b. Larger Areas. The increase in the annual incidence of rabies in France from 1968 to 1975 is shown in Fig. 6. Taking into account the large surface area which is subject to the spread of the enzootic, the overall annual incidence has steadily increased. Yet, the figures of the annual incidence correspond to the sum of the rising and falling incidences of the

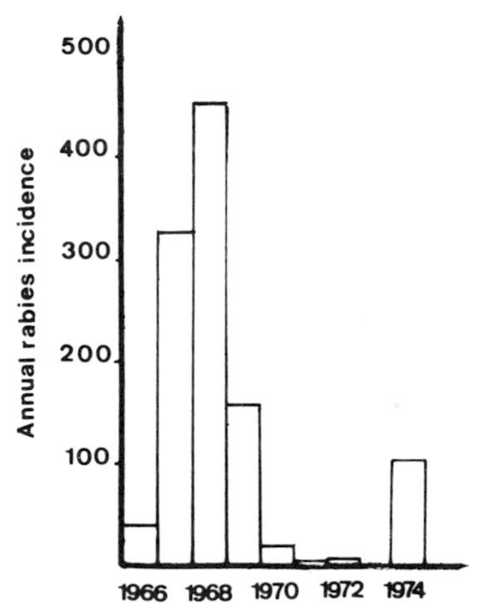

FIG. 4. The annual incidence of rabies in Belgium from 1966 to 1974.

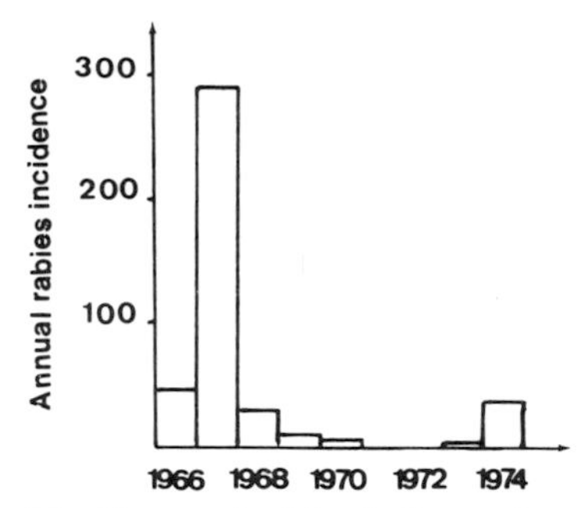

FIG. 5. The annual incidence of rabies in Luxembourg from 1966 to 1974.

different provinces within the infected region. For each zone of smaller surface area, the increase is comparable to those recorded in Belgium and Luxembourg. Thus, for Moselle, the first French province infected, Fig. 7 indicates that fluctuations were as evident in wild animals as in domestic animals. The numbers on which Fig. 7 is based must, however, be considered with a certain degree of reserve for two possible reasons: To begin with, the degree of vigilance for the registration of cases of rabies in wild animals varies according to time, tending to slack off following the first months after the appearance of the disease. Moreover, in the case of cattle, the protection associated with vaccinations has also varied with time because antirabies virus vaccine for cattle has been increasingly used since 1971.

Study of the progression of the annual incidence of vulpine rabies over a limited surface, then, shows a pattern of fluctuations having a varied

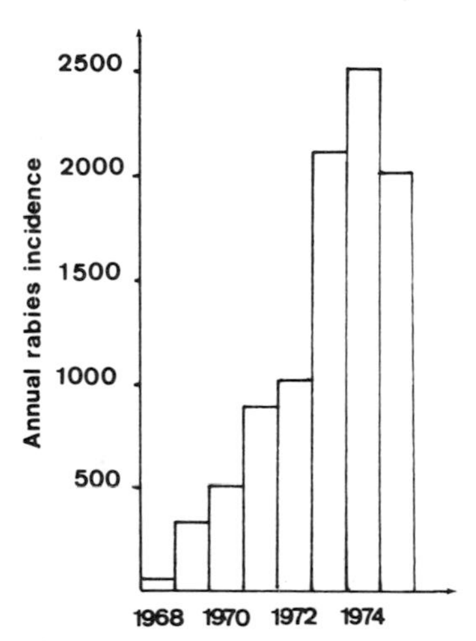

FIG. 6. The annual incidence of rabies in France from 1968 to 1975.

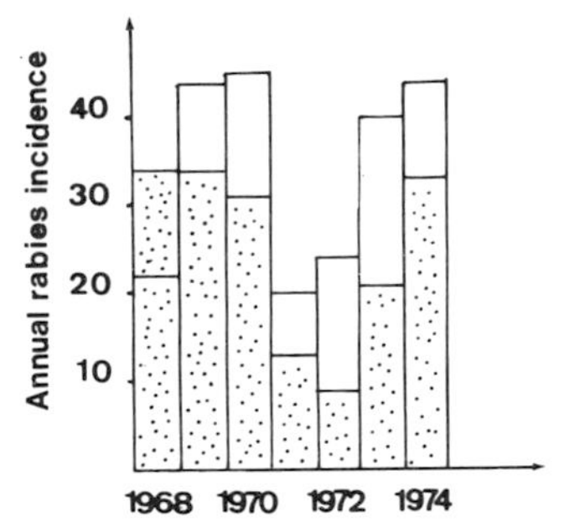

FIG. 7. The annual incidence of rabies in the French province of Moselle from 1968 to 1975. The stippled bars correspond to cases recorded in domestic animals; the rest, to wild animals.

rhythm (periods of 3 to 7 years), corresponding to many successive "waves" of the enzootic; these patterns have been observed in Europe (Kauker and Zettl, 1969) as well as in America (Johnston and Beauregard, 1969).

c. Annual Incidence According to Species. The annual incidence of rabies just alluded to corresponds to the totality of animal species affected. If the increase in the incidence of rabies is studied by individual species, one finds a generally parallel increase in the different animal species. However, the numbers can diverge; for example, in France, the annual incidence of fox, dog, and cat rabies increased in a parallel fashion from 1968 to 1975, whereas that of cattle rabies decreased steadily after having reached a maximum in 1971 (Fig. 8).

2. *Monthly Incidence*

Monthly registration of cases of rabies enables us to monitor the seasonal fluctuations in incidence of the disease in the course of 1 year.

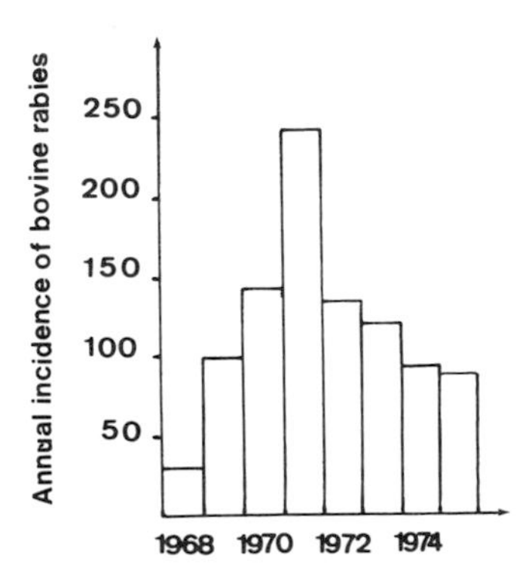

FIG. 8. The annual incidence of bovine rabies in France from 1968 to 1975.

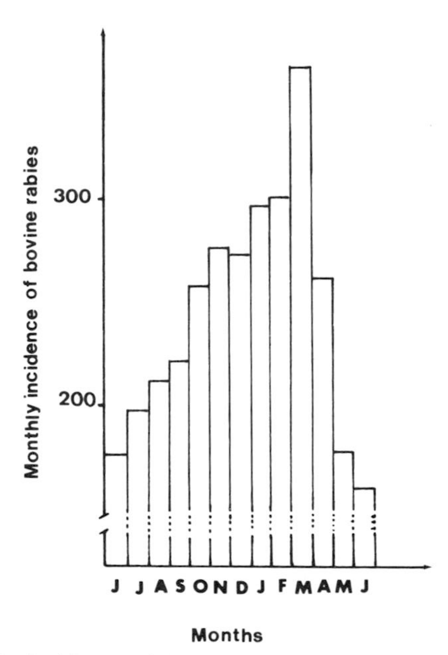

FIG. 9. The monthly incidence of bovine rabies in West Germany from June 1968 to June 1969.

a. In Foxes. Among the countries of Europe there are consistent slight differences in fluctuations in the monthly incidence of fox rabies (as shown in Fig. 9). Generally, the minimum is reached in May or June; then the incidence rises slowly during the second half of the year, reaches a peak in February and March of the following year, and diminishes thereafter. The same progression has been noted in North America (Johnston and Beauregard, 1969).

b. In Cattle. The incidence of rabies rises consistently during the second half of the year, then diminishes after the month of December and during the first third of the year (Fig. 10).

C. Animals Affected

Vulpine rabies is characterized by the great diversity of the infected animal species, both wild (fox, roe deer, badger, stone marten, marten, polecat, wildcat, etc.) and domestic (cow, dog, cat, horse, donkey, sheep, pig, etc.)

1. The Preeminence of Wild Animals

The preeminence of wild animals, especially the fox, in rabies is clearly seen in Table I, which compares the percentage of cases in foxes, other wild animals, and domestic animals in different European countries. The

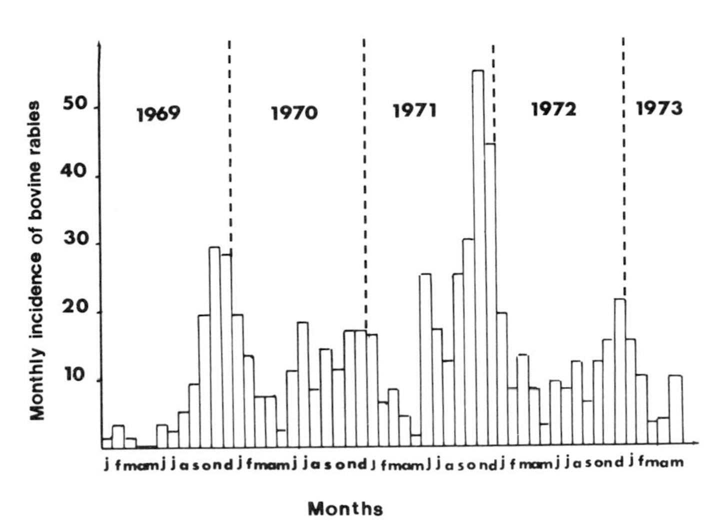

FIG. 10. The monthly incidence of bovine rabies in France from 1969 to 1973.

table shows that 68–85% of the rabid animals in 11 European countries are foxes. In fact, their actual role is certainly much greater. Indeed, while nearly all cases of rabies in domestic animals are registered, the situation regarding wild animals is different. On the one hand, many cases of wild rabies go unnoticed because of the nature of the habitat of the

TABLE I

PERCENTAGE OF RABIES CASES IDENTIFIED IN VARIOUS GROUPS OF ANIMAL SPECIES IN EUROPEAN COUNTRIES[a]

Country	Foxes (%)	Other wild animals (%)	Domestic animals (%)
West Germany	69.0	11.5	18.5
East Germany	75.3	8.5	16.2
Austria	79.0	16.0	5.0
Belgium	68.0	4.2	27.8
Luxembourg	69.4	9.0	21.6
Switzerland	80.6	12.9	6.5
France	75.4	3.8	20.8
Denmark	79.3	5.2	15.5
Poland	70.6	4.4	25.0
Hungary	85.0	2.0	13.0
Czechoslovakia	77.0	5.2	17.8

[a] From Kauker (1975).

affected species. On the other hand, wild animals that are shot or found dead are rarely examined diagnostically except in certain circumstances: to confirm the presence of rabies in an area which is considered free of the disease but which is situated just ahead of the front of the enzootic; or when a wild animal has caused infection in a human or a domestic animal.

For this reason the percentages of fox cases shown in Table I should be considered as minimal figures, considerably below the actual level. In France, for example, the actual number of rabies cases must be two to five times greater than those established by laboratory diagnosis.

In Canada, the wild animal most often infected, next to the fox, is the skunk *(Mephitis mephitis)*, but rabies seems to develop in an almost independent manner in these two species (Tabel *et al.*, 1974). Susceptibility to rabies virus and the level of salivary excretion of rabies virus are different in the two hosts (Sikes, 1962). In the United States, the total number of cases of skunk rabies in 1971 (46%) exceeded that of fox rabies cases (15%).

In Europe, the wild animal most often affected after the fox is the roe deer. This is especially serious in Germany, where it accounted for 14.9% of the rabies cases in Hesse between 1953 and 1971 (Wandeler *et al.*, 1974a). Next come other wild carnivores (badger, stone marten, marten, polecat, etc.), which as a group are apparently of minor importance—except in certain countries [e.g., 6.5% of the rabid animals in Switzerland from 1967 to 1973 were in this group (Wandeler *et al.*, 1974a)].

The infection of wild rodents is rare: an average of one rodent per 1000 cases of rabies reported (Scholz and Weinhold, 1969; Winkler, 1972). Nonetheless, rabies infection in these animals has attracted attention in the past few years because of the special characteristics of the strains isolated (Sodja *et al.*, 1971; Schneider and Schoop, 1972). Forty strains have been isolated in Czechoslovakia, Germany, and Switzerland from Microtidae and from Muridae with about a 2% success rate. The isolation of such strains is difficult: a series of blind intracerebral passages is necessary before the presence of the virus is established. These viruses seem to infect rodents in a latent fashion, since they have often been obtained from animals which were apparently healthy and which had been in captivity for 3 months. The viruses have also been isolated at times from the salivary glands or brown fat when it was not possible to find them in the central nervous system. Inoculated into foxes, these viruses either produce rabies or, if the inoculum is low, produce an inapparent infection. Intrauterine transmission has been reported in experimental conditions, and one can suppose that it exists in natural conditions. We must note the disquieting fact that such strains have been isolated from areas in which no cases of rabies had been recorded in domestic or wild animals

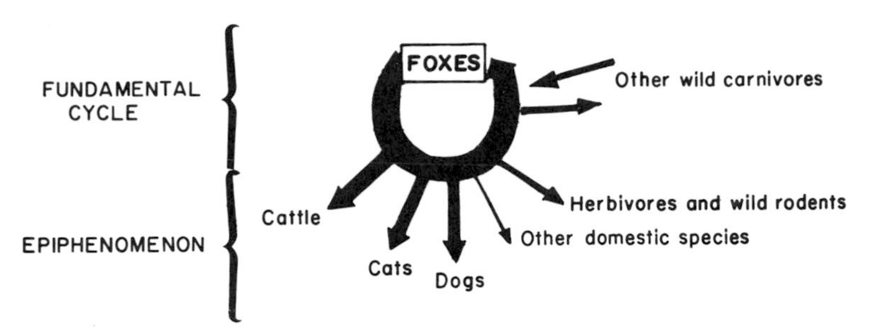

Fig. 11. The epidemiologic cycle of vulpine rabies.

for 7 years. In view of the aforementioned unusual characteristics, it seems that rabies can develop in rodents according to a cycle quite independent of vulpine rabies. More detailed information will doubtless be obtained in this field in the coming years.

2. Percentages of Infection in Various Domestic Animal Species

Percentages of infection in domestic animals vary according to the measures of control used in different countries. In Belgium, the rate of canine rabies infection is minimal thanks to compulsory antirabies vaccination of dogs. In France, the infection in cattle has diminished concomitantly with the development of antirabies vaccination for this species. Cats are still often infected by rabies virus, because it is difficult to restrict their movements in comparison, say, with dogs. In other domestic animals (horse, sheep, pig) the incidence of rabies is low.

From the statistics of animal rabies cases, and the increasing intermediary role of the fox species mentioned earlier, we can depict the epidemiologic cycle of European vulpine rabies as schematized in Fig. 11. Shown here is a basic orbit with a central focus of fox species from which rabies "escapes" into other wild or domestic species. These species are an epidemiological epiphenomenon because their suppression would not check the spread of the rabies enzootic; however, they play the main role in the chain of human infection.

III. Analytical Epidemiology

The basic role of vector and victim played by the fox, which was apparent in the foregoing review of the descriptive epidemiology, leads us to consider the ecology and ethology of this species before analyzing the progress of the virus within the body of the fox and then in the vulpine population.

A. Ecology and Behavior of the Fox

1. Habitat and Food

The ecosystem within which the fox develops is conditioned by various ecological factors that correspond to the physiological requirements of this animal.

The red fox is found most often in the temperate zones in regions with a rather limited range of temperature, but it can also be found in the North, either at altitudes of up to 3,000, depending on the area, or occasionally in desert regions. The most common habitat in the temperate zones is semiopen ground or woodland, with copses, thickets, and hedges furnishing shelter near cultivated areas. The fox prefers to establish its den in loose earth in sloping ground, with a principal entrance facing south and offering an unobstructed view. But some foxes are apparently satisfied with a rather shallow scrape, with a hollow tree trunk, or even with an abandoned building. In a given region, the percentage of foxes living in dens varies according to the nature of the terrain, and this has a bearing on the results of campaigns of gasification of the dens (see Section VI, B).

The study of stomach contents and the examination of feces show that the fox is often polyphagous. The basic food is voles, but the diet can vary a great deal, especially seasonally, and comprises other mammals such as rabbits, rats, squirrels, hares, small domestic animals (lambs), poultry, and birds, as well as frogs, toads, snails, slugs, earthworms, lizards, and (in summer) fruits.

2. Dynamics of the Vulpine Population

Reproduction is annual. The rut takes place in winter with a peak in January–February in the western European countries. Male foxes are not fertile except during the courtship season. Females have an annual period of 3 weeks in heat and can be impregnated during a 3- or 4-day period. Gestation averages 51 to 52 days, with extreme limits of 49 to 56 days, and the dropping of litters usually takes place in March–April. An average of 90% of the females are pregnant in March–April (Lessman, 1971). The average number of cubs per litter is four to five [4.2 according to Pearson and Bassett (1946) and Storm (1972); 4.67 according to Wandeler *et al.* (1974c); 5.4 according to Switzenberg (1950)]; there is an average of 44% females among the cubs. Theoretically, then, if there were no deaths, the fox population could triple each year. The mother nurses the young for 6 to 7 weeks. The young cubs become independent toward their fourth month (August-September) and reach puberty at about the tenth month, or during the winter following their birth.

The lifespan of a captive fox can reach 15 years, with an average of 10 years, but in nature it is shorter because of various factors, especially campaigns against the fox. The age pyramid, therefore, seems always wide at the base, then very steeply pointed. For example, of 522 foxes killed, Jensen and Nielsen (1968) counted 389 individuals (75%) aged between 6 months and 1 year. Of 671 foxes killed in the Arnhem region of Holland, van Haaften (1970) reported that 70% were less than 1 year old.

In the adult fox population, there is usually a surplus of males [1.2 males for every female (Wandeler *et al.*, 1974c)], but reports vary a great deal with authors, regions, and seasons.

Estimation of the fox population in a given area is difficult. Many methods are available: one consists of a census of the inhabited dens in a defined perimeter; another makes use of the capture and marking of foxes. An estimate of the population density can also be based on the average number of foxes killed per year and per square kilometer, provided that hunting activities remain essentially identical. Both of these methods yield more or less unsatisfactory results. Estimates of vulpine population density vary a great deal according to the region being studied.

Bögel and his collaborators (1974) studied seasonal variations in fox populations and propose the values shown in Table II for the numerous zones of western Europe unaffected by rabies. The basic number is one adult fox per square kilometer in the spring; after the birth of cubs, the number reaches three foxes per square kilometer at the beginning of the summer. Hunting, accidents, and sickness reduce this level and bring it back to one fox per square kilometer the following spring.

All accidental reduction of a fox population in an area tends to be annulled within a few years (see Section V, B). In Europe, under present

TABLE II

Annual Increase in the Fox Population per Square Kilometer in an Area Unaffected by Rabies and Having No Control Measures against the Fox[a]

Density before birth of new generation	Density after birth of new generation	Reduction due to hunting	Reduction due to disease and accidents	Total reduction	Density before birth of new generation
1	3	1.2	0.8	2	1

[a] From Bögel *et al.* (1974).

ecological conditions, the fox population seems to be able to conserve a significant potential for recuperation after temporary or more prolonged operation of any factor which reduces the population.

3. Ethology of the Fox

Study of the social life and behavior of the fox is difficult; it can be done by direct observation of the animals, by marking, by the use of radio transmitters, or by examining traces left on the ground. During most of the year, the red fox is a solitary animal that is active during periods of semidarkness. This activity is essentially oriented toward the search for food.

Otherwise, the fox is a sedentary animal: the extent of its home range fluctuates, in general, between 2 and 8 km^2 [2.5–7.8 km^2 (Sargeant, 1972)], depending on the amount of available food. At the time of the rut, the males delimit a territory within the vital domain where they exercise dominance over the total space and the females therein. These territories are marked by vocalization and by the odor of certain of male secretions or excretions (urine, feces, supracaudal glands). The male territory is thus a transitory social phenomenon tied to reproduction and is terminated at the same time as the rut (Tissot Boireau, 1975).

During this period, there is often increased ranging, sometimes over great distances. At other times, the movements of foxes are usually more limited. However, the short distances covered by adult foxes should be distinguished from those of cubs marked when they are still in the den and which may ultimately be recovered many dozens of kilometers from their place of birth. Jensen (1966) indicates that the majority of foxes do not move more than 10 km. However, perhaps one in ten may wander more than 25 km from its place of birth. A later study by this same author (Jensen, 1970), using all of the data available for western Europe, indicates that two-thirds of the animals marked when they were young were found less than 10 km from the spot where they were marked. Displacements up to 50 km were common among the remaining third, and a small percentage reached 100–150 km. According to Behrendt (1955), 64% of 50 marked foxes moved no farther than 8 km. Kauker and Zettl (1963) found that 68% of 25 marked foxes had traveled more than 10 km.

The females seem to move about less than the males, the young as well as the adults: of 65 young males, Phillips and Storm (1968) recorded an average dispersion distance of 42 km (extremes: 0.8 and 336 km) and of 47 young females an average of only 11.4 km (extremes: 0.8 and 77 km). Phillips and his colleagues (1972) have reported that 78% of young marked males left their place of birth during their first year and 95%

during their second year, against, respectively, 33% and 50% of the females. The maximal displacement has been recorded by Ables (1965) as 392 km.

In summary, it is appropriate to point out that the "normal" movements of foxes are limited to a few kilometers, but that a small number of them, notably the young males in search of vital domain, can engage in veritable "migrations," going as far as many dozens of kilometers. These facts are directly relevant to the epidemiology of rabies. Furthermore, the fox is an animal which has a relatively unimportant social life, and the determining factors tend rather toward the dispersion of individuals and the augmentation of social distance than to contacts and to positive social tendencies.

B. Infection of the Fox by Rabies Virus

The fox is very susceptible to the virus of rabies, more susceptible than the skunk (Sikes, 1962). The habitual mode of transmission of the virus from fox to fox is biting. The infection of a fox by rabies usually leads to death, after a progression which includes the stages indicated in Fig. 12.

The mean incubation oscillates around 25 days [an average of 26.5 days according to Tierkel (1966); and 1 month according to Parker and Wilsnack (1966)].

The duration of clinical disease is about 3–4 days [1–3 days (Sikes, 1962); 4–6 days, average (Tierkel, 1966); 7 days (Parker and Wilsnack, 1966)]. Nearly all of the rabid foxes shed rabies virus in their saliva during the disease [93% of 1451 rabid foxes, as reported by Wandeler *et al.* (1974b)]. These last authors found the mean titer of the virus in the supernatant of ground salivary gland tissues to be $10^{3.4}$ LD_{50} per 0.03 ml of supernatant. As in domestic carnivores, the saliva can be virulent before the appearance of the first clinical signs, and virulence increases over time.

It is difficult to specify the percentage of cases of furious rabies in foxes, i.e., the form favorable for the transmission of salivary virus; one could guess that this percentage is around 50. Of a group of 231 rabid foxes, Johnston and Beauregard (1969) have reported wounds on the

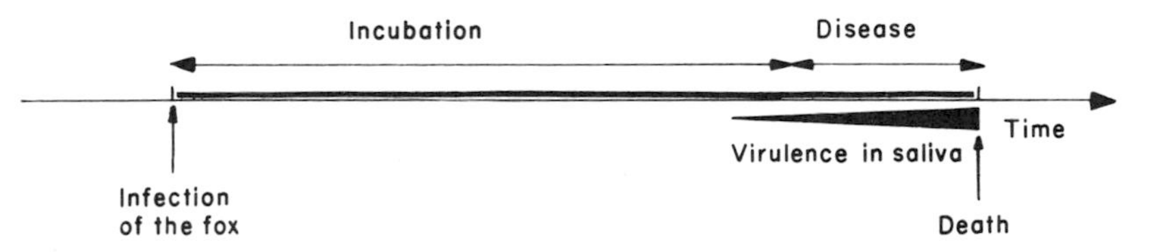

FIG. 12. Progress of rabies virus infection in one fox.

head in 40% of them, lesions which indicate extremely violent behavior.

Another major unknown is the behavior of the rabid fox. Certainly, we know that during the clinical period the animal loses its instinct of self-preservation, enters an open terrain in full daylight, attacks humans, and attacks other domestic animals (e.g., dogs or cattle). But little is known, for example, about the average distance that a fox can cover during the period of virulent salivary excretion, or whether it is in a preferred direction or not. It is difficult to gather this information.

The distance from the point where a fox has been infected to the point where it can infect another fox can vary greatly. Its variation depends, among other things, on two factors: (1) for the fox which has remained on its home range since inoculation, we must determine the mean distance traveled during the period of clinical expression; (2) for wandering foxes, we are interested in the distance they have been able to travel during the incubation period.

In summary, taking into account the period of presymptomatic virulence of the saliva, one may consider that, in the majority of cases, a fox will excrete the virus for about 5–6 days, after having remained noninfectious for 3–4 weeks. Each rabid fox then plays the role of a relay in the transmission of the virus, with a relatively long latency period and an "effective" period of short duration. These elements are essential to establish an analytic scheme of the chain of transmission of rabies virus within a vulpine population.

C. Analysis of Transmission

In a schematic way, the chain of transmission of rabies virus can be represented in a linear fashion:

$$F \rightarrow F \rightarrow F \rightarrow F$$

During incubation each fox serves as a "braking" element in the diffusion, an element which regulates and finally slows down the progression for a few weeks; then, by chance encounters during the period of virulent excretion, usually at distances from its home range of some hundreds of meters to several kilometers, it transmits the virus to another fox. This mechanism is responsible for the apparently regular nature of the progression of the front of a rabies enzootic that lends itself to a mathematical model. The actual chain of transmission in nature is represented schematically in Fig. 13.

In a zone of low vulpine population density (Fig. 13a), the territory of each fox is theoretically larger than in a zone of high population density (Fig. 13b). Each fox then finds itself at a greater distance from a neighbor in a zone of low density than in a zone of high density. The corollary

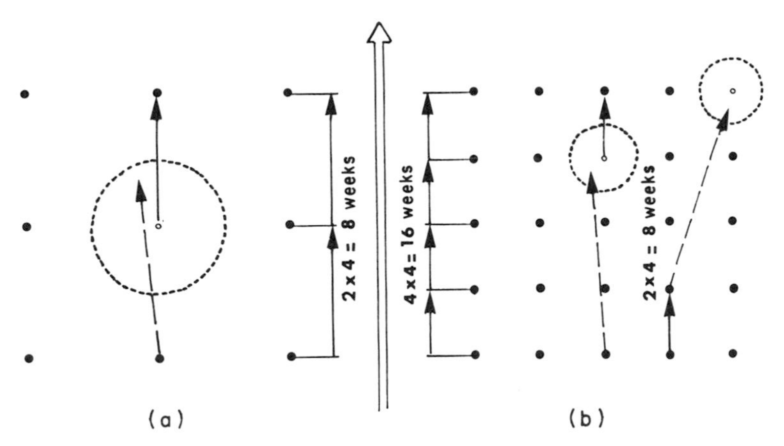

FIG. 13. Comparative transmission of rabies virus within two fox populations of different densities. (a) Low density. (b) High density. KEY: •, center of each fox's territory; *open arrow,* main direction of progression of the rabies enzootic; *solid arrow,* viral transmission between neighbors over a 4-week period; *dashed arrow,* viral transmission from a "migrant" over a 4-week period.

is that a rabid fox transmits the virus, during the period of salivary excretion, to a fox farther away in a low-density zone than to one in a high-density zone. Moreover, as the incubating fox plays an interrupting role in the progression for about 4 weeks, the greater the number of foxes in a given area, the more time the virus will probably take to traverse this area because its progress is impeded by an increased number of relays. In other words, in a high-density zone, the cases of rabies will be numerous but close to one another, whereas in a low-density zone they will be less numerous but farther away from one another.

This chain of reasoning is applicable to perfectly sedentary foxes. The "migrant" foxes come and upset this uniformity, but this phenomenon, doubtless a little more frequent in a zone of high population density, does not permanently disrupt the pattern; however, it can explain some "surprising" progressions during the favorable period, which is during winter.

Without pushing the argument too far, which would require that the progression of the rabies enzootic be slower in a zone of high vulpine population density than in a zone of low density, one may nonetheless understand that there is no direct relation between rapidity of progression and density of vulpine population, i.e., rabies does not progress more rapidly in a zone of high fox population density. This concept carries with it a corollary extremely important to grasp: reduction in density of a fox population in a region, owing to control measures, does not bring about a reduction in the rate of progression. This rate can remain unchanged; indeed, in extreme instances, it can even be augmented. It is then necessary to recognize clearly that the reduction in density of a fox population does not, in the majority of cases, bring about a reduced rate of progression of the front, nor, *a fortiori*, a halt in the progression.

It should be evident that this argument is applicable only when the density of a fox population is above a certain *threshold* corresponding to a good probability that a fox will encounter another fox during the contagious period of virulent excretion. Some statistics have been advanced for such a critical threshold: one fox for 250 hectares (Dean, 1955) ; at least one for 500 hectares (Müller, 1966). In fact, it is difficult to propose a definite figure for any area, and it seems wise to share the opinion offered by the World Health Organization in 1970: "One does not know, at the present time, the threshold at which transmission ceases, even though the figure of one fox or less per 500 hectares has been advanced."

Consequently, and despite the ignorance of the exact value of the density threshold critical for a given region, the essential issue in a region during active invasion by the rabies enzootic is whether the initial level of fox population density is such that the application of available control measures against the fox will or will not allow attainment of the critical threshold. In most cases, the density of the fox population is high, and the use of measures compatible with a moderate financial effort does not bring about a change in the progression of the front of the enzootic unless the latter is narrow (Müller, 1971). A study of Fig. 13 shows, moreover, the existence of a relationship between fox population density and rabies incidence. The greater the number of foxes in a given area, the greater are the chances of encounter and consequently of inoculation by a contagious animal.

The incidence of rabies in a region is then directly proportional to the initial density of the fox population; note, as a corollary, the fact that reduction of fox population density in an area before rabies virus is introduced into that area diminishes the potential incidence of rabies in that area.

One can summarize the practical considerations derived from analytic study of rabies virus transmission by saying that control of the fox in an area contributes to a reduction in the incidence of rabies in that area, but in most cases this approach allows one neither to expect nor to achieve a check or halt in the progression of the enzootic.

IV. Synthetic Epidemiology

The preceding analysis clarifies the findings of descriptive epidemiology at the individual level. We can now turn to epidemiologic phenomena observed at the regional level.

A. Development in Time

1. Seasonal Fluctuations

The regular periodicity of the peak incidence of fox rabies in the first trimester and of its augmentation in the course of the second semester (see Fig. 9) depends directly upon the habits of the fox. Thus, during the rut, wandering increases and encounters multiply, favoring the transmission of the virus during fights among males and, later, biting of females. Considering the relative brevity of the rabies incubation period, this period probably correlates with a maximal incidence of rabies.

The origin of the regular increase in cases of fox rabies found during the second semester is the dispersion of young foxes. The fox population density is artificially elevated by the dissemination of the newly independent young seeking a territory. It is principally the young male foxes which contract rabies during the course of the second semester, as Johnston and Beauregard (1969) have reported; their results show that during this period two-thirds of the rabid foxes are males and that 65% of the males are young. This selective infection of young males is in keeping with their greater precocity and their greater distance of dispersion in comparison with young females.

For bovine rabies, the seasonal fluctuations in incidence are tied to the biological rhythm imposed on these animals by humans, who put them to pasture from about April to November. Nearly all of the cases of bovine rabies seen in Europe are due to contamination by a rabid fox. Consequently, the risks of contamination and subsequently the incidence of bovine rabies are proportional to the incidence of vulpine rabies during the season of pasturage. In general, the cattle are infected at night in fields that border on small woods. The first cases of bovine rabies usually appear 2 months after the first exposure of cattle to fox bites, or in May–June. Then the incidence of bovine rabies rises steadily during the summer, autumn, and early winter up to November, paralleling the rise in incidence of vulpine rabies, but with a lag of 2 to 3 months. The peak is most often reached 1 month after the cattle reenter the barn, during the month of December; then the frequency diminishes, exposure having practically ceased. During the winter, only the cases with relatively long incubation crop up; among these the number decreases during the first trimester of the following year and declines to zero by spring. There may be a few consecutive cases of contamination of cattle in unclosed cattle sheds.

Having reviewed the seasonal fluctuations in the incidence of rabies in foxes and cattle, we will now consider the long-term periodicity of the infection.

2. Long-Term Kinetics

The descriptive epidemiology of the development of rabies in a region of limited area has shown considerable fluctuation over several years (as in Belgium, Luxembourg, etc.). In certain areas, this phenomenon is maintained over time [e.g., Schleswig-Holstein (Kauker and Zettl, 1969)].

The cyclic turnover in rabies incidence depends on the population dynamics of the fox. The appearance of rabies in an area introduces a mechanism of autoregulation of the fox population. After the enzootic has passed, the population of foxes still present in the area is strongly reduced: the mortality due to rabies virus reduces it by about 50% (Steck, 1968; Wandeler *et al.*, 1974c); and an additional proportion is destroyed by humans, since the appearance of rabies in an area induces intensification of the battle against the fox. Thus the number of foxes is stringently reduced, and the following year the incidence of rabies is very low. However, this sharp reduction in fox population density favors increased reproduction, because food is more freely available to the surviving animals; or, rather, it favors reduction of the causes of mortality, since fecundity differs little in areas of high or low density (Wandeler *et al.*, 1974c; Bögel *et al.*, 1974). In two or three generations, i.e., in 2 or 3 years, if the pressure exerted by humans is weak, the initial population can be reestablished and the conditions of vulpine population density are again favorable for a rise in rabies incidence.

Many variations in this simple scheme are possible, depending upon the initial density of the vulpine population, on the intensity and the steadfastness of the control measures applied by humans, on the nature of the milieu, and on other factors [see Figs. 14 to 17, which represent results

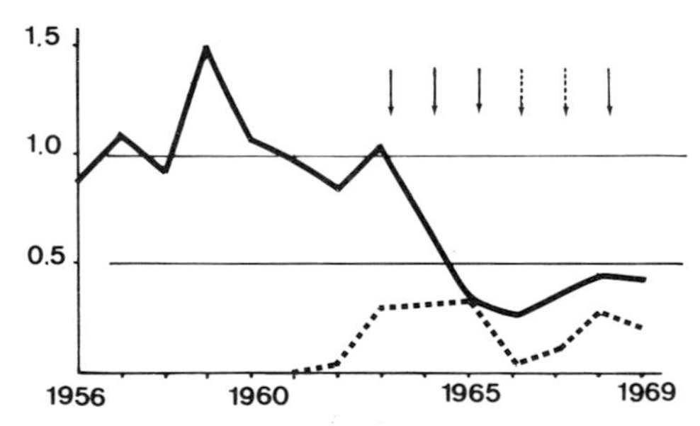

FIG. 14. Changes in fox population density and rabies incidence in the Swabian Jura area (7.191 km²) of Baden-Württemberg, West Germany, from 1956 to 1969. (From World Health Organization, 1972.) The ordinate shows the number of foxes killed per year per square kilometer (*solid line*) and the annual number of rabid foxes per 10 km² (*dotted line*). *Solid arrows* indicate gasification procedures; *dotted arrows*, incomplete operations.

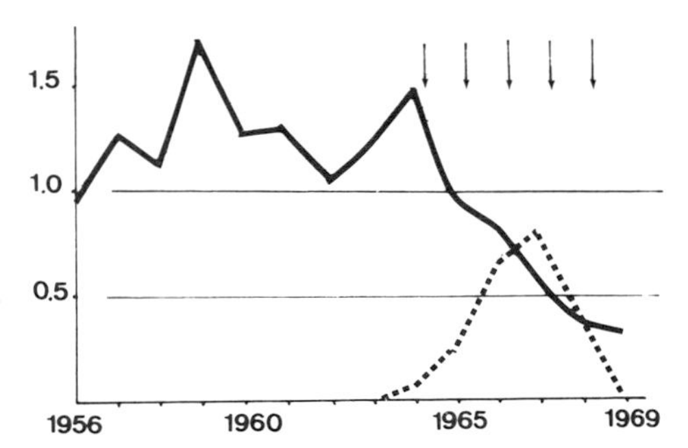

FIG. 15. Changes in fox population density and rabies incidence in the pre-Alpine region (8.744 km^2) of Baden-Württemberg, West Germany. (From World Health Organization, 1972.) Legend as in Fig. 14.

reported in various zones of Baden-Württemberg, West Germany (Moegle *et al.*, 1971; World Health Organization, 1972)].

In Fig. 14, the peak density of the vulpine population before the appearance of rabies was estimated by the number of foxes killed per year and per square kilometer. It is seen that this number clearly fell after 1963 under the combined action of the enzootic of rabies and the use of gasification. After a plateau, the number of cases of rabies fell in 1965, to rise again 2 years later.

Figure 15 demonstrates another kind of progression. In this pre-Alpine region, the density of the vulpine population (estimated according to the same criteria as before) was high, and one can note as a corollary the extent of the ensuing incidence of rabies.

Figure 16 illustrates the reduction in vulpine population density induced by a rabies enzootic, since the use of gas in this particular area did not begin until 1963, i.e., many years after the appearance of rabies. Later, under the combined effects of rabies and gas, the vulpine popula-

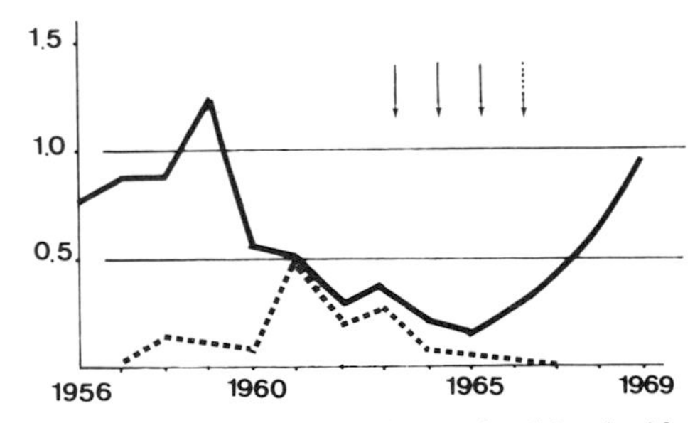

FIG. 16. Changes in fox population density and rabies incidence in the northern part (3.168 km^2) of Baden-Württemberg, West Germany. (From World Health Organization, 1972.) Legend as in Fig. 14.

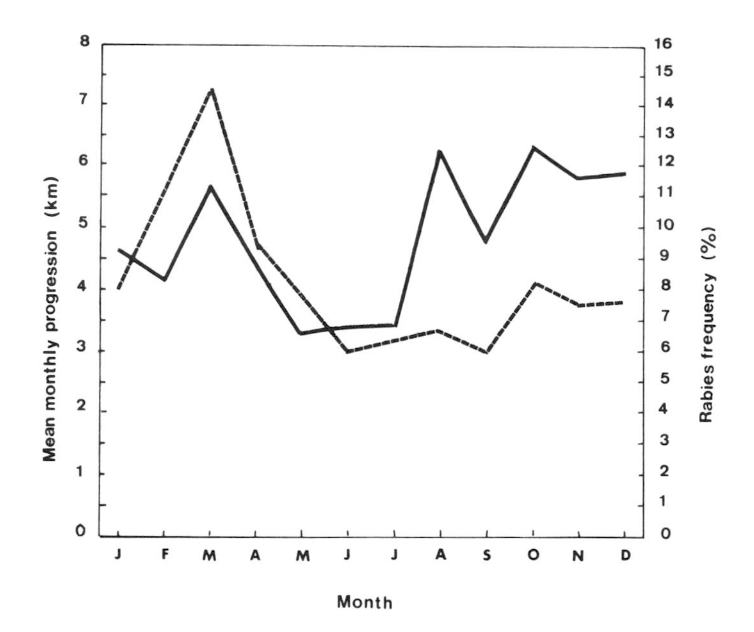

FIG. 18. Comparison between the monthly incidence of rabies (*dashed line,* percentage of cases observed each month) and the rate of progression (*solid line,* average monthly distance of new cases in relation to the front of the enzootic) in the course of the year; cumulative data for 2822 cases recorded from 1963 to 1971. (From Moegle *et al.,* 1974.)

of the enzootic and also relative to the relations among density of fox population, rabies incidence, and rate of progression of the enzootic.

Figure 18 shows the variations, in the course of the year, in rabies incidence and in rate of progression of the enzootic front. A sharp rise in incidence during the first trimester is apparent, as well as two periods of maximal progression separated by a period when the rate is low (May–July). One also sees a dissociation in the fluctuations of the incidence and rate: the rate of progression is accelerated (August) despite a reduced incidence. During the second semester the cases of rabies are fewer but more distant from one another than during the first trimester. This notion is confirmed by Fig. 19, which represents the average rate of progression for each month, as well as the maximal distances recorded. The greatest distances at which new cases appear in relation to the front are recorded in August, September, and October. However, during the period of minimal progression (May–July), one finds no "erratic" cases. In general, the great majority of rabies cases appear less than 10 km from the front of the preceding month (Table III).

Finally, Moegle and his collaborators have compared the data recorded

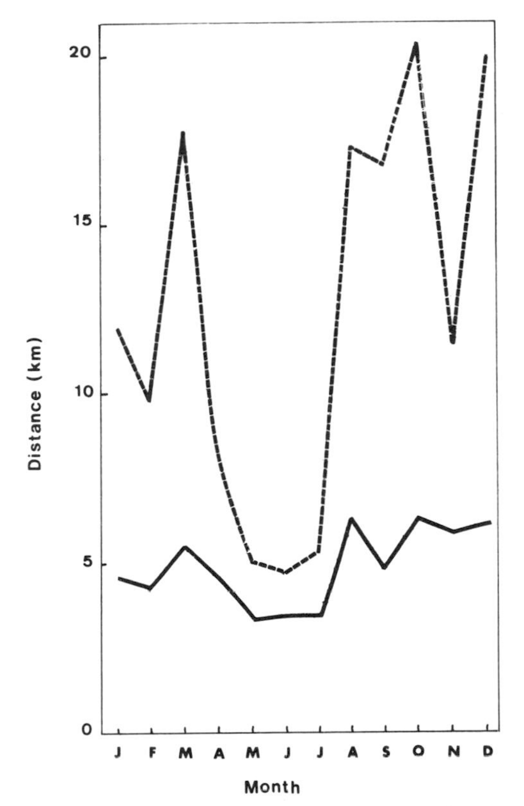

FIG. 19. Average monthly distance of advance of the rabies enzootic (*solid line*) and maximal distances of new cases of rabies in relation to the front of the preceding month (*dashed line*). (From Moegle *et al.*, 1974.)

in three different zones of the region under study. These zones (A, B, and C; cf. Table IV) differ in their fox population density, which rises progressively from A toward C, whereas the mean monthly rate shows little change. These results confirm that there is a much more intense relation-

TABLE III

DISTANCE OF NEW CASES OF RABIES FROM THE ENZOOTIC FRONT OF THE PRECEDING MONTH[a]

Distance (km)	0–5	5–10	10–15	15–20	20
Percentage of new cases	66.4	27	4.7	1.2	0.7
	93.4				

[a] From Moegle *et al.* (1974).

TABLE IV

COMPARISON OF RABIES INCIDENCE AND THE RATE OF PROGRESSION OF THE ENZOOTIC FRONT IN THREE ZONES OF WEST GERMANY HAVING DIFFERENT VULPINE POPULATION DENSITIES[a]

	Zone A	Zone B	Zone C
Number of foxes killed per square kilometer and per year	0.7	1.1	1.5
Rabies incidence per square kilometer and per year	0.044	0.051	0.065
Average monthly progression of enzootic front (km)	4.98	4.71	4.70

[a] From Moegle *et al.* (1974).

ship between vulpine population density and rabies incidence than between vulpine population density and rate of progression, with the consequences already described.

B. *Studies of Bögel and Associates*

The studies of Bögel *et al.* (1974) have essentially affirmed the delay necessary for a vulpine population of a given density to recover its initial level of density after reduction owing to rabies and gasification. This lag is, generally, the time required for reestablishment of a basic number of foxes, whatever the decimation factor may have been.

This study is based on the HIPD index (hunting indicator of population density), which is an indirect estimate of vulpine population density arrived at by recording the number of foxes killed per year and per square kilometer. The index is valid when calculated for surface areas of at least 2000 km^2. Its variations parallel those of the vulpine population density, since this has been controlled in numerous zones of Europe where the density in the spring is on the order of 0.9 to 1.2 adult foxes per square kilometer.

Figure 20 shows the results obtained in different zones of Baden-Württemberg. The ordinate represents percentage values of the HIPD index in relation to the initial level, before the reduction induced by rabies virus and gasification. The time in years is shown along the abscissa. Each line corresponds to one zone and reveals the rise in the HIPD index

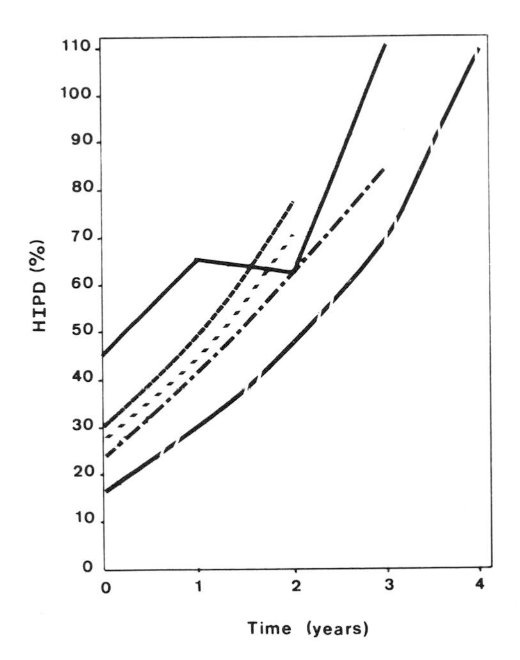

FIG. 20. Change in the estimated fox population density depending upon the HIPD index (number of foxes killed per year and per square kilometer) after the enzootic has passed and gasification procedures have ceased, in different areas of Baden-Württemberg. Each line designates an area. The HIPD percentages on the ordinate are in relation to values before reduction of the fox population. (From Bögel *et al.*, 1974.)

that follows the waning of the rabies enzootic and the end of gasification procedures. Note that the slope of these different lines is quite similar and that reestablishment of the fox population is effected at a comparable rate in the different zones.

So, after a reduction in the density of a fox population brought about by rabies and gasification, it takes about 4 years for the initial population level to be restored from the 20–30% of surviving animals; the same result requires 3 years for 40% survivors, 2 years for 60%, and 1 year for 80%

From these results, Bögel and associates have constructed a diagram (Fig. 21) indicating the period necessary for complete reconstitution of a vulpine population after it has undergone reductions of variable consequence. The ordinate on the left indicates the time necessary in years; that on the right, the percentage of animals remaining after the reduction.

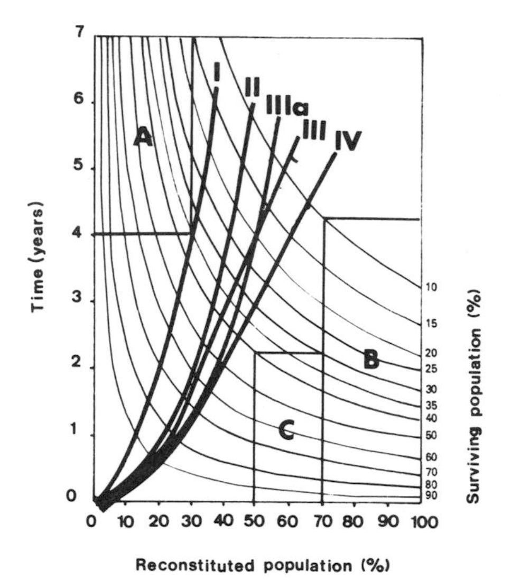

FIG. 21. Diagram indicating the time necessary for complete reestablishment of a fox population depending upon the initial reduction and the average percentage of reconstitution. Zones A, B, and C are not compatible with the rate of reconstitution of fox populations. Curves I, II, III, IIIa, and IV correspond to possible gradients of total population reconstitution. *On the right:* percentages of the original population remaining after the reduction. (From Bögel *et al.*, 1974.)

The rate of reconstitution of the vulpine population is shown on the abscissa, i.e., percentage of the original population stored per year. Certain zones of the diagram are excluded because they correspond to delays which differ from those recorded under natural conditions. Thus, zone A would imply a rate of reconstitution clearly lower than the recorded rates, whereas zones B and C suggest rates which are too high. Within the remaining area, many curves can be drawn and studied (I, II, III, IIIa, IV). Curve II is proposed by Bögel's group as the correct one according to the field results. Using this diagram it is possible to predict the growth of a fox population which has been reduced by both a rabies enzootic and by control measures against rabies; one can also determine, with satisfactory approximation, the delay necessary for the full reconstitution of the population. Bögel and co-workers insist on the necessity of applying this diagram only for the study of populations located in relatively large areas. However, many factors (e.g., climatic variations) intervene in the dynamics of fox populations, and the fact that it is impossible to specify them all in advance introduces a margin of error which cannot be suppressed, but which does not lessen the interest of the results of Bögel and his colleagues.

In the fight against rabies, it is well to keep these concepts in mind. They imply that the campaign against foxes must be permanently applied, despite the cost, because relaxation of control measures will lead ineluctably to the situation predicted by Fig. 21, hence to conditions propitious for a new outburst of rabies.

C. Computer Models

Preston (1973) has studied by computer a mathematical model representing the epidemiology of vulpine rabies after first defining the characteristics of a vulpine population: number of animals per surface area, age and sex of animals, reproductive characteristics of the fox, movements, etc. The only source of mortality introduced was that of the rabies virus. The simulation accomplished by varying the rate of transmission yielded interesting results, which we shall discuss after having briefly presenting the various data introduced into the model.

The total vulpine population was distributed over a surface area comprising 121 zones, each of 65 km^2. To begin with, four males and four females were "placed" in each zone. The size of the litter was fixed as 4.59 ± 1.56 cubs for the adult females and 4.16 ± 1.52 for the young females, with an equal number of male and female offspring. For the dispersion of the young foxes in autumn, the recorded differences between the males (which migrate farther) and the females were taken into account. The percentage of foxes with the furious form of rabies was fixed at 50%. It was considered that each fox which developed furious rabies could cover 32–48 km during the period of clinical expression and that a fox could cover 16 km a day in its own territory or when roving. The mean incubation period chosen was 1 month with a maximum of 7 months.

The virus of rabies was "introduced" into the area in the form of infection of half the foxes in nine zones (3.7% of the total vulpine population). Ultimately, the dynamics of rabies and of the vulpine population were simulated by varying the index P1, which is the proportion of animals contracting rabies after contact with a fox that had furious rabies.

Depending on the value of P1, different results were obtained: for one P1 value of 0.005 or less, the disease disappeared in a few months (Fig. 22a). For values of 0.025 or more, a severe epizootic struck down 90% of the vulpine population by the end of the first year and the disease disappeared (Fig. 22d). For values between 0.005 and 0.025 (Fig. 22b and 22c), there were cyclic oscillations of rabies and of vulpine population density, but the disease seemed established indefinitely. For a value of 0.015 (Fig. 22b), there was a 1-year cycle, and the seasonal fluctuations in incidence are similar to those reported in the United States and Canada.

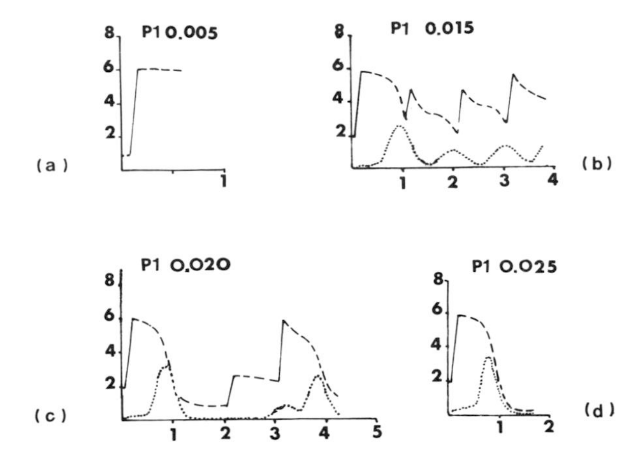

Fig. 22. Computer simulation of fox population dynamics and rabies incidence based on the index P1 corresponding to the proportion of animals contracting rabies in each area crossed by a fox affected with the furious form. The ordinates express the total number of foxes in thousands (*dashed line*) and the number of rabies cases (*dotted line*); the abscissas, the time in years. (a) Disappearance of rabies. (b) An annual cycle of rabies. (c) A self-maintained rabies cycle of 3 years. (d) Progress of a severe epizootic and disappearance of rabies. (From Preston, 1973.)

With a value of 0.020 (Fig. 22c), a self-sustaining cycle of 3 years was established, within which periodic reintroduction of the virus was not necessary.

In this model, in which rabies constitutes the only cause of mortality, it was not possible to obtain a self-sustaining cycle of 4 or 5 years, which tends to confirm that when such cycles are observed in nature they depend upon a control campaign against the fox.

Finally, these computer simulations indicate that to eliminate rabies from a region where it follows a 3-year cycle, it would be necessary to eliminate or to immunize 75% of the vulpine population.

The results of this study are interesting inasmuch as they confirm the concepts put forward by the epidemiologic analysis of rabies virus and also in that they contribute precision. It can thus be proposed: (1) that in the absence of an unknown reservoir, the disease in a given region may become perennial in the form of a 3-year cycle; (2) that measures of control against fox populations which are apparently responsible for the prolongation of the rabies cycle are justified in certain regions; and (3) that it is necessary to destroy or vaccinate a large proportion of foxes in order to eliminate the disease.

VI. Prophylaxis Against Fox Rabies

In a region of enzootic vulpine rabies it is advisable to distinguish explicitly between measures designed to prevent rabies in domestic animals and those aimed at rabies in wild animals. The first, provided that they are correctly applied, yield excellent results.

A. Rabies in Domestic Animals

Prophylaxis includes both veterinary public health and medical measures. The jusification of vaccination will be all the greater as the species of animals which need protection cannot be protected from fox bites by the simple application of veterinary health measures.

1. Preventive Measures

Preventive measures consist of avoiding as often as feasible all possible contact of domestic animals with wild animals in an enzootic area, especially keeping dogs closed up or leashed. Such measures are not applicable to certain categories of dogs (hunting dogs, sheep dogs, etc.) or to herbivores.

2. Medical Measures

Various live or inactivated vaccines can effectively protect domestic animal species. In particular, mixed vaccines (antirabies and antiaphthous for ruminants; antirabies, anti-Carré's disease, anti-Rubarth's disease, and antileptospirosis for dogs; antirabies and anti-infectious leucopenia for cats) simplify the vaccination procedures. Wide application of animal vaccination produces excellent results that are demonstrable, for instance, by the very low incidence of canine rabies in Belgium or the diminution of the incidence of bovine rabies in France.

B. Rabies in Wild Animals

1. Preventive Measures

Among the techniques which assure a reduction in vulpine population density (shooting, trapping, gasification of dens, distribution of poisoned bait, and the like), the one which has been found most effective is lethal gasification of dens by hydrocyanic acid during the period of sojourn of the females and cubs in the den. Its application over wide areas raises many problems, not only concerning the precautions necessary at every stage of its use to avoid accidents to humans, but the opposition which may be engendered by conservationists and hunters. Moreover, in the regions where a significant proportion of foxes does not live in dens because of the nature of the terrain, the results of gasification are greatly limited. For example, in the Swabian Jura and in the Swiss Oberland, 45% and 55% of the litters, respectively, are found elsewhere than in dens (Wandeler *et al.,* 1974c). Figure 14 showed that rabies has persisted in the Swabian Jura in spite of gasification of dens. Another obstacle limits the efficacy of gasification: in a region where it has been used, part of the dens are no longer inhabited and instead other shelters are occupied, shelters impossible to treat with gas. Thus, in one region of Switzerland the proportion of litters of cubs found in dens has gone from 63 to 49% in a single year, doubtless because of gasification (Wandeler *et al.,* 1974c). Gasification of dens also risks increased displacement of foxes, with the consequences that this implies for transmission of the virus. Finally, gasification procedures should be perfectly coordinated among different regions of a country in order to avoid favoring colonization of zones of very low vulpine population density from zones where measures of population control are moderate or weak.

At any rate, in general, the gasification of dens on a large scale at the moment of the progression of a rabies enzootic contributes toward lowering the vulpine population density to a level such that the disease may disappear. Moreover, as we have seen previously, this action must subsequently be fully maintained for some years to avoid having the vulpine population reconstitute itself and rabies reappear. Experience shows that after passage of the enzootic front and virtual disappearance of rabies, it is very difficult, if not impossible, to maintain a campaign of intense gasification over a large surface area. When the strip of land to be protected is relatively small owing to favorable geographic features, such as the "neck" of Denmark, such an effort is feasible and is crowned with success (Müller, 1971).

For regions situated in advance of the enzootic front, one may hope for a good outcome on condition that there is intensive reduction of the vulpine population in a belt at least 80 km wide in advance of the front, with progressively less intensive control measures beyond this zone.

Bögel and Arata (1972) studied the theoretical effects of various measures of control of fox populations and arrived at certain conclusions. The objective of measures to control fox populations should be to reduce the population density by 75–80% [which corresponds to the figure proposed by Preston's (1973) computer stimulation] in a zone where the initial density is one adult fox per square kilometer, before the period of births.

Observations made in Germany, Denmark, and Switzerland have indeed shown that rabies can completely disappear when the HIPD index falls from 1.2 to 0.3 or from 1 to 0.2.

a. Prophylaxis in a Noninfected Area Threatened by Rabies. In a threatened area, if the density can be reduced below the critical threshold, gasification of dens must be initiated and hunting intensified; should this not be feasible, especially because the majority of litters is outside of dens, one can foresee that the density will stay relatively high until rabies, if it appears, brings about a great reduction in that density.

Even very intensive hunting, if used as the exclusive means of control of fox population, is incapable of reducing the density of a vulpine population below the critical threshold.

Similarly, the exclusive use of gasification of dens in a region where a maximum of 60% of the litters is born in dens precludes the possibility of arriving at a threshold of 0.25 foxes per square kilometer. In an area where 80% of litters are born in dens, and providing that there is an intense and steady gasification effort, the critical threshold could be attained at the end of the third year of gasification. Should a major hunting effort be associated with gasification, this time limit might be reduced to 2 years.

b. Prophylaxis in an Area of Enzootic Rabies. In an enzootic area, the combined effects of rabies and intensification of hunting are not usually enough to reduce the density below the critical threshold. Similarly, the effects of rabies plus a single campaign of gasification at the peak of the enzootic are not generally sufficient.

In a region where 80% of foxes live in dens, regular use of gasification in the years preceding the peak of the enzootic, and one gasification following it, can bring about the elimination of rabies. If less than 60% of the foxes live in dens, other measures (say, hunting) must be combined with gasification.

Bögel and Arata (1972) have constructed graphs showing the results of the anticipated reduction of vulpine population density in an area when the reduction is due to rabies itself, gasification procedures, and hunting efforts. Such constructs may seem quite theoretical and quite far removed from the exigencies of action in the field; but they were devised on the basis of results obtained in the field. It thus seems desirable to take into account both positive field action, with its large variations and its unknowns, and those theoretical proposals which, though imperfect, it would be unwise to ignore merely on the pretext that they are insufficiently specific for a given region.

2. Medical Measures

The idea of immunization of foxes instead of their elimination is attractive (Black and Lawson, 1970; Debbie *et al.*, 1972; Mayr *et al.*, 1972; Debbie, 1974). Theoretically, and on condition that the vaccine could be applied to 75–80% of the vulpine population in a region, vaccination would be useful. However, its realization faces numerous difficulties, stemming as much from problems of presentation of the vaccine as from how it is to be conserved. There is also the question of residual pathogenic effects of vaccine strains for various rodents (Bijlenga and Joubert, 1974). Techniques of field vaccination are currently under study, and the investigators who convened in Frankfurt in September 1975 concluded that these must be thoroughly worked out before field use can be envisaged.

VII. Conclusion

The perennial establishment of rabies virus in numerous areas of Europe and North America, as well as its progressive extension into neighboring areas, have led to the accumulation of a great deal of data, to considerable research, and to the program coordinated by the WHO and the FAO that provides current epidemiologic information on this disease. Vulpine population density and the biology of the fox are the major factors which affect the overall epidemiology of vulpine rabies.

At present, only preventive measures are prescribed for efforts to reduce the incidence and the spread of rabies in wild animals. The two measures chiefly recommended are gasification of dens and bounty hunting. Their application usually results in only a reduction in rabies incidence. To halt the progression of the virus in a noncontaminated area or to make it disappear from infected areas, the theoretically necessary procedures are clear: reduction of the vulpine population density to the general critical threshold of about 0.25 foxes per square kilometer (which is one fox per 400 hectares); and, eventually, to keep the population density at this level. This implies a continuous intense effort against the natural dynamics of fox populations, since these exert constant pressure for reestablishment of a quite high density. These intensive measures necessary for control of fox populations are extremely burdensome, and when one takes into account the numerous other public and animal health imperatives which arise in each country, it seems unlikely that they will be consistently applied over vast areas. One is consequently obliged to be reconciled to the existence of enzootic rabies, to be accustomed to the return of successive waves every 3 to 7 years, depending on the region, and to set as an accessible objective, in default of the disappearance of vulpine rabies, the absence of human rabies.

This implies concomitant control of vulpine population density to reduce rabies incidence in wild animals, an extremely wide vaccine coverage of those domestic animals which are usual conduits of transmission of rabies virus to humans, and energetic public information campaigns.

ACKNOWLEDGMENT

The authors wish to thank Mrs. Frederik B. Bang for translating the text from the original French.

REFERENCES

Ables, E. D. (1965) *J. Mammol.* **46,** 102.
Andral, L., and Toma, B. (1973). *Cah. Méd. Vét.* **42,** 203.
Behrendt, G. (1955). *Osterwald Ith. Z. Jagdk.* **1,** 161.
Bijlenga, G., and Joubert, L. (1974). *Bull. Acad. Vét.* **47,** 423.
Black, J. G., and Lawson, K. F. (1970). *Can. J. Comp. Med.* **34,** 309.
Bögel, K., and Arata, A. A. (1972). *O.M.S. Doc., Geneva* **VPH/R/72.9.**
Bögel, K., Arata, A. A., Moegle, H., and Knorpp, F. (1974). *Zentralbl. Veterinaermed., Reihe B* **21,** 401.
Chalmers, A. V., and Scott, G. R. (1969). *Trop. Anim. Health Prod.* **1,** 33.
Costy-Berger, F., and Marchal, A. (1975). *Ann. Méd. Vét.* **119,** 241.
Dean, D. J. (1955). *Public Health Rep.* **70,** 567.
Debbie, J. G. (1974). *Infect. Immun.* **9,** 681.
Debbie, J. G., Abelseth, M. K., and Baer, G. M. (1972). *Am. J. Epidemiol.* **96,** 231.
Grégoire, M. (1969). *Bull. Acad. R. Méd. Belg.* **9,** 229.
Haagsma, J. (1975). *Tijdschr. Diergeneeskd.* **100,** 516.
Irvin, A. D. (1970). *Vet. Rec.* **86,** 333.
Jensen B. (1966). *Beitr. Jagd Wildforsch.* 187.
Jensen. B. (1970). *WHO/FAO Res. Programme Wildl. Rabies Cent. Eur., 2nd Reunion, Munich, working document.*
Jensen, B., and Nielsen, L. (1968). *Dan. Rev. Game Biol.* **5,** No. 6.
Johnston, D. H., and Beauregard, M. (1969). *Bull. Wildl. Dis. Assoc.* **5,** 357.
Kaplan, M. (1969). *Nature (London)* **221,** 421.
Kauker, E. (1975). *Sitzungsber. Heidelb. Akad. Wiss., Math.-Naturwiss. Kl.* No. 2, p. 49.
Kauker, E., and Zettl, K. (1963). *Veterinaermed. Nachr.* **96,** 106.
Kauker, E., and Zettl, K. (1969). *Berl. Münch. Tierärztl. Wochenschr.* **16,** 301.
Kohn, A., and Maar, L. (1968). *Bull. Soc. Sci. Méd. Grand-Duché Luxemb.* **105,** 9.
Lessman, F. G. (1971). Dissertation, Univ. of Munich, Munich.
Mayr, A., Kraft, H., Jaeger, O., and Haacke, H. (1972). *Zentralbl. Veterinaermed., Reihe B* **19,** 615.
Moegle, H., Knorpp, F., and Bögel, K. (1971). *Berl. Münch. Tierärztl. Wochenschr.* **84,** 437.
Moegle, H., Knorpp, F., Bögel, K., Arata, A., Dietz, K., and Diethelm, P. (1974). *Zentralbl. Veterinaermed. Reihe B* **21,** 647.

Mol, H. (1971). *Bull. Off. Int. Epizoot.* **75,** 808.
Müller, J. (1966). *Bull. Off. Int. Epizoot.* **65,** 21.
Müller, J. (1971). *Bull. Off. Int. Epizoot.* **75,** 763.
Parker, R. L., and Wilsnack, R. E. (1966). *Am. J. Vet. Res.* **27,** 33.
Pearson, O. P., and Bassett, C. F. (1946). *Am. Nat.* **80,** 45.
Phillips, R. L., and Storm, G. L. (1968). *Midwest Fish Wildl. Conf., Columbus, Ohio* 10 pp.
Phillips, R. L., Andrews, R. D., Storm, G. L., and Bishop, R. A. (1972). *J. Wildl. Manage.* **36,** 237.
Preston, E. M. (1973). *J. Wildl. Manage.* **37,** 501.
Sansyzbaev, B. K. (1975). *Zh. Mikrobiol., Epidemiol. Immunobiol.* No. 5, p. 139.
Sargeant, A. B. (1972). *J. Wildl. Manage.* **36,** 225.
Schneider, L. G., and Schoop, U. (1972). *Ann. Inst. Pasteur, Paris* **123,** 469.
Scholz, M., and Weinhold, E. (1969). *Berl. Münch. Tierärztl. Wochenschr.* **82,** 255.
Shigailo, V. T., Vetchinin, V. V., Potapova, V. G., Klimenko, A. F., and Sukhopyatkina, A. A. (1975). *Zh. Mikrobiol., Epidemiol. Immunobiol.* No. 5, p. 138.
Sikes, P. K. (1962). *Am. J. Vet. Res.* **23,** 1041.
Sodja, I., Lim, D., and Matouch, O. (1971). *J. Hyg., Epidemiol., Microbiol., Immunol.* **15,** 271.
Steck, F. (1968). *Vet. Rec.* **83,** No. 11. (Clin. Suppl.)
Storm, G. L. (1972). Ph.D. Thesis, Univ. of Minnesota, Minneapolis.
Switzenberg, D. F. (1950). *J. Mammal.* **31,** 194.
Tabel, H., Corner, A. H., Webster, W. A., and Casey, C. A. (1974). *Can. Vet. J.* **15,** 271.
Tierkel, E. (1966). *Symp. Ser. Immunobiol. Stand.* **1,** 245.
Tissot Boireau, D. (1975). D.V.M. Thesis, Ecole Nat. Vét. d'Alfort, Alfort, France.
Toma, B., and Andral, L. (1970). *Cah. Méd. Vét.* **39,** 99.
van Haaften, J. L. (1970). Cited in Bögel and Arata (1972).
Wandeler, A., Wachendörfer, G., Förster, U., Krekel, H., Schale, W., Müller, J., and Steck, F. (1974a). *Zentrabl. Veterinaermed., Reihe B* **21,** 735.
Wandeler, A., Wachendörfer, G., Förster, U., Krekel, H., Müller, J., and Steck, F. (1974b). *Zentralbl. Veterinaermed., Reihe B* **21,** 757.
Wandeler, A., Müller, J., Wachendörfer, G., Schale, W., Förster, U., and Steck, F. (1974c). *Zentralbl. Veterinaermed., Reihe B* **21,** 765.
Winkler, W. G. (1972). *J. Infect. Dis.* **126,** 565.
World Health Organization (1972). *Weekly Epidemiol. Rec.* No. 5, p. 49.
World Health Organization (1970). *Chronique O.M.S.* (French version), **24,** 54.

ECOLOGY OF WESTERN EQUINE ENCEPHALOMYELITIS IN THE EASTERN UNITED STATES*

C. G. Hayes† and R. C. Wallis

Department of Epidemiology and Public Health,
Yale University, New Haven, Connecticut

I. Introduction

Western equine encephalomyelitis (WEE) has been recognized as a serious public health problem in western North America for more than 30 years. WEE appears to exist endemically in numerous foci in that region, with a low incidence rate among humans. Severe outbreaks, however, have occurred periodically. For example, during the summer of 1941 a severe epidemic involving more than 3000 cases in humans occurred in North Dakota, Minnesota, and adjacent areas of Canada

* In this review the term *eastern United States* refers to the following states: Maine, New Hampshire, Vermont, Massachusetts, Rhode Island, Connecticut, New York, New Jersey, Maryland, Delaware, Virginia, North Carolina, South Carolina, Georgia, Florida, Alabama, Mississippi, and Louisiana.

† Present address: Pakistan Medical Research Center, 6 Birdwood Road, Lahore, Pakistan.

(Olitsky and Casals, 1952). The case fatality rate ranged from 8 to 15%. Epizootics among horses are more common. More than 600 cases of WEE were diagnosed among horses in the central and western United States during the summer of 1975 (Foreign Animal Diseases Report, 1975).

More recently, the virus of WEE has been found to be widely distributed in the eastern United States in regions along the Atlantic and Gulf seaboard; however, WEE has not been recognized as a public or veterinary health problem in the East. This situation presents an interesting epidemiological problem: a virus present in different regions is causing disease in one area but not another area even though susceptible hosts are present. The reasons for the absence of WEE in humans and equines in the East are not clearly understood but probably are a reflection of parallel variations in the enzootic transmission cycles of this virus in different regions of the United States.

Several reviews are available on the ecology of WEE virus in areas of the western United States (Cockburn *et al.*, 1957; Longshore, 1960; Longshore *et al.*, 1960; Loomis and Meyers, 1960; Meyers *et al.*, 1960; Reeves and Hammon, 1962; Hess and Hayes, 1967). This article will examine data from studies on the ecology of WEE virus conducted in the eastern United States over the past two decades since the virus was first recognized in this region.

II. Classification of WEE Virus

A. *Ecological Classification*

WEE virus is maintained naturally in an enzootic cycle involving mosquitoes and wild birds. On the basis of this cycle, it is classified as an arbovirus (arthropod-borne virus). This classification refers to a virus being maintained in nature "through biological transmission between susceptible vertebrate hosts by haematophagous arthropods" (W.H.O. Scientific Group, 1967). Inherent in the definition is the production of viremia in the vertebrate host of sufficient magnitude to initiate infection in the arthropod taking a blood meal from the vertebrate. The term biological transmission means that the virus ingested by the arthropod must escape from the alimentary tract and undergo replication, particularly in the salivary glands, before transmission by bite to a susceptible vertebrate host can occur. This period of reproduction in the arthropod between the moment of initiation of infection and development of the ability to transmit virus by bite is termed the *extrinsic incubation period.*

B. Physicochemical Classification

Classification of an agent as an arbovirus serves as a useful concept for the epidemiologist because it is based on the mode of virus transmission between hosts; however, it has little taxonomic value in the current universal classification scheme for viruses, which is based primarily on physicochemical characteristics (Casals, 1971; Fenner, 1974). This is reflected by the fact that, of the 234 viruses listed as definite, probable, or possible arboviruses in the *International Catalogue of Arboviruses* (1975), those which have been characterized by physicochemical properties can be divided into at least six distinct taxa. Based on this scheme, WEE virus has been placed in the family Togaviridae. Virions in the Togaviridae "contain single-stranded RNA, 3×10^6 to 4×10^6 daltons, have isometric, probably icosahedral, nucleocapsids surrounded by a lipoprotein envelope containing host cell lipid and virus-specified polypeptids including one or more glycopeptides" (Fenner *et al.*, 1974). The virions appear to undergo replication in the cytoplasm and mature by budding either from the plasma membrane or intracellular membranes. Infectious RNA can also be extracted from virions in this family, suggesting that the virion RNA has the same base sequence as messenger RNA (Baltimore, 1971; Fenner *et al.*, 1974). Details of the structure and replication of the Togaviridae in various host systems have been covered in recent reviews by Horzinek (1973), Murphy *et al.* (1975), Mussgay *et al.* (1975), and Pfefferkorn and Shapiro (1975).

C. Serological Classification

At the present time, two genera, *Alphavirus* and *Flavivirus,* have been defined within the Togaviridae. The genus *Alphavirus*, which includes WEE virus, is identical to the serologically defined arbovirus Group A originally described by Casals and Brown (1954). All members of this group are antigenically interrelated but unrelated to any other members of the Togaviridae, and all members have mosquitoes as invertebrate hosts (Casals, 1957; Casals and Clarke, 1965).

Within the genus *Alphavirus*, several complexes have been recognized on the basis of antigenic relationships (Porterfield, 1961). WEE virus is included in one such complex with the viruses of Sindbis, Highlands J, Whataroa, Aura, and Y62–33 (Karabatsos, 1975). In addition, recent evidence with radioimmune precipitation techniques suggests that all Alphaviruses contain three antigenic components: (1) a type-specific and (2) a complex-specific determinant residing in the viral envelope, and (3) a broadly reacting group antigen common to all members of the genus residing in the nucleocapsid (Dalrymple *et al.*, 1972a, 1973).

III. Distribution

Numerous studies involving virus isolations and serological surveys indicate that WEE virus is present in practically all areas of the continental United States. The virus is also widely distributed in southwestern Canada from the provinces of British Columbia to Manitoba (Dillenberg, 1965; McKay *et al.*, 1968; Shemanchuk and Morgante, 1968; Kettylis *et al.*, 1972). An isolation from sentinel chickens in Grande Prairie, Alberta, is the most northerly report of WEE virus in North America (Brown and Marken, 1966). WEE virus is probably widely distributed in Central and South America as well, but isolations have only been reported from a few areas. Virus has been isolated in Argentina (Meyer *et al.*, 1935; Mettler, 1962), and recent epizootics of equine encephalomyelitis in that country and adjacent areas of Uruguay have been attributed to WEE virus (*Vigilancia Epidemiologica*, 1973). Isolations have also been reported from Brazil (Bruno-Lobo *et al.*, 1961; Shope *et al.*, 1966) and Guyana [formerly Brtish Guiana (Spence *et al.*, 1961)]. Sudia and co-workers (1975a) recently reported the first isolation of WEE virus from Mexico.

Earlier reports from Czechoslovakia of the isolation of WEE virus now appear to be in error (von Sprockhoff and Ising, 1971).

IV. Chronological History

The first isolations of WEE virus were made by Meyer and co-workers (1931) from the brain tissues of two horses during an epidemic of equine encephalomyelitis in the San Joaquin Valley of California during the summer of 1930. These workers were able to reproduce the disease in experimentally inoculated horses and demonstrated the usefulness of guinea pigs as a laboratory animal for virus isolation and experimental work. Shortly after these initial isolations, Meyer (1932) made the suggestion that the equine encephalomyelitis virus may be responsible for producing a similar disease in man. His suggestion was based on the observation of three cases of encephalitis in persons who had been in close contact with infected horses prior to the onset of their illnesses. Eklund and Blumstein (1938) subsequently made observations on six human cases of encephalitis that occurred during an epidemic of equine encephalomyelitis in northwestern Minnesota in 1937. Five of the six cases were in individuals who had been in close contact with sick horses.

In three of the patients who recovered, blood samples were collected approximately 4 months after the date of onset of illness. The presence of neutralizing antibodies to WEE virus was demonstrated in the serum of one of these patients.

The relationship of WEE virus to human disease was firmly established by Howitt (1938) with the isolation of virus from the brain of a 20-month-old boy who died of acute encephalitis in California. In the same paper, she also reported finding neutralizing antibodies to WEE virus in the sera of three children who had been hospitalized with an encephalitic illness. Later she reported the isolation of virus from the blood of an adult male who died of acute encephalitis during the same outbreak (Howitt, 1939).

The observation that epidemics of equine encephalomyelitis were restricted to the warmer periods of the year, along with observations made during early studies in the Yakima Valley of Washington and Kern County, California, that cases were largely limited to irrigated areas with large mosquito populations, suggested the involvement of an arthropod vector in the spread of WEE virus (Buss and Howitt, 1941; Hammon and Howitt, 1942). Experimental transmission work in the laboratory, particularly with mosquitoes, supported this concept. Kelser (1933) was able to achieve transmission from infected to normal guinea pigs using *Aedes aegypti.* The mosquitoes could transmit as early as six days after feeding on an infected guinea pig. More importantly, in the same study, he also demonstrated that *Ae. aegypti* could transmit WEE virus from an infected guinea pig to a horse, resulting in the production of clinical equine encephalomyelitis. Since *Ae. aegypti* did not occur in the areas of the United States where epidemics of equine encephalomyelitis caused by WEE virus had been reported, it had no importance as a natural vector; however, a number of other species of mosquitoes in the genus *Aedes* were present in these epidemic areas. Kelser's observations led to studies with many of these species as experimental vectors (see review by Ferguson, 1954).

The first field isolation of WEE virus from mosquitoes was reported by Hammon and co-workers (1941a) from the Yakima Valley, Washington. The virus was obtained from a pool of *Culex tarsalis* collected in July of 1941. Following this initial report, a number of additional virus isolations were made from this species during 1941 and 1942 in the same region (Hammon *et al.*, 1942a). Experimental transmission work had not been done with this species prior to the isolation of virus, since it was considered to be primarily a bird-feeding mosquito and not an important feeder on large mammals such as humans and equines. However, subsequent studies on the feeding habits of *C. tarsalis* in the

Yakima Valley revealed that this mosquito did frequently feed on large domestic mammals, thereby demonstrating that it had the potential for serving as a vector of WEE virus to equines and humans (Bang and Reeves, 1942; Reeves and Hammon, 1944).

Following these epidemiological findings, Hammon and Reeves (1943) conducted laboratory experiments on the vector capability of *C. tarsalis* for WEE virus. Mosquitoes were exposed to virus by feeding on an infected guinea pig, duck, and on virus–blood suspensions. Mosquitoes that became infected transmitted virus to chickens and a guinea pig. Since this work, several other investigators have confirmed the vector efficiency of this species under experimental conditions (Barnett, 1956; Chamberlain and Sudia, 1957; Hayles *et al.*, 1972). The aforementioned epidemiological and experimental evidence along with correlations on mosquito population dynamics and virus activity led to the conclusion that *C. tarsalis* was the most important vector of WEE virus in the Yakima Valley of Washington (Hammon *et al.*, 1942b). Studies in other areas of the central and western United States have generally supported this finding.

Along with the first isolations of WEE virus from field-collected mosquitoes, the early studies in the Yakima Valley also produced evidence of the involvement of vertebrates other than horses in the epidemiology of WEE. Neutralizing antibodies to WEE virus were found in both domestic and wild mammals and birds (Hammon, 1941; Hammon *et al.*, 1941b, 1942c). Howitt and van Herick (1941) reported similar findings from Kern County, California. In these studies the prevalence of WEE antibodies in domestic mammals and fowl was greater than that found for wild mammals. Certain species of wild birds also were shown to possess a high antibody rate similar to that found for domestic animals.

This information on antibody rates combined with data indicating that domestic fowl developed a high-level viremia after infection whereas mammals generally did not, and other epidemiological evidence, such as the availability of a large susceptible population each year in close proximity to man, led to the hypothesis that domestic fowl were the most important reservoir hosts for WEE virus (Hammon *et al.*, 1945). However, when studies designed to reduce vector contact with domestic fowl by spraying of residual insecticides failed to affect virus-transmission and vector-infection rates, attention was shifted to the investigation of wild birds as reservoir hosts of WEE virus (Reeves *et al.*, 1948; Hammon and Reeves, 1948).

Prior to 1950, only a single isolation of WEE virus had been obtained from wild birds (Cox *et al.*, 1941). In that instance, virus was recovered from brain and spleen tissues taken from a prairie chicken (*Tympanuchus cupido*) collected in North Dakota. Three additional isolations of

WEE virus were made from the blood of nestling wild birds collected in Colorado during the summer of 1950 (Sooter *et al.*, 1951). Since these early reports, many additional virus isolations have been obtained from these hosts. In addition, experimental studies have shown that the viremia developed by wild birds is comparable in titer to that found with domestic fowl. On the basis of this evidence, wild birds have been incriminated as the most important enzootic vertebrate hosts for WEE virus (Hammon *et al.*, 1951; Kissling *et al.*, 1957).

The evidence presented above, mainly from early studies in Washington and California, pointed to a basic cycle of WEE virus in nature involving transmission between mosquitoes, primarily *C. tarsalis,* and wild birds. Humans and equines were shown to be involved only as secondary hosts that probably played no role in the spread or maintenance of virus. This evidence has generally been substantiated in numerous more recent investigations undertaken in areas of the central and western United States.

The first evidence that WEE virus was active in the eastern part of the United States came from studies conducted by the Public Health Service in southeastern Louisiana (Kissling *et al.*, 1954a). During the late spring and summer of 1950, a serological survey of wild birds in this area revealed the presence of neutralizing antibodies to WEE virus in the sera of 15 of 226 birds tested. More significantly, 12 of the 15 immune birds were juveniles, indicating that infection had probably occurred locally. Following this preliminary survey, more detailed studies on virus activity were conducted in the same area of Louisiana during 1952 and 1953 (Kissling *et al.*, 1955). In addition to confirming the presence of wild birds immune to WEE in the study area, virus was recovered from three birds collected in June of 1952. Three additional isolations of WEE virus were obtained from mosquitoes, further strengthening the evidence for the existence of an enzootic focus of virus activity. The initial discovery of WEE virus in the Northeast was made in 1953 in New Jersey (Holden, 1955). In this study, isolations were obtained from "apparently normal" immature house sparrows (*Passer domesticus*) collected during September. In this same study, the presence of neutralizing antibody to WEE virus was found in 12 of 65 plasma specimens obtained from penned ring-necked pheasants (*Phasianus colchicus*). Immature as well as 2-year-old pheasants were surveyed, and the greater percentage of immunity found in older birds compared to that in immature birds suggested that WEE virus had been active in the area for at least 2 years.

Since these early studies, WEE virus has been discovered to be present in a number of additional states in the East. Isolations were first reported from North Carolina in 1955 (Chamberlain *et al.*, 1958); Alabama in 1957 (Stamm *et al.*, 1962); Massachusetts in 1959 (Hayes *et al.*, 1961); Florida in 1960 (Henderson *et al.*, 1962); Georgia in 1963 (Chamberlain *et al.*, 1969); Maryland in 1964 (Moussa *et al.*, 1966); Virginia in 1965 (Lord and Calisher, 1970); Connecticut in 1972 (Yale Arbovirus Research Unit, unpublished data); and New York in 1973 (J. Woodall, unpublished data). These studies demonstrate that WEE virus also is present in the East in an enzootic cycle involving transmission between wild birds and mosquitoes, similar to the cycle found in the western United States.

V. Enzootic Cycle

A. *Vertebrate Hosts*

1. *Wild Birds*

Results from field studies have implicated wild birds as important reservoir hosts for WEE virus in the eastern United States. Numerous virus isolations have been made from these hosts in various regions of the East (Table I), and virus has been recovered from the blood of as many as 7–14% of the birds collected during a given period in some study areas (Stamm, 1963; Wheeler, 1966). Results from experimental work also have shown wild birds to be excellent hosts for this virus.

TABLE I

Isolations of WEE Virus from Wild Birds in Several Areas of the United States

State	Study period	No. isolations/ total	Percentage positive	Reference
Louisiana	April 1952–June 1953	3/1421	0.2	Kissling *et al.* (1955)
Alabama	May 1957–June 1958	3/1014	0.3	Stamm *et al.* (1962)
Alabama	Nov. 1959–Dec. 1960	23/3020	0.8	Stamm (1968)
Alabama	Sept.–Oct. 1960	21/649	3.2	Stamm and Newman (1963)
Florida	May–July 1960; Aug. 1961	2/801	0.3	Henderson *et al.* (1962)
Maryland	May–Oct. 1968	2/991	0.2	Williams *et al.* (1974)
Maryland	1969	3/2866	0.1	Dalrymple *et al.* (1972b)
New Jersey	1960–1969	127/55,000	0.2	Goldfield and Sussman (1970)
Connecticut	Sept.–Oct. 1974	1/292	0.3	Yale Arbovirus Research Unit (unpublished)

Kissling *et al.* (1957) tested six species of wild birds captured in Alabama. These birds were infected with WEE virus by subcutaneous inoculations and by the bite of infected mosquitoes. All six species developed viremia of sufficient titer to infect *Ae. aegypti* and *Ae. triseriatus* mosquitoes. Transmission of virus from infected to normal birds was also accomplished using these vectors.

Experimental studies also have revealed that virus persists at detectable levels in the blood of wild birds for only 3–5 days following infection (Hammon *et al.*, 1951; Kissling *et al.*, 1957; Holden *et al.*, 1973a). Because of the transient nature of viremia, the virus recovery rate from wild birds collected in an area may not accurately reflect the extent of virus activity that has occurred over a season. This problem has been partially overcome by obtaining indirect evidence of infection by serological surveys of wild bird populations. Since antibodies to WEE virus have been found to persist for months and even years in some species of birds, the proportion of immune birds collected in a locale usually exceeds the proportion of birds from which virus can be recovered during any given period (Kissling *et al.*, 1957; Wheeler, 1966; Holden *et al.*, 1973a). Between 60 and 70% of some species of wild birds have been found to possess antibody to WEE virus by the end of the transmission season (Stamm and Newman, 1963; Dalrymple *et al.*, 1972a).

However, because of the persistence of antibodies, crude WEE immune rates in birds cannot be used as a direct measure of current virus activity; i.e., individuals might have been infected in a previous season. In order to obtain more accurate data on the level of virus activity during a season, wild birds have been classified according to age as "immatures" (less than 1 year old) or "matures" (greater than 1 year old). Antibodies to WEE virus in immature birds indicates that infection has occurred during the present season. In studies in Alabama, Stamm (1968) measured the neutralizing antibody rate for WEE virus in immature birds throughout the transmission season. Only 2 of 147 immature birds sampled during July had WEE antibodies, but by the end of the transmission season, 52% of the immatures showed evidence of prior infection. A similar study in the Pocomoke Swamp area of Maryland revealed no antibodies to WEE virus in immature birds prior to August but almost equal rates of immunity for adult and immature birds collected in August and September (Williams *et al.*, 1974). Twelve of 59 immatures and 15 of 77 mature birds sampled during this period had antibodies to WEE virus.

One complicating factor in interpreting the significance of antibodies present in immature birds results from the transovarian passage of antibodies from immune females to their offspring. Maternal antibodies to

WEE virus have been demonstrated in doves (*Streptopelia* spp.) up to 12 weeks and in chickens up to 4 weeks after hatching (Reeves *et al.*, 1954). Neutralizing substances also have been detected in 1- and 2-day-old nestling house sparrows born of WEE virus-immune parents (Holden *et al.*, 1973a). However, because maternal antibodies are of short duration, their presence probably has been unimportant in most studies in the eastern United States, as immature wild birds generally have been collected by netting after they have fledged. Furthermore, results from several studies in the East indicate that WEE virus activity among nestling birds is minimal since most immatures collected early in the summer have been found antibody negative (Stamm, 1968; Dalrymple *et al.*, 1972b; Williams *et al.*, 1974). The apparent lack of exposure of nestling birds to virus activity in the eastern United States appears contrary to the situation in some areas of the West where nestling house sparrows play an important role as amplifying hosts for WEE virus early in the transmission season (Cockburn *et al.*, 1957; Reeves *et al.*, 1964; Burton *et al.*, 1966a; R. O. Hayes *et al.*, 1967; Holden *et al.*, 1973b).

The mobility patterns of many species of wild birds also complicate the interpretation of antibody data obtained from adults. Individuals sampled in one area may have been infected in another locale and subsequently moved into the study area. By classifying birds according to their migratory status as permanent, summer, winter, or transient and unstable species, investigators have been able to obtain a better index of local transmission activity (Table II). Because virus transmission among wild birds occurs primarily during the warmer parts of the year when adult mosquitoes are active (See Section V, D), the finding of a higher proportion of antibody-positive birds among the species present in an area during this period (permanent and summer residents) compared to the winter residents and transient species indicates that active transmission is occurring in that area. Moreover, an increase in the proportion of antibody-positive summer and permanent resident adults sampled over the course of the season can provide an indication of the level of virus activity that has occurred. For example in Stamm's (1968) Alabama study, the immune rate among the adult summer and permanent resident species captured increased from 26 to 47% by the end of the transmission season. However, even among species classified as permanent or summer residents, considerable fall wandering, nondirection migration, and east–west migration may occur over a large geographical region (Sussman *et al.*, 1966; Anderson *et al.*, 1967). In addition, individuals of some species classified as permanent residents of an area may actually be breeding farther north and spending only the winter

TABLE II

PROPORTION OF WEE ANTIBODY POSITIVE BIRDS, BY MIGRATORY STATUS, CAPTURED IN SEVERAL STUDY AREAS OF THE EASTERN UNITED STATES[a]

State	Permanent residents	Summer residents	Winter residents	Transient and unstable species	References
Maryland	238/743 (32)	220/784 (28)	50/707 (7)	13/632 (2)	Dalrymple *et al.* (1972b)
Connecticut	34/118 (29)	8/31 (26)	0/30	0/110	Yale Arbovirus Research Unit (unpublished data)
Louisiana	39/501 (8)	45/300 (15)	3/209 (1)	—	Kissling *et al.* (1955)
Alabama	150/498 (30)	150/716 (21)	143/534 (27)	12/175 (7)	Stamm (1968)
Alabama	35/88 (40)	41/255 (16)	2/10 (20)	9/213 (4)	Stamm and Newman (1963)

[a] Parenthesized numbers indicate the percentage positive.

in the study area. Likewise, individuals of some species classified as summer residents may also breed to the north and be captured only when they pass through the study area in the spring on their way north to breed or in the fall when they are passing through on their way south (Stamm *et al.*, 1962; Stamm, 1968).

Recaptured birds probably give the most accurate indication of the resident population in an area. As pointed out by Stamm *et al.* (1962) a bird captured on one date and recaptured in the same vicinity after a period which is less than the time needed for the bird to have undergone a complete cycle of migration can be assumed to have spent the interval between captures in the local area. Therefore, birds that are negative for virus antibody when first sampled but which convert to antibody positive or are viremic at a later recapture date within an appropriate time span can be considered to have been exposed to virus in the vicinity of the collection site. In Florida, Henderson *et al.* (1962) found a conversion rate from WEE antibody negative to positive of 20.7% among 58 recaptured birds bled at an interval of approximately 1 month. Wheeler (1966) applied "life table" methods to recapture data in calculating the prevalence and incidence of WEE among three species of wild birds studied over a 6-year period in the Hockomoke Swamp in Massachusetts. In addition to providing a reliable method of assessing the trend of virus transmission in the study area over a number of years, important interspecific differences in regard to virus activity and antibody retention were revealed among the three species studied. For example, during 1962, a year of intense virus activity among wild birds in the area, approximately 11% of the catbirds (*Dumatella carolinensis*) and 6% of the black-capped chickadees (*Parus atricapillus*) were shown to have been exposed to WEE virus during a particular sampling period. In this same period, less than 1% of recaptured swamp sparrows (*Melospiza georgiana*) revealed evidence of infection. Wheeler also found that catbirds and swamp sparrows retained detectable antibody for long periods, whereas chickadees were found capable of reverting from antibody positive to negative within as short a period as 2 weeks.

The evidence gathered from the field and laboratory studies in the East does not allow any particular species of wild bird to be singled out as being more important than other species in the enzootic cycle of WEE virus. However, birds in the order Passeriformes have been consistently found to be involved to a greater extent than are birds in other orders sampled. This may simply be a reflection of the population density of passerines in the type of habitat where WEE virus activity characteristically is found in the eastern United States (See Section V, C). On occasion, birds in other orders have been shown to have a high rate of exposure to WEE virus. In studies in Louisiana in 1952–1953, where substantial numbers of wading birds in the order Ciconiiformes were sampled, 27 of 197 white ibis (*Eudocimus albus*) and 7 of 27 yellow-crowned night herons (*Nyctamassa violacea*) were found to possess antibodies to WEE virus (Kissling *et al.*, 1955). In a later study in the same area, 28 of 49 yellow-crowned night herons were positive for WEE antibodies (Stamm, 1958).

In general, wild birds appear universally susceptible to infection with WEE virus. Over 50 species in at least eight orders are known to be naturally or experimentally susceptible to this virus. No species has been found refractory to infection (Stamm, 1958, 1966). In fact, most wild birds develop inapparent infections of WEE, although individuals of some species such as the house sparrow and purple finch (*Carpodacus purpureus*) appear to experience occasional mortality (Hammon *et al.*, 1951; Kissling *et al.*, 1957; Holden *et al.*, 1973a). A number of factors such as viremia titer and duration, antimosquito behavior, and longevity of particular species of birds as well as biases inherent in commonly used sampling methods may all influence the results obtained in studies on the dynamics of bird–virus interaction (Stamm, 1963; Edman and Kale, 1971).

2. Nonavian Vertebrates

Virus isolation and antibody data for wild nonavian vertebrates sampled in WEE enzootic areas of the eastern United States indicate these animals are not as extensively involved in the transmission cycle of WEE virus as are wild birds. Goldfield and Sussman (1967) reported the isolation of this virus from wild mammals, reptiles, and amphibians collected in New Jersey. Workers trapped and processed 11,728 nonavian vertebrates between June of 1962 and the end of 1966. WEE virus was isolated from the blood of 22 of 8523 and from the brains of 12 of 7571 animals. Virus was recovered from eight species of mammals, two species of reptiles, and one amphibian species. Virus isolation attempts from nonavian vertebrates collected in other areas of the East have all been negative (Henderson *et al.*, 1962; Stamm *et al.*, 1962; Wellings *et al.*, 1972; Massachusetts Department of Public Health, unpublished data; Yale Arbovirus Research Unit, unpublished data). However, the number of animals tested in these studies has been considerably less than the number sampled in New Jersey. An interesting aspect of the New Jersey study is that 25 of the 34 isolations of WEE virus reported from nonavian vertebrates were obtained from animals captured between the months of November and April. Enzootic activity among wild birds and mosquitoes is minimal or nonexistent during this period because of cold weather (c.f. Section V, D). This suggests that these animals may be more involved in the maintenance of WEE virus through the winter in the East rather than in the summer transmission cycle (Section VI, B).

In addition to the virus isolation data from New Jersey, results from serological surveys conducted in Florida (Henderson *et al.*, 1962), New York (Whitney, 1963; Whitney *et al.*, 1968), New Jersey (Goldfield and Sussman, 1967), and Maryland (Dalrymple *et al.*, 1972b) also confirm that wild mammals, reptiles, and amphibians do occasionally become infected with WEE virus in the eastern United States. Virus recoveries and antibody data from studies in which wild nonavian vertebrates have been sampled in several regions of western North America also show that these animals become exposed to WEE virus. (Lennette *et al.*, 1956; Cockburn *et al.*, 1957; Reeves and Hammon, 1962; Gebhardt *et al.*, 1964; Spalatin *et al.*, 1964; Yuill and Hanson, 1964; Burton *et al.*, 1966a; Rueger *et al.*, 1966; Bowers *et al.*, 1969; Emmons and Lennette, 1969; Yuill *et al.*, 1969; Burton and McLintock, 1970; Hardy, 1970; Hoff *et al.*, 1970; Iversen *et al.*, 1971; Prior and Agnew, 1971; Hardy *et al.*, 1974a; Leung *et al.*, 1975; Sudia *et al.*, (1975b). However, as pointed out by Hardy *et al.* (1974a), their role in the ecology of WEE virus in the West is uncertain and might vary between different geographic regions.

Experimental studies with wild mammals have shown a number of species to be susceptible to infection with WEE virus. Some species readily succumb to disease after infection, whereas others develop only inapparent infections (Syverton and Berry, 1936a, 1940; Howitt, 1940; Mitchell and Walker, 1941; Grundmann *et al.*, 1943). Several species of rodents and the snowshoe hare (*Lepus americanus*) have been found to develop viremia of high titer after experimental infection (Hardy *et al.*, 1974b; Kiorpes and Yuill, 1975), but no mosquito feeding or transmission studies have been reported using these animals as hosts. Several species of cold-blooded vertebrates also have been found to be susceptible to infection with WEE virus and capable of developing a viremia of sufficient titer to infect mosquitoes feeding on them. Some investigators think that poikilotherms may play an important role in the overwintering of WEE virus (See Section VI, B).

B. Invertebrate Hosts

Insects are the only invertebrates from which WEE virus has been isolated in the eastern United States. Most isolations have been obtained from mosquitoes, although workers in New Jersey (M. Goldfield, unpublished data) have also recovered virus from fleas and lice collected from white-footed mice (*Peromyscus leucopus*). Because of the consistency and frequency of virus isolations reported from various study areas, *Culiseta melanura* has been incriminated as the most important mosquito vector to wild birds in this region.

Kissling *et al.* (1955) made the initial isolation of WEE virus from *C. melanura* in Louisiana in May of 1952. In addition, they reported two recoveries of virus from another mosquito, *Aedes infirmatus*, in June of the same year. However, the isolations from the latter species came at a time of high population level and when virus activity was at a peak in the study area. The minimum infection ratio (MIR) for *Ae. infirmatus* for the study period was greater than 1:2000. By contrast, *C. melanura*, which constituted less than 1.0% of the total mosquitoes collected, had an MIR of 1:387. Chamberlain *et al.* (1958) recorded the next isolation of WEE virus from *C. melanura* during studies in New Jersey. Based on this and the previously reported Louisiana isolation, these workers suggested that *C. melanura* might be serving as an enzootic vector for the virus in the eastern part of the country. Their suggestion has been substantiated by subsequent studies in which the virus recovery rate from *C. melanura* has been consistently higher than that for other tested mosquitoes.

For example, during studies in Alabama from the spring of 1957 to the summer of 1958, Stamm *et al.* (1962) made eight isolations of WEE virus. All were from *C. melanura*, which had an MIR of 1:336. No virus (WEE) was recovered from an additional 6500 specimens representing 29 other mosquito species collected during the study. Sudia *et al.* (1968), working in the same general vicinity of Alabama from 1959 to 1963, made 11 isolations of WEE virus from mosquitoes; nine of these were from *C. melanura*. This mosquito had an MIR of 1:240, whereas the other two species from which virus was recovered both had an MIR greater than 1:2000. An additional 32,435 mosquitoes representing 29 species were negative for virus. In the Pocomoke Swamp area of Maryland, 50 isolations of WEE virus were recorded from mosquitoes over a 6-year period from 1964 to 1969 (Moussa *et al.*, 1966; Le Duc *et al.*, 1972; Williams *et al.*, 1972; Muul *et al.*, 1975). All virus strains were recovered from *C. melanura*. In addition to the above examples, virus isolation reports from studies in Georgia (Chamberlain *et al.*, 1969), Florida (Wellings *et al.*, 1972), New Jersey (Kandle, 1961, 1963, 1964), Massachusetts (Massachusetts Department of Public Health, unpublished data) and New York (J. Woodall, unpublished data) also support the hypothesis that *C. melanura* is the primary enzootic vector of WEE virus in the eastern region of the country.

In addition to virus isolation data, ecological observations on the feeding behavior of *C. melanura* have strengthened the concept of involvement of this mosquito in the enzootic cycle of WEE virus. As outlined in a previous section, wild birds, especially in the order Passeriformes, have been implicated as the main enzootic vertebrate hosts of WEE virus. Host attraction and blood meal identification studies have shown *C. melanura* to be closely associated with wild birds.

R. O. Hayes (1961a) conducted host preference tests for *C. melanura* in Massachusetts using animal-baited traps. Seven species of birds, eight species of mammals, six species of reptiles, and four species of amphibians were used in this study. Although the results did not reveal any particular species within a vertebrate class to be significantly more attractive than other species in that class, birds as a group were shown to be more attractive for *C. melanura* than mammals, reptiles, or amphibians.

An average of 153 *C. melanura* per trap were attracted to a bird, whereas 48, 10, and 5 mosquitoes per trap night were attracted to a mammal, amphibian, and reptile, respectively. Engorgement rates also were substantially higher for the mosquitoes attracted to birds than for those attracted to mammals or reptiles. None of the *C. melanura* attracted to the amphibians took a blood meal. Means (1968) obtained similar results in New York using "host-preference traps." This latter study was also designed to compare the relative attractiveness of the different taxonomic classes of vertebrates. Wild birds attracted the greatest numbers of *C. melanura*. Decreasing numbers of females were attracted to amphibians, reptiles, and mammals. Of 401 *C. melanura* that had the opportunity to be attracted to birds versus mammals, reptiles, or amphibians, 348 (87%) were attracted to the birds. Engorgement rates of the females attracted to the various hosts reflected the same trend. Ninety-two percent of those attracted to birds were blooded, whereas 81%, 74%, and 40% of the females attracted to reptiles, amphibians, and mammals, respectively, were engorged.

Other workers have used precipitin testing to determine the blood source of engorged specimens of *C. melanura* collected in the field. A high percentage of meals determined by this method have come from birds, with only occasional feedings positive for other classes of animals. In New Jersey, Jobbins *et al.* (1961) tested 157 engorged *C. melanura* for blood-meal identification, and 137 were positive for avians. In subsequent studies in New Jersey, Crans (1962, 1964, 1966) confirmed that *C. melanura* was primarily an avian feeder. Of the 2893 engorged *C. melanura* in which the blood meal source could be determined, 2812 had fed on birds. An exception to this trend was reported by Moussa and coworkers (1966). Of 130 blooded *C. melanura* collected in the Pocomoke

Swamp area of Maryland, the 48 specimens giving a positive precipitin reaction had all fed on nonavian vertebrates. Their results indicated that *C. melanura* fed on pig, deer, racoon, reptile, cow, dog, horse, and man in descending order of preference. However, the large number of engorged females whose blood-meal source could not be identified (82 of the 130 tested) casts doubt on the accuracy of these results as a reflection of *C. melanura's* feeding behavior. Subsequent studies in the same area have failed to confirm the 1966 report. Joseph and Bickley (1969) working in the Pocomoke Swamp tested 1776 blood-fed *C. melanura* to determine the source of feeding, and 87.6% were positive for birds. Muul *et al.* (1975) and Le Duc *et al.* (1972) both reported that more than 90% of the engorged *C. melanura* from the Pocomoke Swamp contained bird blood. Serological studies on blood-fed *C. melanura* collected in Georgia (Chamberlain *et al.*, 1969) and in Florida (Edman *et al.*, 1972) also show this species to be ornithophilic in its feeding behavior. In those studies where the order of bird has been determined, Passeriformes generally have accounted for 80–90% of the total bird blood meals (Crans, 1964; Joseph and Bickley, 1969; Edman *et al.*, 1972; Le Duc *et al.*, 1972; Muul *et al.*, 1975). However, Edman *et al.* (1972) suggested that the feeding pattern of *C. melanura* among different avian hosts reflected an opportunistic rather than a fixed behavior pattern. Their studies showed that 80–90% of the nocturnal avian residents were passerines in the two study areas in Florida where blood-engorged *C. melanura* were collected. Anderson and Maxfield (1962), working in an area where WEE virus activity had been reported in southeastern Massachusetts, found that 12 of the 15 avian species breeding in the area were passerines.

The small proportion of engorged *C. melanura* found positive for mammal, reptile, or amphibian blood is in agreement with the virus isolation and serological data from these animals showing that they do not become involved in the enzootic cycle of WEE virus as extensively as do wild birds. However, the fact that *C. melanura* has been shown to feed occasionally on these animals and the finding of double-fed specimens containing both avian and nonavian blood indicate that the opportunity exists for virus exchange between these hosts (Jobbins *et al.*, 1961).

C. Habitat Association

In the East, most strains of WEE virus have been isolated from wild birds and mosquitoes collected in freshwater swamps and bogs. In fact, enzootic activity of this virus appears to be confined to this type of habitat. Such swamps often contain a mixed growth of deciduous trees such as maple, gum, and oak and evergreens such as cypress, cedar, and pine. These areas are usually in close proximity to rivers or streams and characteristically have a high water table, so that even when surface water is scarce, water is abundant a short distance below the substrate (Beaven and Oosting, 1939; Kissling *et al.*, 1955; Stamm *et al.*, 1962; Saugstad *et al.*, 1972). The restriction of enzootic activity of WEE virus to the wooded-swamp habitat appears to be related more to the presence of the proper mosquito vector rather than to the availability of suitable vertebrate hosts.

Wild birds, particularly the small passerines, are very numerous in the wooded swamp habitat (See Section V, B). However, passerines are also abundant in many other types of sylvan habitats where WEE virus activity has not been recorded. Although the species composition varies between the swamp and other types of wooded habitats, the wide range of wild birds demonstrated to be susceptible to WEE virus (Stamm, 1963), as well as the mobility patterns of many species between various habitat types, indicates that the localization of the enzootic WEE virus activity to the wooded swamp is not attributable to a lack of suitable avian hosts occurring outside of such areas. This conclusion is reinforced by the existence of enzootic bird–mosquito cycles of WEE virus in parts of the central and western United States in habitats which differ ecologically from the wooded-swamp area characteristic of the East (Reeves and Hammon, 1962; Hayes *et al.*, 1967; Hess *et al.*, 1970). As with wild birds, many species of nonavian vertebrates that occur in wooded swamps are also found in a variety of other forested as well as grassland areas. Isolations of WEE virus have not been reported in the East from wild nonavian vertebrates, particularly small mammals, which have been collected in areas away from the typical swamp habitat.

On the other hand, *C. melanura*, the enzootic vector of WEE virus in the East, is very selective in its breeding site preference and is found almost exclusively in wooded freshwater swamps. Burbutis and Lake (1956) found this species to be present in many areas of New Jersey, but occurrence was rather localized. They collected larvae from sphagnum bogs, large permanent cedar-type swamps, rock piles, and rarely in tussock swamps and partially dried stream beds. All sites were well shaded and contained cold water. R. O. Hayes *et al.* (1962) also reported *C. melanura* in freshwater swamps adjacent to coastal salt marshes and from inland swamp areas in New Jersey. Jamnback (1961) described this species on Long Island as breeding in restricted habitats, usually in sphagnum bogs. Bickley and Byrne (1960) considered *C. melanura* to be rare in Maryland; however, subsequent surveys by Moussa *et al.* (1966) showed large numbers of this species to be present in the Pocomoke Swamp area of that state. Joseph and Bickley (1969) also working in Maryland found *C. melanura* to be abundant in swamps along the Pocomoke and Nanticoke rivers in the southeastern part of the state, but the species was scarce in northern areas of Maryland where there was less swampland. Several investigators have studied the distribution of breeding sites of *C. melanura* within the Pocomoke Swamp. Larvae were found most commonly in depressions beneath uprooted trees, in rotted out stump holes, and in various other holes in the root matt on the floor of the swamp. The greatest number of larvae per unit area sampled were collected in the swamp interior, as compared to upland forest sites within the swamp or areas on the swamp periphery (Joseph and Bickley, 1969; Williams *et al.*, 1971, 1974; Saugstad *et al.*, 1972; Muul *et al.*, 1975). Studies on *C. melanura* in Georgia (Siverly and Schoof, 1962), Massachusetts (A. J. Main, unpublished data), and Connecticut (Yale Arbovirus Research Unit, unpublished data) also have shown that the larvae of this species occur principally in wooded freshwater swamps.

Occasionally, *C. melanura* larvae and adults have been reported in areas distant from the typical swamp habitat, but the number of specimens collected has been small. Several investigators have reported the collection of larvae from water in artificial containers (Dorsey, 1944; Wallis and Whitman, 1967; Joseph and Bickley, 1969). Spielman (1964) reported the occurrence of adults and larvae of *C. melanura* in an urban area 11 km from the nearest freshwater swamps. Over a 4-year period in Boston, Massachusetts, adult specimens were collected from an air shaft of a building, and egg rafts and larvae were found in a pool formed in an eroded portion of a river bank sheltered by thick brush. MacCreary and Stearns (1937) collected adult *C. melanura* from light traps setup on a lighthouse in the Delaware Bay 13 km from the nearest point of land.

A number of other mosquito species besides *C. melanura* also breed in wooded freshwater swamps, and larvae of several species such as *Aedes canadensis, Culex restuans, C. territans,* and *C. salinarius* are often found in direct association with the larvae of *Culiseta melanura* (Burbutis and Lake, 1956; Joseph and Bickley, 1969; Williams *et al.*, 1971; Saugstad *et al.*, 1972). Some of these species feed readily on birds (R. O. Hayes, 1961a; Crans, 1964, 1966; Murphy *et al.*, 1967; Means, 1968). WEE virus also has been isolated from some of these species (Chamberlain *et al.*, 1958; Hayes, 1961b; Kandle, 1961; Sudia *et al.*, 1968; M. Goldfield, unpublished data). However, all of these species are more catholic in their selection of breeding sites than is *C. melanura* and occur in areas outside of freshwater swamps (Carpenter and La Casse, 1955; Horsefall, 1955; Wallis, 1960). Adults as well as larvae are frequently collected in abundance in habitat types where enzootic activity of WEE virus has not been reported, and they appear to play no major role in the enzootic bird–mosquito cycle of this virus.

The main ecological factor limiting enzootic activity of WEE virus to the wooded swamp in the eastern United States appears to be the breeding habitat preference exhibited by *C. melanura*. A wider distribution would be anticipated based on the dispersal patterns of wild birds, mammals, and other mosquito species that occur in this type of habitat.

D. Seasonal Pattern of Virus Activity

The eastern United States lies within the North Temperate Zone, and the occurrence of low temperatures is normally associated with the onset of the winter season. Because of this, enzootic activity of WEE virus is primarily limited to the warmer periods of the year when adult mosquitoes are active, even though wild birds are present during all seasons. The length of the active transmission period varies in regions of different latitude within the East.

In some areas of the Southeast, adult *C. melanura* can be collected throughout the year. Stamm *et al.* (1962) found evidence of feeding or breeding activity in Alabama during every month except February. Love and Goodwin (1961) collected *C. melanura* in southwest Georgia from light traps throughout the year. Likewise, Taylor *et al.* (1966) found adults of this species active throughout the year in the Tampa Bay area of Florida. Although these studies show that *C. melanura* can be active during the winter in the Southeast, the abundance and activity of this vector are greatly reduced as a result of cooler weather, and most studies have failed to find evidence of virus activity during this period.

Kissling *et al.* (1955) found WEE virus activity in Louisiana to be at a peak during May and June. No evidence of transmission during the winter months was detected (Kissling *et al.*, 1955, 1957). Researchers in Alabama also failed to detect continuous virus activity. All of their isolations came from wild birds and mosquitoes captured from July to December (Stamm *et al.*, 1962; Stamm, 1968; Sudia *et al.*, 1968). Wellings *et al.* (1972) did report the isolation of several strains of WEE virus from *C. melanura* collected during the winter in Florida; however, the intensity of virus activity was much lower during this period than in the spring and early summer.

In the Northeast, adult *C. melanura* are not present throughout the winter. Emergence of adults begins during the spring, and substantial populations are present in most areas by late May or early June (Burbutis and Lake, 1956; Moussa *et al.*, 1966; Main *et al.*, 1968; Joseph and

Bickley, 1969; Saugstad *et al.*, 1972; Muul *et al.*, 1975). Because this species is multivoltine, adults are present during the entire summer, and large numbers may be collected as late as October in the Northeast. In fact, Joseph and Bickley (1969), studying the emergence patterns of this mosquito from man-made breeding sites in Maryland, recorded the highest emergence peak of the season during October. Adult activity as measured by light trap collections usually terminates in late October or early November when the mean temperature falls below 10°C (Joseph and Bickley, 1969; Muul *et al.*, 1975). However, adults have on occasion been captured in man-made and natural resting sites as late as the end of November to mid-December (Wallis, 1953; Joseph and Bickley, 1969; Gusciora *et al.*, 1972).

Only one isolation of WEE virus has been recorded from mosquitoes collected before July in the Northeast. M. Goldfield (unpublished data) reported the isolation of virus from *C. melanura* collected in New Jersey during June of 1974. Workers in New Jersey (Kandle, 1963, 1967) have recovered virus from wild birds captured during June. Workers in New York (Bast *et al.*, 1973) and Maryland (Dalrymple *et al.*, 1972b) also have reported the conversion of sentinel quail from negative to positive for WEE virus antibody during June. The Pocomoke Swamp studies in Maryland indicate that the peak period of WEE virus activity occurs from mid-July to mid-September and that transmission seldom continues past October (Williams *et al.*, 1971, 1972, 1974; Dalrymple *et al.*, 1972b; Saugstad *et al.*, 1972; Muul *et al.*, 1975).

The factors responsible for the reported absence of virus activity in the late spring and early summer in the Northeast are not clear, as large populations of *C. melanura* and wild birds are present before July. Females of this mosquito have been shown to be feeding actively during this period. Wallis (1959a) reported as many as 50% of the females sampled during June in Connecticut were freshly blooded or gravid. Joseph and Bickley (1969) and Muul *et al.* (1975) also found high rates of engorgement during May and June among specimens collected in Maryland.

Wild birds are also numerous during May and June in the Northeast. However, the population is composed mainly of mature birds such as permanent residents and recently returned summer residents. At this time the WEE antibody rate in this group may be fairly high from virus exposure in past seasons (cf. Section V, A). This could produce a herd immunity effect which might dampen a buildup of virus activity early in the season. Immature birds constitute a small portion of the total birds collected by mist-netting prior to July in the Northeast but may account for more than 50% of the population sampled by the latter part of that month (Anderson and Maxfield, 1962; Williams *et al.,* 1974). Thus a large number of susceptible hosts are entering the population by midsummer, and this may serve as an amplifying mechanism to increase the virus activity to a level where it can be more readily detected. For example, Dalrymple *et al.* (1972b) studied the kinetics of virus transmission among immature white-eyed vireos (*Vireo griseus*) collected in the Pocomoke Swamp area of Maryland (Figs. 1 and 2). Immatures of this species were first collected in mid-June, but at that time they constituted less than 20% of the sampled population of this species. No immune birds were detected among the immatures until early July. At this time, the immature group comprised approximately 50% of the population collected, and 20% of them had WEE antibodies. By mid-July, when the initial isolation of WEE virus was made in the study area, close to 100% of the white-eyed vireos collected were immatures. Most of the immatures sample had WEE antibody by this date, indicating that an extremely rapid spread of virus occurred among this population during July and early August. This is the period when the greatest number of virus isolations were made in this study area.

The length of the extrinsic incubation period (Section I, A) of WEE

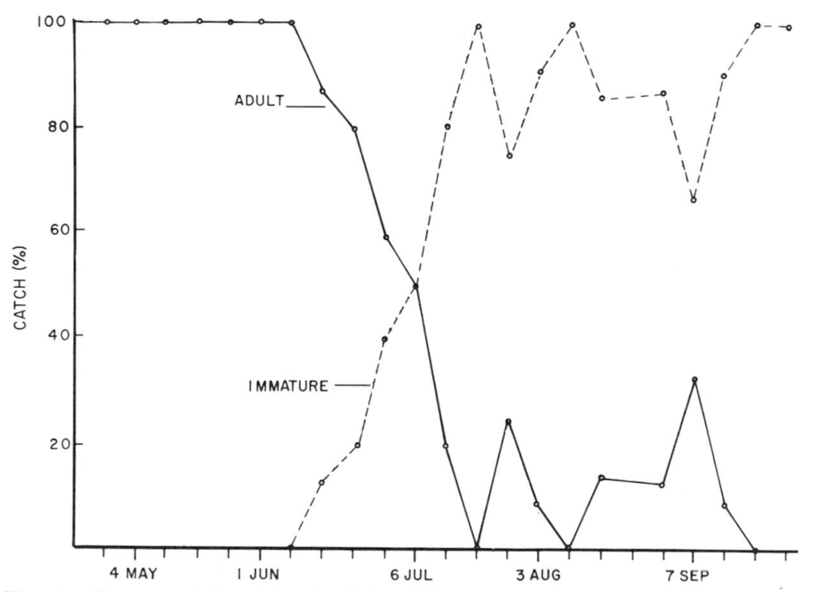

Fig. 1. Season relationship of adult and immature white-eyed vireo population in the Pocomoke Swamp expressed as a percentage of the total number of birds of this species captured. (From Dalrymple *et al.,* 1972b.)

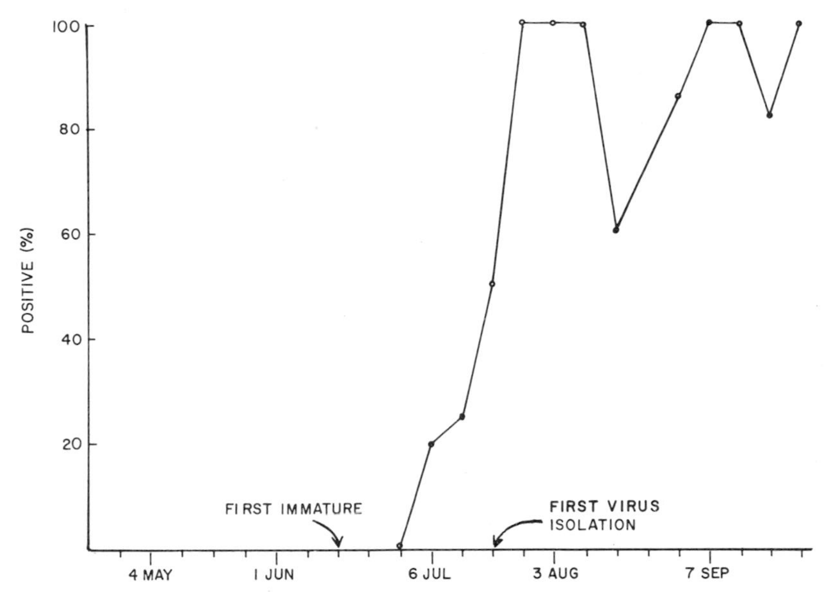

FIG. 2. Rate of infection with WEE virus of the immature white-eyed vireo population in the Pocomoke Swamp. (From Dalrymple *et al.*, 1972b.)

virus in *C. melanura* at temperatures prevailing during spring and early summer in the Northeast may also be an important factor delaying onset of transmission activity. Several investigators have shown that the extrinsic incubation period of arboviruses in mosquitoes is lengthened as temperatures are reduced (Davis, 1932; Bates and Roca-Garcia, 1946; Chamberlain and Sudia, 1955; La Motte, 1963; Rush *et al.*, 1963a; Hurlburt, 1973; McLean *et al.*, 1974). C. G. Hayes (unpublished data) has shown that *C. melanura* females fed on viremic chicks were able to transmit WEE virus after 7 days incubation at 27°C. Mean temperatures in the wooded-swamp habitat preferred by *C. melanura* are somewhat lower than this during the spring (Saugstad *et al.*, 1972); however, the kinetics of WEE virus replication in this mosquito have not been determined at reduced temperatures. *Culiseta melanura* has been shown to retain virus for life once it becomes infected (C. G. Hayes, unpublished data), and this mosquito has been found capable of surviving for more than 3 months under laboratory conditions (Love and Goodwin, 1961). These two factors probably contribute to the maintenance of a high rate of infection in this species throughout the summer and early fall once virus activity has commenced.

VI. Overwintering

Since there is no evidence of year-round transmission of WEE virus among wild birds in the Northeast (as discussed in Section IV, D), this agent must either survive the winter by some alternate mechanism or be reintroduced into the enzootic foci each season from areas where activity is occurring. This situation probably holds true for most parts of the Southeast as well, although adult mosquitoes may retain virus throughout the year in some of the warmer areas (Section IV, D). To explain the phenomenon of continued appearance of virus activity in enzootic foci in the eastern United States each summer in the absence of demonstrable year-round transmission, the following hypotheses have been proposed: (1) Virus is introduced yearly by migrating vertebrate hosts from other regions where transmission is occurring. (2) Virus survives the winter in the enzootic foci in chronically infected vertebrate hosts or in chronically infected invertebrate hosts.

A. Reintroduction by Migrating Vertebrate Hosts

1. Spring Migration of Birds

The Louisiana studies of Kissling *et al.* (1955) showed that wild birds migrating into the tropics during the winter had a higher antibody-positive rate for WEE virus than that of species which were permanent residents of the study area. From this, these authors concluded that migrants were probably exposed to virus for a greater portion of the year than were nonmigrant birds; in addition, they suggested that birds arriving in the study area from the tropics during the spring migration could transport virus from over-wintering sites and serve as a source of infection for mosquitoes. In subsequent studies, however, conducted in Louisiana and in Florida, Kissling *et al.* (1957) failed to recover virus from spring migrants from South and Central America. The permanent and winter residents of these areas were found to have an antibody rate as high as or higher than those species that overwintered in the tropics. Workers in the Northeast have failed to isolate WEE virus from spring migrants, even though large numbers have been tested (Anderson, 1961; Wheeler, 1966; Goldfield and Sussman, 1970; Dalrymple *et al.*, 1972b). In addition, summer resident birds in the Northeast that overwinter in the southern United States or in the tropics and have been captured returning to the swamp habitat the following spring have shown little change in their antibody profiles for WEE virus from the previous season (Wheeler, 1966; Dalrymple *et al.*, 1972b). These data indicate that summer residents are at minimal risk for acquiring WEE in the regions where they overwinter.

2. *Fall Migration of Birds*

Workers in Alabama showed that birds migrating from the north during the fall could introduce WEE virus into an area (Stamm *et al.*, 1962; Stamm and Newman, 1963; Stamm, 1968). For example, during studies in Baldwin County, Alabama, in 1957, Stamm *et al.* (1962) were unable to detect virus activity from May through September; but concurrent with the arrival of migrant birds, isolations of virus were obtained from wild birds and mosquitoes. In an expanded study on the role of migrating birds in disseminating virus, Lord and Calisher (1970) sampled migrant birds at seven sites along the Atlantic flyway from Maryland to southern Florida during the fall of 1965. Isolations of WEE virus were made from actively migrating wild birds captured in Virginia and North Carolina. In the Northeast, Anderson *et al.* (1966) sampled fall migrants in Massachusetts but were unable to recover virus. Although the results from these studies show that migrating birds are capable of spreading virus during the fall from one region to another in the East, there is no evidence to support the hypothesis that virus is being introduced from areas north of the United States. To the contrary, the fact that WEE virus activity has been detected in the Northeast and Southeast during the spring and early summer (cf. Section IV, D) before fall migration begins indicates that this is not a necessary mechanism for the annual occurrence of virus activity in enzootic foci in the East.

B. Survival in Enzootic Foci

1. *Chronic Infection of Vertebrate Hosts*

a. Birds. Investigators in the eastern United States have failed to detect virus activity among wild birds collected during the winter (Section V, D)—with the exception of Kandle (1967), who reported the isolation of several strains of WEE virus from birds captured during the winter months in New Jersey in 1966 and 1967. WEE virus has also been recovered from birds collected during the winter or early spring in several areas of western North America (Burton *et al.*, 1966a; reviewed in Reeves, 1974).

In addition to these field isolates, results from experimental studies have shown that birds can develop persistent infections with WEE virus. Reeves *et al.* (1958a) isolated virus from various tissues of infected birds at intervals of 1–10 months after the original exposure. An isolation made 10 months after the initial infection came from the spleen of a tricolored blackbird (*Agelaius tricolor*), but more significantly, two isolations were made from blood at intervals of 198 days from a cowbird (*Molothrus ater*) and 234 days from a house sparrow. In later studies, Reeves (1974) attempted to infect mosquitoes by feeding them on birds that had been inoculated with WEE virus 1 or more months earlier. No infections were demonstrated in 14,022 mosquitoes that survived 14 days postfeeding. Until chronically infected birds can be shown capable of recirculating enough virus in their blood to infect mosquitoes or other ectoparasites feeding on them, their role in the overwintering of WEE virus remains speculative.

b. Mammals. Goldfield and Sussman (1967) reported the isolation of WEE virus from rodents collected in New Jersey during the winter. Over a 5-year period, 10 strains of WEE virus were recovered from the blood of Norway rats (*Rattus norvegicus*) and white-footed mice captured between the months of October and March. No virus was detected in the blood of rodents collected during the warmer months from April through September. WEE virus was also recovered from the brains of eight rodents. There are no other reports of WEE virus isolations from rodents captured in the East (cf. Section V, A), although several isolations have been obtained from rodents collected during the winter and early spring in parts of western North America (Burton *et al.*, 1966a; Hardy *et al.*, 1974a; Leung *et al.*, 1975).

In experimental studies, Hardy *et al.* (1974b) infected six species of rodents with WEE virus. Organs from nine animals that survived 58 weeks postinoculation were tested for persistence of virus, but no isolations were made. However, factors such as physiological state and environmental conditions have been shown to be important variables influencing an animal's response to virus infection. For example, Kiorpes and Yuill (1975) demonstrated that viremia duration, titer, and antibody response were increased in snowshoe hares (*Lepus americanus*) experimentally infected with WEE virus that were exposed to variable climatic conditions compared to those held at constant temperature and humidity. The course of infection was also influenced by the time of year in which the hares were captured for experimental work. Animals captured during the winter, before reproductive activity had started, experienced a greater viremia when exposed to fluctuating temperatures than did those collected during the summer in the midseason of reproductive activity. Studies with another arbovirus, Colorado tick fever, in hibernating and nonhibernating ground squirrels, *Spermophilus lateralis*, have shown that viremia may be more prolonged in the hibernating animals (Emmons, 1966). Bats have also been found to develop persistent infections with several arboviruses when held at hibernating temperatures (reviewed by Sulkin and Allen, 1974). Field studies, however, have not implicated bats as natural hosts of WEE virus in the eastern United States. In studies in Massachusetts, A. J. Main (unpublished data) tested over 500 bats but failed to find virus or neutralizing antibodies to WEE virus. Workers in New Jersey reported a single isolation from 2425 bats that were submitted for rabies examination from 1962 through 1967 (Goldfield *et al.*, 1968; Goldfield and Sussman, 1970). They recovered virus from the salivary gland of a big brown bat (*Eptesius fuscus*). Reeves (1974) suggested that the development of immunological tolerance, similar to that demonstrated with the arenaviruses in their mammalian hosts, may be a factor in the development of chronic arboviral infections by vertebrate hosts. In experimental studies on immunological tolerance, Gresikova and Zavada (1966) infected pregnant rabbits, 2–7 days prior to delivery, and 1-day-old rabbits with WEE virus. Complete tolerance was not demonstrated, but both congenitally and postnatally infected rabbits developed a protracted viremia lasting up to 32 and 20 days, respectively. The congenitally infected rabbits also failed to develop antibodies to WEE virus.

c. Poikilotherms. Workers in New Jersey (Goldfield and Sussman, 1967) reported the isolation of WEE virus from the blood of a northern diamondback terrapin (*Malaclemys terrapin terrapin*) and an eastern box turtle (*Terrapine carolina carolina*) and from the brain tissue of two northern spring peepers (*Hyla crucifer*). Viremia in turtles was detected only during July, August, and September. In this same study, WEE virus antibody was found in turtles, frogs, and snakes. Workers in Massachusetts (Massachusetts Department of Public Health, unpublished data) failed to recover virus or detect antibody in 194 Massachusetts reptiles tested from 1961 to 1965. Other workers in Florida (Henderson *et al.*, 1962) and Maryland (Dalrymple *et al.*, 1972b) were also unsuccessful in attempts to isolate virus from cold-blooded vertebrates, but antibody to WEE virus was found in turtles, frogs, and snakes in the Maryland studies and in turtles captured in New York (Whitney *et al.*, 1968).

Snakes and frogs also have been implicated as possible overwintering hosts for WEE virus in several areas of western North America. Gebhardt *et al.* (1964) isolated virus from the blood of wild snakes captured in Utah in 1962 and 1963. Some of these animals were captured during the spring before the population of *Culex tarsalis*, the primary vector in the West, had started to build up. A number of the naturally infected snakes which these investigators collected gave birth while being held in the laboratory. WEE virus was isolated from 4 of 15 pools of blood tested from the offspring of these females. WEE antibody and virus also has been demonstrated in the blood of snakes and frogs collected in Saskatchewan, Canada (Spalatin *et al.*, 1964; Burton *et al.*, 1966b; Prior and Agnew, 1971), and Sudia *et al.* (1975b) recently reported the isolation of WEE virus from tortoises collected in Texas. Workers in other WEE endemic areas of western North America that have tested large numbers of snake and frog blood specimens have failed, however, to confirm the involvement of these animals in the ecology of this virus (reviewed in Reeves, 1974).

Snakes, frogs, and turtles have been found to be susceptible to infection with WEE virus in experimental studies. Similar to hibernating mammals, these cold-blooded vertebrates are capable of maintaining virus for prolonged periods at reduced temperatures and can circulate enough virus to infect mosquitoes feeding on them after they have been removed from hibernation (Rosenbusch, 1939; Thomas *et al.*, 1958; Thomas and Eklund, 1960, 1961; Gebhardt and Hill, 1960; Massachusetts Department of Public Health, unpublished data). In detailed studies on the kinetics of WEE virus replication in hibernating garter snakes (*Thamnophis* spp.), Gebhardt *et al.* (1973) found that persistence of virus was related to the length of time following initial infection before the snakes entered hibernation. Their results indicated that the snakes had to enter hibernation between 18 hours and 11 days after WEE infection in order for virus to persist through the hibernation period of 50–95 days at 5°C. Snakes placed at 5°C later than this 11-day limit lost their infection during the hibernation period. Snakes maintained at room temperature (23°C) had a demonstrable viremia for about 3 weeks only.

2. Chronic Infection of Invertebrate Hosts

a. Mosquitoes. Since WEE virus is vectored by mosquitoes, winter survival in these hosts is one of the more attractive hypotheses. In the western United States, the primary vector, *Culex tarsalis,* overwinters as an adult female. Isolations of WEE virus have been reported from females of this mosquito collected during the winter in California and Colorado (Blackmore and Winn, 1956; Reeves *et al.*, 1958b). This species also has been shown capable of carrying WEE virus through the winter under experimental conditions (Bellamy *et al.*, 1958, 1967). However, because of the scarcity of known natural infections in this species during the winter, together with the evidence that most female *C. tarsalis* mosquitoes that survive the winter rarely take a blood meal before entering hibernation, most investigators have concluded that this mechanism is probably not important in the overwintering of WEE virus in the West

(Bennington *et al.*, 1958; Blackmore and Dow, 1962; Bellamy and Reeves, 1963; Burdick and Kardos, 1963; Rush *et al.*, 1963a,b).

In the Northeast, WEE virus has been isolated during the summer from some of the *Culex* species that overwinter as adults (Goldfield, unpublished data; Massachusetts Department of Public Health, unpublished data); however, attempts to isolate virus from hibernating *Culex* adults in this region have been unsuccessful (Wallis *et al.*, 1958; Hayes *et al.*, 1962; Gusciora *et al.*, 1972). Like *C. tarsalis* in the West, evidence suggests that most species that overwinter as adults in the Northeast also fail to take a blood meal before going into diapause (Wallis, 1959b; Eldridge *et al.*, 1972; Gusciora *et al.*, 1972). Gusciora and co-workers (1972) presented evidence, however, that some *C. restuans* females surviving the winter may take a prediapausal blood meal. Gonotrophic dissociation has also been demonstrated to occur in some species of *Culex* (Eldridge *et al.*, 1972). The possibility that these species play an important role in the overwinter survival of this virus cannot be ruled out.

Culiseta melanura, the primary vector of WEE virus in the eastern United States, survives the winter in the larval stage (Burbutis and Lake, 1956; Joseph and Bickley, 1969). Adults are not found during this period, except in some areas of the Southeast where year-round breeding occurs (cf. Section V, D). Therefore, for this species to play an important role in the winter survival of WEE virus, the immature stage would have to become infected. This could occur via transovarial transmission of virus from infected females to their progeny. In an early study on the replication of WEE virus in mosquitoes, Trager (1938) demonstrated the ability of ovaries from *Aedes aegypti* to support replication of WEE virus *in vitro*. In a later study, Thomas (1963) found that WEE virus persisted in the ovaries of experimentally infected *Culex tarsalis* for as long as 25 days. WEE virus also has been isolated from eggs laid by experimentally infected *Aedes triseriatus*, *C. tarsalis* and *C. melanura* (Kissling *et al.*, 1957; Thomas, 1963; C. Hayes, unpublished). However, this virus has not been recovered from progeny reared from experimentally infected females (Hammon and Reeves, 1945; Barnett, 1956).

Isolations of WEE virus have not been reported from field-collected mosquito larvae nor from adults reared from field-collected larvae in either the eastern or western United States. *C. melanura* larvae have been sampled for virus in several areas of the East with negative results (Dougherty and Price, 1960; Hayes *et al.*, 1962; Sudia *et al.*, 1968; Muul *et al.*, 1975). However, an isolation of another *Alphavirus*, eastern equine encephalomyelitis (EEE), was reported from larvae of this species collected during December in Massachusetts, and an isolation of WEE virus has been reported from a pool of 15 *C. melanura* adult males collected in

Alabama during October (Stamm *et al.*, 1962; Massachusetts Department of Public Health, unpublished data). In the latter case, the authors did not rule out the possibility of contamination, however, since infected females were taken in the same collection with the males.

Observations made by Reeves and co-workers (1958b) on the properties of WEE virus strains isolated from *C. tarsalis* collected in California during the winter may be relevant to the scarcity of winter isolations. Of 19 virus isolations from adult mosquitoes made during the months of January–April in California, many strains exhibited very low pathogenicity for adult mice. These strains could only be propagated in a chick embryo system. Several such isolates required three to six sucessive passages in embryos before producing consistent mortality in adult mice. Many of these strains also failed to produce a significant antibody response when inoculated into chickens. On the other hand, typical summer isolates from mosquitoes usually had a high pathogenicity for mice as well as chick embryos and produced a marked immunogenic response in chickens. One experiment designed to study the overwintering of WEE virus in laboratory-infected *C. tarsalis* also resulted in demonstrating a similar pattern of low mouse pathogenicity and reduced immunogenicity of virus; however, subsequent experiments under similar conditions failed to produce significant attenuation of the virus (Bellamy *et al.*, 1958). Since *C. melanura* larvae in the Northeast experience a long period of diapause at low winter temperatures, WEE virus in these larvae might undergo changes similar to those observed by Reeves *et al.* (1958b), just discussed. It is of interest to note that the isolate of EEE virus from overwintering *C. melanura* larvae mentioned earlier was initially classified as an unknown agent because of atypical behavior in tissue culture (Massachusetts Department of Public Health, unpublished data). It was only recognized as EEE virus after a change to a typical cytopathogenic effect (CPE) during the course of subculturing.

b. Other Arthropods. In the East, only three isolations of WEE virus have been reported from arthropods other than mosquitoes (Section V, B). However, several strains of WEE virus have been recovered from such sources in areas of the central and western United States. In fact, the first reported isolation of WEE virus from a naturally infected arthropod was from a cone-nose bug (*Triatoma sanguisuga*) collected in Kansas (Kitselman and Grundmann, 1940). A number of isolations have been made from mites in the West (Reeves *et al.*, 1947, 1955; Sulkin, 1945; Sulkin and Izumi, 1947; Hammon *et al.*, 1948; Miles *et al.*, 1951; Cockburn *et al.*, 1957), but results from experimental studies have shown these arachnids to be poor hosts for WEE virus (Chamberlain and Sikes, 1955; Reeves *et al.*, 1955; Sulkin *et al.*, 1955; Winn and Bennington, 1959).

Experimental infection of a few other arthropods also has been reported, but at present none of these have been incriminated as important hosts for the overwintering of WEE virus (Herms *et al.*, 1934; Syverton and Berry, 1936b, 1941; Gwatkin, 1939; Grundmann *et al.*, 1943; Hammon *et al.*, 1942b; Hurlburt and Thomas, 1960; Mangiafico *et al.*, 1968).

VII. Epizootic Expression

A. Vector Biology

One of the more puzzling aspects of the epidemiology of WEE is the variation in disease associated with geographical distribution of the virus. Illness in humans and equines appears to be virtually absent in the East but is frequently recognized in the central and western parts of the country. No confirmed human case has been reported from the eastern United States. Isolation of WEE virus has been reported from an ill horse in Hillsborough County, Florida (Jennings *et al.*, 1966). Workers in New Jersey have reported the isolation of virus from domestic pets (Goldfield *et al.*, 1968). Three isolations were obtained from 1653 dogs and two isolations from 1179 cats submitted for rabies examination over a 6-year period. The only animals in which outbreaks of disease attributable to infection with WEE virus have been reported in the eastern United States are pen-raised chukars (*Alectorsis graeca*) and pheasants (Ranck *et al.*, 1965; Lord and Calisher, 1970).

This variation in disease pattern is probably a reflection of the operation of certain causative factors produced by differences in the biology of WEE virus in each region. An obvious factor that differs between the transmission chains of WEE virus in the eastern and western United States is the species of mosquito incriminated as the primary vector. *Culex tarsalis* serves as the vector in the central and western part of the country (Reeves and Hammon, 1962). *Culiseta melanura* has been shown to be the primary enzootic vector in the East (Section V, B). The distribution of these two species has been reported to overlap in 22 states (Fig. 3): *C. melanura* has been collected as far west as Colorado; *C. tarsalis* occurs in the southern United States and has been reported as far north as Pennsylvania in the East (Carpenter and La Casse, 1955; Carpenter, 1968, 1970; Sollers-Riedel, 1972). The bionomics of these species are distinct, however, and as a result their abundance differs in areas where overlap occurs.

Culex tarsalis breeds in sunlit grassy marshes, open ground pools containing emergent vegetation, and pools in stream beds. This mosquito is particularly abundant in irrigated areas where seepage and improper

FIG. 3. Distribution of *Culex tarsalis* and *Culiseta melanura* in the United States: O, *C. tarsalis*; □, *C. melanura*; △, *C. tarsalis* and *C. melanura*. Black circles, squares, or triangles represent areas where clinical cases of WEE have been reported in either humans or equines.

flooding of pasture land result in extensive aquatic habitats favorable to the production of mosquitoes. It is in such irrigated areas of the Great Plains and western North America that WEE virus has commonly been recognized as a disease threat to humans and equines (Hess *et al.*, 1970). *C. tarsalis* has not been reported to be abundant in the eastern United States (Snow and Packard, 1956), and WEE virus has not been isolated from this species in this region.

By contrast, *C. melanura* has a more limited breeding habitat, being restricted almost entirely to wooded freshwater swamps (See Section V, C). Areas where this mosquito occurs in abundance have been reported only from the Atlantic or eastern Gulf seaboard states. The feeding habits of these two species also vary. *C. melanura* feeds almost exclusively on birds (Section V, B). Although predominantly a bird feeder, *C. tarsalis* appears to be more opportunistic in its choice of host. Precipitin testing has revealed that this species will readily feed on mammals, including horses and occasionally humans (Tempelis *et al.*, 1965; Tempelis and Washino, 1967). *C. melanura* has been observed feeding on humans in nature on only three separate occasions (Stabler, 1948; Hayes and Doane, 1958; Schober, 1964). Precipitin testing has revealed a few additional cases of this mosquito feeding on humans (Jobbins *et al.*, 1961; Moussa *et al.*, 1966). This species also has been observed to feed on humans under laboratory conditions but only at a very low rate (Wallis, 1959a; D. E. Hayes, 1968; C. G. Hayes, unpublished data). Reports of *C. melanura* feeding on equines are even rarer. Chamberlain and co-workers (1969) reported the collection of three specimens feeding on a mule in a pasture in North Carolina. Two separate studies in Maryland using precipitin testing have each resulted in the collection of a single specimen positive for horse blood (Moussa *et al.*, 1966; Joseph and Bickley, 1969).

These behavioral differences suggest that the absence of disease attributable to WEE virus in humans and equines in the eastern United States may be related to a lack of exposure of these hosts to infected *C. melanura*. Results of serological surveys also support this conclusion. Serological evidence of human infection has not been detected in this region (Schaeffer *et al.*, 1954; Stamm *et al.*, 1962; Work, 1964; Sudia *et al.*, 1968; Whitney *et al.*, 1968; Yale Arbovirus Research Unit, unpublished data). Surveys of horses indicate occasional infections may occur; however, the interpretation of serological data is complicated by incomplete knowledge of vaccination and history of previous residence (Goldfield and Sussman, 1967; Morris *et al.*, 1973).

Before the explanation of lack of exposure to infected *C. melanura* can be accepted as a reason for the absence of WEE in humans and equines, the occurrence of cases of EEE in these hosts in the East has to be reconciled with the aforementioned data. *C. melanura* also has been found to be the primary vector for this virus. As with WEE, wild birds serve as the reservoir vertebrate hosts for EEE virus, and enzootic activity appears to be limited primarily to the wooded freshwater swamp habitat (Wallis, 1965; Wallis and Main, 1974). Between 1955 and 1971, 116 cases of EEE in humans were reported (McGowan *et al.*, 1973). Equine deaths occur annually from this disease, and several large epizootics involving over 100 horses have occurred in the East (Schaeffer *et al.*, 1954). The variation in pathobiological expression of these agents suggests the possibilities that either *C. melanura* is a less efficient vector of WEE virus than of EEE or that other mosquito species are serving as epidemic vectors for EEE virus but not for WEE virus.

In areas of the eastern United States where long-term ecological studies have been conducted, the recovery rate of virus from *C. melanura* does not indicate that this vector becomes infected with EEE virus more frequently than with WEE virus. Workers in Alabama isolated 19 strains of WEE virus and 23 strains of EEE virus from *C. melanura* over a 6-year period (Stamm *et al.*, 1962; Sudia *et al.*, 1968). In studies in Maryland from 1966 to 1969, 24 isolations of EEE virus and 49 isolations of WEE virus were reported from this mosquito (Muul *et al.*, 1975; Saugstad *et al.*, 1972; Williams *et al.*, 1974). In Massachusetts during the summers of 1973, 1974, and 1975, outbreaks of arthropod-borne encephalitis involving equine or human cases of EEE were reported. Cases of WEE were not observed; however, 410 isolations of WEE virus were made from *C. melanura* during these periods, whereas only 174 isolations of EEE virus were reported from this vector (Massachusetts Department of Public Health, unpublished data). In addition, experimental studies on the vector capabilities of *C. melanura* have not shown this mosquito to be more susceptible to infection or better able to transmit EEE virus than WEE virus. Howard and Wallis (1974) reported the ID_{50} of EEE virus for *C. melanura* by feeding on viremia chicks to be ca. $10^{3.8}$ suckling mouse intracerebral (ic) LD_{50}. C. G. Hayes (unpublished data) has demonstrated the ID_{50} of WEE virus for *C. melanura* by feeding on viremic chicks to be ca. $10^{1.5}$ suckling mouse (ic) LD_{50}. Both investigators found that approximately 90% of the infected mosquitoes were capable of transmitting to susceptible chicks after an appropriate incubation period.

Several *Aedes* species have been suggested as epidemic vectors of EEE virus. Because of their abundance, feeding habitat, and distribution, *Ae. sollicitans* and *Ae. vexans* have a high index of suspicion in the East (Feemster and Getting, 1941; Wallis, 1960; Goldfield and Sussman, 1970). L. D. Beadle (unpublished data) listed 16 isolations of EEE virus from *Ae. sollicitans* and three from *Ae. vexans* in this region. *Ae. vexans* also has been found naturally infected with WEE virus in this area (M. Goldfield, unpublished data), but no isolations have been reported from *Ae. sollicitans.* However, isolations of WEE virus have been reported from *Ae. sollicitans* collected in Texas (Sudia *et al.*, 1975c). Both species have been demonstrated capable of transmitting the viruses of WEE and EEE experimentally (Kelser, 1938; Davis, 1940; Chamberlain *et al.*, 1954). In a laboratory study by Chamberlain and co-workers (1954), *Ae. sollicitans* was found to be equally susceptible to infection with both viruses. In subsequent transmission attempts, 75% of the EEE-infected mosquitoes could transmit but only 3% of the WEE-infected females transmitted. These investigators were working with too few specimens to draw firm conclusions from their data. *Ae. sollicitans* also has been shown capable of vectoring EEE virus from an infected horse to a susceptible horse on two separate occasions (Sudia *et al.*, 1956; Ten Broeck and Merrill, 1935). Another species, *Ae. canadensis,* has been implicated as a vector of EEE virus to horses in New York State (J. Woodall, unpublished data). WEE virus also has been isolated from this mosquito

(R. O. Hayes *et al.*, 1961). No experimental studies have been reported with *Ae. canadensis*.

Other studies have shown that viremia titers reached in wild birds infected with a strain of WEE virus isolated in Louisiana were lower than those reported in an earlier study for birds infected with EEE virus (Kissling *et al.*, 1954a, 1957). Stamm and Kissling (1957) also demonstrated that EEE-immune house sparrows infected with WEE virus developed a lower titered viremia of shorter duration than nonimmune birds. However, the titer and length of viremia did not differ significantly between WEE-immune and -nonimmune birds infected with EEE virus. These factors might favor a higher infection rate of potential epidemic vectors with EEE virus compared to WEE virus if such vectors also had a relatively high threshold of infection.

An apparent difference in seasonal expression of WEE and EEE viruses in the enzootic focus may also favor the escape of EEE virus from the swamp habitat. EEE virus activity appears to peak later in the transmission season than WEE virus activity—particularly in the Northeast (Williams *et al.*, 1971, 1972). Outbreaks of EEE in humans and equines also occur most frequently in late summer and early fall (Feemster *et al.*, 1958; Byrne *et al.*, 1959; Goldfield and Sussman, 1968; Bryant *et al.*, 1973). Several studies in the Pocomoke Swamp in which the seasonal distribution of *C. melanura* has been recorded indicate that this species may occasionally migrate to areas on the periphery of the swamp habitat in late summer and fall (Saugstad *et al.*, 1972; Muul *et al.*, 1975). A study of Le Duc *et al.* (1972) in the same area also suggests that *C. melanura* may undergo a shift in feeding behavior toward mammals late in the season when EEE virus activity is at a peak and WEE virus activity is declining. All of the engorged *C. melanura* found positive for mammalian blood by these investigators were collected during the late summer and fall. Joseph and Bickley (1969) also reported that the percentage of bird-fed specimens collected in the Pocomoke Swamp reached a seasonal low during September. In addition, both of the observations of *C. melanura* feeding on humans in nature in which dates were reported occurred during September (Hayes and Doane, 1968; Schober, 1964). Although these observations are only suggestive, other mosquito species that serve as vectors of arboviruses have been shown to have a seasonal biphasic pattern of feeding on birds and mammals (Hess and Hayes, 1967; Edman and Taylor, 1968). Edman and Taylor (1968) suggest that for a species to function both as an enzootic vector for avian hosts and as an epidemic vector such a shift in host feeding behavior may be a basic requirement.

Even if both the viruses of WEE and EEE were escaping from the enzootic focus at a low rate the occurrence of EEE in humans and equines would probably be more readily detected than would WEE since the former produces a more severe illness in these hosts. For example, in summarizing data on the occurrence of arboviral encephalitis in the United States over a 17-year period, McGowan *et al.* (1973) reported a case fatality rate in humans of 2.9% for WEE and 50.0% for EEE. Experimental studies suggest that a similar mortality pattern also occurs in horses (Kissling *et al.*, 1954b; Schaeffer and Arnold, 1954; Reeves and Hammon, 1962).

B. Variation in Virus Strains

Sponseller *et al.* (1966) conducted experimental studies on the infection of ponies with several strains of WEE virus. These investigators used four low-passage strains of virus originally isolated from various areas of North America. Strains isolated from a nestling sparrow and from *C. tarsalis* collected in California and a strain recovered from a sparrow collected in Saskatchewan were used. One strain from the eastern United States which had been isolated from *C. melanura* collected in Florida also was used. Six ponies inoculated intracerebrally (ic) with the viruses from western North America developed elevated body temperatures. Five of these ponies also showed other clinical signs of encephalitis. The single pony inoculated ic with the Florida strain did not develop fever or encephalitis. This pony did develop a low-titered viremia, but viremia was not detected in another pony inoculated subcutaneously (sc) with this virus. All of the ponies inoculated ic or sc with the viruses from western North America developed a higher-titered viremia than did the pony inoculated ic with the virus isolated in Florida. These experimental results combined with the epidemiological evidence that WEE seldom occurs in humans and equines in the East suggest that virus strains from this region may be less virulent for these hosts than are strains from the western and central United States.

Other workers have shown that strains of WEE virus from the eastern and western United States can be distinguished antigenically. Karabatsos and co-workers (1963), studying seven strains of WEE virus, were able to show that four strains isolated in the western United States and Canada were indistinguishable from each other in complement-fixation (CF), hemagglutination-inhibition (HI), and neutralization tests but differed significantly from three eastern United States strains. The three strains from the East also showed slight variation among themselves. In an elaboration of these studies, Henderson (1964) compared 100 strains of WEE virus. Henderson tested these strains in plaque-neutralization tests using antisera prepared against antigenically distinct plaque derivatives selected from a standard strain of WEE virus (MacMillan strains). The strains were classified into three separate antigenic groups labeled aphasic, phase I, and phase II. All 33 strains from the Atlantic and Gulf Coast region formed a single group (phase II). The strains from the central and western United States fell into two categories, (aphasic and phase I) with overlapping geographical distribution. The only exception to this general pattern was a single strain from Colorado that was placed in the phase II group.

Henderson *et al.* (1965) also reported that strains of WEE virus from the East underwent changes in antigenic status from phase II to aphasic after serial ic passage in infant mice. In addition, when mixtures containing varying proportions of purified aphasic and phase II virus types were inoculated into infant mice, the aphasic virus was selected over the phase II at a rate of 1:1000 after a single passage. In similar experiments in which *Ae. aegypti* were used, no preferential selection of one virus type over the other was apparent after a single oral passage. Such observation led these workers to suggest that each strain of WEE virus was composed of a number of subpopulations of virions, each with distinct properties, and that qualitative and quantitative differences in the several subpopulations comprised in each strain are responsible for antigenic differences. They further suggested that the preferential selection of subpopulations by different hosts encountered by the virus in various geographic areas was responsible for the antigenic variation found among the virus strains (Henderson, 1964; Saturno and Henderson, 1965; Henderson *et al.*, 1965).

Antigenic variation among strains of virus isolated in different geographical regions has important epidemiological implications for virus isolation and serological survey work conducted in a locale. For example, serological studies conducted by Henderson *et al.* (1962) on wild birds captured in Florida demonstrated that a greater percentage of birds were positive for WEE HI antibody when tested using a local strain of virus as an antigen than when a strain of virus from California was used as the antigen. Similar results also have been reported for wild birds tested by workers in Connecticut (Table III) and New York (A. J. Main, unpublished data; J. Woodall, unpublished data). The experimental studies of Sponseller *et al.* (1966) on the infection of ponies with WEE virus revealed this same phenomenon. All of the surviving ponies except the two inoculated with the Florida isolate of virus had detectable WEE neutralizing antibodies when tested using an antigen prepared from a strain of virus isolated from the brain of a horse in Iowa. The sera of

TABLE III

HI Titers of Sera from Wild Birds Netted in Connecticut in 1974 Tested with a Local and Western North American Strain of Virus

Specimen No.[a]	Virus strain: Western North American (Macmillan)	Virus strain: Connecticut (B–8–74 or Spec. No. 72666)	Titer difference
B–6–74	1/10	1/40	4×
7	1/10	1/160	16×
9	<1/10	1/80	>8×
10	<1/10	1/20	>2×
11	1/10	1/20	2×
12	1/10	1/20	2×
13	1/10	1/80	8×
16	<1/10	1/20	>2×
21	<1/10	1/20	>2×
22	<1/10	1/80	>8×
23	1/10	1/80	8×
25	1/40	>1/320	>8×
26	1/10	1/80	8×
29	<1/10	1/160	>16×
32	<1/10	1/10	>1×
33	1/10	>1/320	>32×
36	<1/10	1/10	>1×
40	<1/10	1/40	>4×
44	1/10	1/40	4×
52	1/20	>1/80	>4×
53	1/10	>1/80	>4×
55	<1/10	>1/80	>8×
76	<1/10	1/10	>1×
78	<1/10	1/10	>1×
80	<1/10	1/20	>2×
90	<1/10	1/40	>4×
91	<1/10	>1/80	>8×
92	1/20	1/80	4×
99	1/20	>1/80	>4×
104	1/10	>1/80	>8×
106	<1/10	1/40	>4×
107	1/40	>1/80	>2×
108	<1/10	1/20	>2×
147	1/20	>1/80	>4×
154	1/20	>1/80	>4×
183	1/10	>1/40	>4×
201	1/10	>1/80	>8×
218	1/20	>1/80	>8×
224	1/20	>1/80	>4×
225	<1/10	>1/80	>8×
294	<1/10	>1/80	>8×
295	1/10	1/80	8×

(*continued*)

TABLE III—*continued*

Specimen No.[a]	Virus strain: Western North American (McMillan)	Virus strain: Connecticut (B-8-74 or Spec. No. 72666)	Titer difference
Total positive	23	42	
Total negative	19	0	
Total	42	42	
GMT (including negative)	8.6	72.9	8.4×
GMT (excluding negative)	15.7	72.9	4.6×

[a] Only positive sera are included (A. J. Main, unpublished data).

the ponies inoculated with the virus isolated in Florida gave a positive reaction only when tested with an antigen made from the homologous virus strain. Difficulties also have been encountered in identifying strains of WEE virus isolated in the eastern United States when tested by CF when antisera prepared against virus isolates from western North America were used (Yale Arbovirus Research Unit, unpublished data; J. Woodall, unpublished data).

VIII. Concluding Remarks

Studies in the eastern United States have shown that WEE virus activity occurs in this area during the warmer parts of the year. For the most part, transmission appears to be limited to the wooded freshwater swamp habitat. Wild birds, particularly in the order Passeriformes, serve as the main vertebrate reservoir, and *C. melanura* has been demonstrated to be the primary vector of virus to these hosts. Other vertebrates such as small mammals, reptiles, and amphibians occasionally become infected with WEE virus in the East; however, their importance for the maintenance of virus is unknown.

Although most investigators feel that WEE virus survives throughout the year in enzootic foci in the East, the mechanism of overwintering remains unknown. Of interest is the recent recovery of strains of WEE virus which are avirulent for baby mice from bird-nest–occupying bedbugs *(Oeciacus vicarius)* collected during the winter in eastern Colorado. These bugs appear to be serving as an overwintering source of virus in that area and transmitting virus to nestling birds in the spring to initiate the wild bird–mosquito cycle (Vector-Borne Diseases Division, Center for Disease Control, unpublished data).

WEE virus does not appear to pose a public or veterinary health problem in the East at the present time. This is partially attributable to the habitat restriction and host-feeding preference of *C. melanura,* the primary vector in this region. Some evidence also suggests that eastern strains of WEE virus may be less virulent for humans and equines than are some strains that occur in the western United States.

In view of the large overlap in the geographical range of *Culex tarsalis* and *Culiseta melanura,* the introduction of a more virulent strain of virus into the eastern United States appears possible. *C. melanura* has been shown to be capable of transmitting a strain of WEE virus originally isolated in California (C. G. Hayes, unpublished data). An analogy can be drawn with another *Alphavirus,* Venezuelan equine encephalitis (VEE). Prior to 1969, VEE apparently existed endemically for at least 50 years in parts of Central America without a major outbreak of human disease having been recorded. However, starting in June of 1969, an epidemic of VEE started on the Pacific Coast of Central America and eventually spread into Texas during the summer of 1971. This outbreak was probably caused by an "epidemic" strain of virus that was introduced into Central America from Ecuador during 1969 (Franck, 1972).

References

Anderson, K. S. (1961). *Mass. Audubon* Winter.

Anderson, K. S., and Maxfield, H. K. (1962). *Wilson Bull.* **31,** 381–385.

Anderson, K. S., Randall, E. J., Main, A. J., and Tonn, R. J. (1966). *Mass. Audubon* Summer,

Anderson, K. S., Randall, E. J., Main, A. J., and Tonn, R. J. (1967). *Bird-Banding* **38,** 135–138.

Baltimore, D. (1971). *Bacteriol. Rev.* **35,** 235–241.

Bang, F. B., and Reeves, W. C. (1942). *J. Infect. Dis.* **70,** 272–274.

Barnett, H. C. (1956). *Am. J. Trop. Med. Hyg.* **5,** 86–98.

Bast, T. F., Whitney, E., and Benach, J. L. (1973). *Am. J. Trop. Med. Hyg.* **22,** 109–115.

Bates, M., and Roca-Garcia, M. (1946). *Am. J. Trop. Med.* **26,** 585–605.

Beaven, G. F., and Oosting, H. J. (1939). *Bull. Torrey Bot. Club* **66,** 367–389.

Bellamy, R. E., and Reeves, W. C. (1963). *Ann. Entomol. Soc. Am.* **56,** 314–323.

Bellamy, R. E., Reeves, W. C., and Scrivani, R. P. (1958). *Am. J. Hyg.* **67,** 90–100.

Bellamy, R. E., Reeves, W. C., and Scrivani, R. P. (1967). *Am. J. Epidemiol.* **85,** 282–296.

Bennington, E. E., Scooter, C. A., and Baer, H. (1958). *Mosq. News* **18,** 299–304.

Bickley, W. E., and Byrne, R. J. (1960). *Mosq. News* **20,** 93–94.

Blackmore, J. S., and Dow, R. P. (1962). *Mosq. News* **22,** 291–294.

Blackmore, J. S., and Winn, J. F. (1956). *Proc. Soc. Exp. Biol. Med.* **91,** 146–148.

Bowers, J. H., Hayes, R. O., and Hughes, T. B. (1969). *J. Med. Entomol.* **6,** 175–178.
Brown, J. H., and Marken, J. A. (1966). *Can. J. Public Health* **57,** 366–367.
Bruno-Lobo, G., Bruno-Lobo, M., Travassos, J., Pinheiro, F., and Pazin, I. P. (1961). *An. Microbiol.* **9A,** 183–195.
Bryant, E. S., Anderson, C. R., and van der Heide, L. (1973). *Avain Dis.* **17,** 861–867.
Burbutis, P. P., and Lake, R. W. (1956). *Proc. N.J. Mosq. Exterm. Assoc.* **43,** 155–161.
Burdick, D. J., and Kardos, E. H. (1963). *Ann. Entomol. Soc. Am.* **56,** 527–535.
Burton, A. N., and McLintock, J. (1970). *Can. Vet. J.* **11,** 232–235.
Burton, A. N., McLintock, J. R., Spalatin, J., and Rempel, J. G. (1966a). *Can. J. Microbiol.* **12,** 133–141.
Burton, A. N., McLintock, J. R., and Rempel, J. G. (1966b). *Science* **154,** 1029–1031.
Buss, W. C., and Howitt, B. F. (1941). *Am. J. Public Health* **31,** 935–944.
Byrne, R. J., Yancey, F. S., Bickley, W. E., and Finney, G. (1959). *J. Am. Vet. Med. Assoc.* **135,** 211–215.
Carpenter, S. J. (1968). *Calif. Vector Views* **15,** 72–97.
Carpenter, S. J. (1970). *Calif. Vector Views* **17,** 40–65.
Carpenter, S. J., and La Casse, W. J. (1955). "Mosquitoes of North America." Univ. of California Press, Berkeley.
Casals, J. (1957). *Trans. N.Y. Acad. Sci.* **19,** 219–235.
Casals, J. (1971). *In* "Comparative Virology" (K. Maramorosch and H. Kurstak, eds.), pp. 307–333. Academic Press, New York.
Casals, J., and Brown, L. V. (1954). *J. Exp. Med.* **99,** 429–449.
Casals, J., and Clarke, D. H. (1965). *In* "Viral and Rickettsial Infections of Man" (F. L. Horsefall and I. Tamm, eds.), pp. 583–605.
Chamberlain, R. W., and Sikes, R. K. (1955). *Am. J. Trop. Med. Hyg.* **4,** 106–118.
Chamberlain, R. W., and Sudia, W. D. (1955). *Am. J. Hyg.* **62,** 295–305.
Chamberlain, R. W., and Sudia, W. D. (1957). *Am. J. Hyg.* **66,** 151–159.
Chamberlain, R. W., Sikes, R. K., Nelson, D. B., and Sudia, W. D. (1954). *Am. J. Hyg.* **60,** 278–285.
Chamberlain, R. W., Sudia, W. D., Burbutis, P. P., and Bogue, M. D. (1958). *Mosq. News* **18,** 305–308.
Chamberlain, R. W., Sudia, W. D., Coleman, P. H., Johnson, J. G., and Work, T. H. (1969). *Am. J. Epidemiol.* **89,** 82–88.
Cockburn, T. A., Sooter, C. A., and Langmuir, A. D. (1957). *Am. J. Hyg.* **65,** 130–146.
Cox, H. R., Jellison, W. L., and Hughes, L. E. (1941). *Public Health Rep.* **56,** 1905–1906.
Crans, W. J. (1962). *Proc. N.J. Mosq. Exterm. Assoc.* **49,** 120–126.
Crans, W. J. (1964). *Proc. N.J. Mosq. Exterm. Assoc.* **51,** 50–58.
Crans, W. J. (1966). *Proc. Northeast. Mosq. Control Assoc.* **12,** 4–8.
Dalrymple, J. M., Teramoto, A. Y., Cardiff, R. D., and Russell, P. K. (1972a). *J. Immunol.* **109,** 426–433.
Dalrymple, J. M., Young, O. P., Eldridge, B. F., and Russell, P. K. (1972b). *Am J. Epidemiol.* **96,** 129–140.
Dalrymple, J. M., Vogel, S. N., Teramoto, A. Y., and Russell, P. K. (1973). *J. Virol.* **12,** 1034–1042.
Davis, N. C. (1932). *Am. J. Hyg.* **16,** 163–176.
Davis, W. A. (1940). *Am. J. Hyg.* **32,** 45–49.
Dillenberg, H. (1965). *Can. J. Public Health* **56,** 17–20.
Dorsey, C. K. (1944). *Ann. Entomol. Soc. Am.* **37,** 376–387.
Dougherty, E., and Price, J. (1960). *Avian Dis.* **4,** 247–258.

Edman, J. D., and Kale, H. W. (1971). *Ann. Entomol. Soc. Am.* **64,** 513–516.
Edman, J. D., and Taylor, D. J. (1968). *Science* **161,** 67–68.
Edman, J. D., Webber, L. A., and Kale, H. W. (1972). *J. Med. Entomol.* **9,** 429–434.
Eklund, C. M., and Blumstein, A. (1938). *J. Am. Med. Assoc.* **111,** 1734–1735.
Eldridge, B. F., Bailey, C. L., and Johnson, M. D. (1972). *J. Med. Entomol.* **9,** 233–238.
Emmons, R. W., (1966). *Am. J. Trop. Med. Hyg.* **15,** 428–433.
Emmons, R. W., and Lennette, E. H. (1969). *Science* **163,** 945–946.
Feemster, R. F., and Getting, U. A. (1941). *Am. J. Public Health* **31,** 791–802.
Feemster, R. F., Wheeler, R. E., Daniels, J. B., Rose, H. D., Schaeffer, M., Kissling, R. E., Hayes, R. O., Alexander, R. O., and Murray, W. A. (1958). *New Engl. J. Med.* **259,** 107–113.
Fenner, F. (1974). *Aust. J. Exp. Biol. Med. Sci.* **52,** Part 2, 223–240.
Fenner, F., Pereira, H. G., Porterfield, J. S., Joklik, W. K., and Downie, A. W. (1974). *Intervirology* **3,** 193–198.
Ferguson, F. F. (1954). *Public Health Monogr. No. 23.*
Foreign Animal Diseases Report (1975). "United States Encephalitis—1975," Sept.–Oct. U.S. Gov. Printing Office, Washington, D.C.
Franck, P. T. (1972). *Symp. Venez. Encephalitis Virus, Washington, D.C., 1971* pp. 322–328.
Gebhardt, L. P., and Hill, D. W. (1960). *Proc. Soc. Exp. Biol. Med.* **104,** 695–698.
Gebhardt, L. P., Stanton, G. J., Hill, D. W., and Collett, G. C. (1964). *New Engl. J. Med.* **271,** 172–177.
Gebhardt, L. P., St. Jeor, S. C., Stanton, G. J., and Stringfellow, D. A. (1973). *Proc. Soc. Exp. Biol. Med.* **142,** 731–733.
Goldfield, M., and Sussman, O. (1967). *Wildl. Dis. Assoc. Meet., Urbana, Ill.*
Goldfield, M., and Sussman, O. (1968). *Am. J. Hyg.* **87,** 1–10.
Goldfield, M., and Sussman, O. (1970). *Proc. N.J. Mosq. Exterm. Assoc.* **57,**
Goldfield, M., Sussman, O., Black, H. C., Horman, J. T., Taylor, B. F., Carter, W. C., Kerlin, R. E., and Birne, H. (1968). *J. Am. Vet. Med. Assoc.* **153,** 1780–1787.
Gresikova, M., and Zavada, J. (1966). *Acta Virol. (Engl. Ed.)* **10,** 75–77.
Grundmann, A. W., Kitselman, C. M., Roderick, L. M., and Smith, R. C. (1943). *J. Infect. Dis.* **72,** 163–171.
Gusciora, W., Adam, D., Bordash, G., Sussman, O., and Goldfield, M. (1972). *Proc. N.J. Mosq. Exterm. Assoc.* **59,** 147–164.
Gwatkin, R. (1939). *Can. J. Comp. Med.* **3,** 131–133.
Hammon, W. McD. (1941). *J. Am. Med. Assoc.* **117,** 161–167.
Hammon, W. McD., and Howitt, B. F. (1942). *Am. J. Hyg.* **35,** 163–185.
Hammon, W. McD., and Reeves, W. C. (1943). *J. Exp. Med.* **78,** 425–434.
Hammon, W. McD., and Reeves, W. (1945). *Am. J. Public Health* **35,** 994–1004.
Hammon, W. McD., and Reeves, W. C. (1948). *Am. J. Hyg.* **47,** 93–102.
Hammon, W. McD., Reeves, W. C., Brookman, B., and Izumi, E. M. (1941a). *Science* **94,** 328–330.
Hammon, W. McD., Gray, J. A., Evans, F. C., Izumi, E. M., and Lundy, H. W. (1941b). *Science* **94,** 305–307.
Hammon, W. McD., Reeves, W. C., Brookman, B., and Izumi, E. M. (1942a). *J. Infect. Dis.* **70,** 263–266.
Hammon, W. McD., Reeves, W. C., Brookman, B., and Gjullin, C. M. (1942b). *J. Infect. Dis.* **70,** 278–283.
Hammon, W. McD., Lundy, H. W., Gray, J. A., Evans, F. C., Bang, F., and Izjumi, E. M. (1942c). *J. Immunol.* **44,** 75–86.

Hammon, W. McD., Reeves, W. C., Benner, S. R., and Brookman, B. (1945). *J. Am. Med. Assoc.* **128,** 1133–1139.
Hammon, W. McD., Reeves, W. C., Cunha, R., Espana, C., and Sather, G. (1948). *Science* **107,** 92–93.
Hammon, W. McD., Reeves, W. C., and Sather, G. E. (1951). *J. Immunol.* **67,** 357–367.
Hardy, J. L. (1970). *Proc. Pap. Annu. Conf. Calif. Mosq. Control Assoc.* **38,** 31–34.
Hardy, J. L., Reeves, W. C., Scrivani, R. P., and Roberts, D. R. (1974a). *Am. J. Trop. Med. Hyg.* **23,** 1165–1177.
Hardy, J. L., Reeves, W. C., Rush, W. A., and Nir, Y. D. (1974b). *Infect. Immun.* **10,** 553–564.
Hayes, D. E. (1968). *Mosq. News* **28,** 109–110.
Hayes, R. O. (1961a). *Mosq. News* **21,** 179–187.
Hayes, R. O. (1961b). *Proc. N.J. Mosq. Exterm. Assoc.* **48,** 59–62.
Hayes, R. O., and Doane, O. W. (1958). *Mosq. News* **18,** 216–217.
Hayes, R. O., Daniels, J. B., and Mac Cready, R. A. (1961). *Proc. Soc. Exp. Biol. Med.* **108,** 805–808.
Hayes, R. O., Beadle, L. D., Hess, A. D., Sussman, O., and Bonese, M. J. (1962). *Am. J. Trop. Med. Hyg.* **11,** 115–121
Hayes, R. O., La Motte, L. C., and Holden, P. (1967). *Am. J. Trop. Med. Hyg.* **16,** 675–687.
Hayles, L. B., McLintock, J., and Saunders, J. R. (1972). *Can. J. Comp. Med.* **36,** 83–88.
Henderson, J. R. (1964). *J. Immunol.* **93,** 452–461.
Henderson, J. R., Karabatsos, N., Bourke, A. T. C., Wallis, R. C., and Taylor, R. M. (1962). *Am. J. Trop. Med. Hyg.* **11,** 800–810.
Henderson, J. R., Shah, H. H., and Wallis, R. C. (1965). *Virology* **26,** 326–332.
Herms, W. B., Wheeler, C. M., and Herms, H. P. (1934). *J. Econ. Entomol.* **27,** 987–998.
Hess, A. D., and Hayes, R. O. (1967). *Am. J. Med Sci.* **253,** 333–348.
Hess, A. D., Harmston, F. C., and Hayes, R. O. (1970). *Crit. Rev. Environ. Control* **1,** 443–465.
Hoff, G. L., Yuill, T. M., Iversen, J. O., and Hanson, R. P. (1970). *J. Wildl. Dis.* **6,** 472–478.
Holden, P. (1955). *Proc. Soc. Exp. Biol. Med.* **88,** 490–492.
Holden, P., Francey, D. B., Mitchell, C. J., Hayes, R. O., Lazuick, J. S., and Hughes, T. B. (1973a). *Am. J. Trop. Med. Hyg.* **22,** 254–262.
Holden, P., Hayes, R. O., Mitchell, C. J., Francy, D. B., Lazuick, J. S., and Hughes, T. B. (1973b). *Am. J. Trop. Med. Hyg.* **22,** 244–253.
Horsefall, W. R. (1955). "Mosquitoes, Their Bionomics and Relation to Disease." Ronald Press, New York.
Horzinek, M. C., (1973). *Prog. Med. Virol.* **16,** 109–156.
Howard, J. J., and Wallis, R. C. (1974). *Am. J. Trop. Med. Hyg.* **23,** 522–525.
Howitt, B. (1938). *Science* **88,** 455–456.
Howitt, B. (1939). *Science* **89,** 541–542.
Howitt, B. F. (1940). *J. Infect. Dis.* **67,** 177–187.
Howitt, B. F., and van Herick, W. (1941). *J. Infect. Dis.* **71,** 179–191.
Hurlbut, H. S. (1973). *J. Med. Entomol.* **10,** 1–12.
Hurlbut, H. S., and Thomas, J. I. (1960). *Virology* **12,** 391–407.
"International Catalogue of Arboviruses" (1975). (T. O. Berge, ed.). U.S. Dep. Health, Educ. Welfare, Washington, D.C.

Iversen, J. O., Seawright, G. I., and Hanson, R. P. (1971). *Can. J. Public Health* **62,** 125–132.

Jamnback, H. (1961). *Mosq. News* **21,** 140–141.

Jennings, W. L., Allen, R. H., and Lewis, A. L. (1966). *Am. J. Trop. Med. Hyg.* **15,** 96–97.

Jobbins, D. M., Burbutis, P. P., and Crans, W. J. (1961). *Proc. N.J. Mosq. Exterm. Assoc.* **48,** 211–215.

Joseph, S. R., and Bickley, W. E. (1969). *Mr., Agric. Stn., Bull.* **A–161.**

Kandle, R. P. (1961). *Proc. N.J. Mosq. Exterm. Assoc.* **48,** 15–20.

Kandle, R. P. (1963). *Proc. N.J. Mosq. Exterm. Assoc.* **50,** 16–21.

Kandle, R. P. (1964). *Proc. N.J. Mosq. Exterm. Assoc.* **51,** 1–5.

Kandle, R. P. (1967). *Proc. N.J. Mosq. Exterm. Assoc.* **54,** 15–19.

Karabatsos, N. (1975). *Am. J. Trop. Med. Hyg.* **24,** 527–532.

Karabatsos, N., Bourke, A. T. C., and Henderson, J. R. (1963). *Am. J. Trop. Med. Hyg.* **12,** 408–412.

Kelser, R. A. (1933). *J. Am. Vet. Med. Assoc.* **82,** 767–771.

Kelser, R. A. (1938). *J. Med. Vet. Med. Assoc.* **92,** 195–203.

Kettylis, G. D., Verral, V. M., Wilton, L. D., Clapp, J. B., Clarke, D. A., and Rublee, J. D. (1972). *J. Can. Med. Assoc.* **106,** 1175–1179.

Kiorpes, A. L., and Yuill, T. M. (1975). *Infect. Immun.* **11,** 986–990.

Kissling, R. E., Chamberlain, R. W., Sikes, R. K., and Eidson, M. E. (1954a). *Am. J. Hyg.* **60,** 251–265.

Kissling, R. E., Chamberlain, R. W., Edison, M. E., Sikes, R. K., and Bucca, M. A. (1954b). *Am. J. Hyg.* **60,** 237–250.

Kissling, R. E., Chamberlain, R. W., Nelson, D. B., and Stamm, D. D. (1955). *Am. J. Hyg.* **62,** 233–254.

Kissling, R. E., Chamberlain, R. W., Sudia, W. D., and Stamm, D. D. (1957). *Am. J. Hyg.* **66,** 48–55.

Kitselman, C. H., and Grundmann, A. W. (1940). *Kans., Agr. Exp. Stn., Tech. Bull.* **50.**

La Motte, L. C. (1963). *Mosq. News* **23,** 330–335.

Le Duc, J. W., Suyemoto, W., Eldridge, B. F., and Saugstad, E. S. (1972). *Am. J. Epidemiol.* **96,** 123–128.

Lennette, E. H., Ota, M. I., Dobbs, M. E., and Browne, A. S. (1956). *Am. J. Hyg.* **64,** 276–280.

Leung, M.-K., Burton, A., Iversen, J., and McLintock, J. (1975). *Can. J. Microbiol.* **21,** 954–958.

Longshore, W. A., Jr. (1960). *Am. J. Hyg.* **71,** 363–367.

Longshore, W. A., Jr., Lennette, E. H., Peters, R. F., Loomis, E. C., and Meyers, E. G. (1960). *Am. J. Hyg.* **71,** 389–400.

Loomis, E. C., and Meyers, E. G. (1960). *Am. J. Hyg.* **71,** 378–388.

Lord, R. D., and Calisher, C. H. (1970). *Am. J. Epidemiol.* **92,** 73–78.

Love, G. J., and Goodwin, M. H. (1961). *Mosq. News* **21,** 195–215.

MacCreary, D., and Stearns, L. A. (1937). *Proc. N.J. Mosq. Exterm. Assoc.* **24,** 188–197.

McGowan, J. E., Bryan, J. A., and Gregg, M. B. (1973). *Am. J. Epidemiol.* **97,** 199–207.

McKay, J. F. W., Stockiw, W., and Brust, R. A. (1968). *Manit. Med. Rev.* **48,** 56–57.

McLean, D. M., Clarke, A. M., Coleman, J. C., Montalbetti, C. A., Skidmore, A. G., Walters, T. E., and Wise, R. (1974). *Can. J. Microbiol.* **20,** 255–262.

Main, A. J., Hayes, R. O., and Tonn, R. J. (1968). *Mosq. News* **28,** 619–626.
Mangiafico, J. A., Whitman, L., and Wallis, R. C. (1968). *J. Med. Entomol.* **5,** 469–473.
Means, R. G. (1968). *Ann. Entomol. Soc. Am.* **61,** 116–120.
Mettler, N. E. (1962). *Rev. Soc. Argent. Biol.* **48,** 55–62.
Meyer, K. F. (1932). *Ann. Intern. Med.* **6,** 645–654.
Meyer, K. F., Haring, C. M., and Howitt, B. (1931). *Science* **74,** 227–228.
Meyer, K. F., Wood, F., Haring, C. M., and Howitt, B. (1935). *Proc. Soc. Exp. Biol. Med.* **32,** 56–58.
Meyers, E. G., Loomis, E. C., Fujimoto, F. Y., Ota, M. I., and Lennette, E. H. (1960). *Am. J. Hyg.* **71,** 368–377.
Miles, V. I., Howitt, B. F., Gorrie, R., and Cockburn, T. A. (1951). *Proc. Soc. Exp. Biol. Med.* **77,** 395–396.
Mitchell, C. A., and Walker, R. V. I. (1941). *Can. J. Comp. Med.* **5,** 314–319.
Morris, C. D., Whitney, E., Bast, T. F., and Deibel, R. (1973). *Am. J. Trop. Med. Hyg.* **22,** 561–566.
Moussa, M. A., Gould, D. J., Nolan, M. P., and Hayes, D. E. (1966). *Mosq. News* **26,** 385–393.
Murphy, F. A., Whitfield, S. G., Sudia, W. D., and Chamberlain, R. W. (1975). *In* "Invertebrate Immunity" (K. Maramorosch and R. E. Shope, eds.), pp. 25–53. Academic Press, New York.
Murphy, F. J., Burbitis, P. P., and Bray, D. F. (1967). *Mosq. News* **27,** 366–374.
Mussgay, M., Enzmann, P. J., Horzinek, M. C., and Weiland, E. (1975). *Prog. Med. Virol.* **19,** 257–323.
Muul, I., Johnson, B. K., and Harrison, B. A. (1975). *J. Med. Entomol.* **11,** 739–748.
Olitsky, P. K., and Casals, J. (1952). *In* "Viral and Rickettsial Infections of Man" (M. Rivers, ed.), pp. 214–266. Lippincott, Philadelphia, Pennsylvania.
Pfefferkorn, E. R., and Shapiro, R. (1975). *In* "Comprehensive Virology" (H. Fraenkel-Conrat and R. R. Wagner, eds.), Vol. 2, pp. 171–230. Plenum, New York.
Porterfield, J. S. (1961). *Bull. W.H.O.* **24,** 735–741.
Prior, M. G., and Agnew, R. M. (1971). *Can. J. Comp. Med.* **35,** 40–43.
Ranck, F. M., Gainer, J. H., Hanley, J. E., and Nelson, S. L. (1965). *Avian Dis.* **9,** 8–20.
Reeves, W. C. (1974). *Prog. Med. Virol.* **17,** 193–220.
Reeves, W. C., and Hammon, W. McD. (1944). *Am. J. Trop. Med.* **24,** 131–134.
Reeves, W. C., and Hammon, W. McD. (1962). *Univ. Calif., Berkeley, Publ. Public Health* **4.**
Reeves, W. C., Hammon, W. McD., Furman, D. P., McClure, H. E., and Brookman, B. (1947). *Science* **105,** 411–412.
Reeves, W. C., Washburn, G. E., and Hammon, W. McD. (1948). *Am. J. Hyg.* **47,** 82–92.
Reeves, W. C., Sturgeon, J. M., French, E. M., and Brookman, B. (1954). *J. Infect. Dis.* **95,** 168–178.
Reeves, W. C., Hammon, W. McD., Doetchman, W. H., McClure, H. E., and Sather, G. (1955). *Am. J. Trop. Med. Hyg.* **4,** 90–105.
Reeves, W. C., Hutson, G. A., Bellamy, R. E., and Scrivani, R. P. (1958a). *Proc. Soc. Exp. Biol. Med.* **97,** 733–736.
Reeves, W. C., Bellamy, R. E., and Scrivani, R. P. (1958b). *Am. J. Hyg.* **67,** 78–89.
Reeves, W. C., Bellamy, R. E., Geib, A. F., and Scrivani, R. P. (1964). *Am. J. Hyg.* **80,** 205–220.

Rosenbusch, F. (1939). *Proc. Pac. Sci. Congr., 6th* pp. 209–214.
Rueger, M. E., Olson, T. A., and Price, R. D. (1966). *Am. J. Epidemiol.* **83,** 33–37.
Rush, W. A., Kennedy, R. C., and Eklund, C. M. (1963a). *Am. J. Hyg.* **77,** 258–264.
Rush, W. A., Kennedy, R. C., and Eklund, C. M. (1963b). *Mosq. News* **23,** 285–286.
Saturno, A., and Henderson, J. R. (1965). *J. Immunol.* **94,** 365–370.
Saugstad, E. S., Dalrymple, J. M., and Eldridge, B. F. (1972). *Am. J. Epidemiol.* **96,** 114–122.
Schaeffer, M., and Arnold, E. H. (1954). *Am. J. Hyg.* **60,** 231–236.
Schaeffer, M., Kissling, R. E., Chamberlain, R. W., and Vanella, J. M. (1954). *Am. J. Hyg.* **60,** 266–268.
Schemanchuk, J. A., and Morgante, O. (1968). *Can. J. Microbiol.* **14,** 1–5.
Schober, H. (1964). *Mosq. News* **24,** 67.
Shope, R. E., de Andrade, A. P. H., Bensabath, G., Causey, O. R., and Humphrey, P. S. (1966). *Am. J. Epidemiol.* **84,** 467–477.
Siverly, R. E., and Schoof, H. F. (1962). *Mosq. News* **22,** 274–282.
Snow, W. E., and Packard, E. (1956). *Mosq. News* **16,** 143–148.
Sollers-Riedel, H. (1972). *Proc. N.J. Mosq. Control Exterm. Assoc.* **48** (Suppl.: "1970 World Studies on Mosquitoes and Diseases Carried by Them").
Sooter, C. A., Howitt, B. F., Gorrie, R., and Cockburn, T. A. (1951). *Proc. Soc. Exp. Biol. Med.* **77,** 393–394.
Spalatin, J., Connell, R., Burton, A. N., and Gollop, B. J. (1964). *Can. J. Comp. Med. Vet. Sci.* **28,** 131–142.
Spence, L., Belle, E. A., McWatt, E. M., Downs, W. G., and Aitken, T. H. G. (1961). *Am. J. Hyg.* **73,** 173–181.
Spielman, A. (1964). *Science* **141,** 361–362.
Sponseller, M. L., Binn, L. N., Wooding, W. L., and Yager, R. H. (1966). *Am. J. Vet. Res.* **27,** 1591–1598.
Stabler, R. M. (1948). *Mosq. News* **8,** 17–19.
Stamm, D. D. (1958). *Am. J. Public Health* **48,** 328–335.
Stamm, D. D. (1963). *Proc. Int. Ornithol. Congr.,* **13,** 591–603.
Stamm, D. D. (1966). *Auk* **83,** 84–97.
Stamm, D. D. (1968). *Am. J. Epidemiol.* **87,** 127–137.
Stamm, D. D., and Kissling, R. E. (1957). *J. Immunol.* **79,** 342–347.
Stamm, D. D., and Newman, R. J. (1963). *Ann. Microbiol. Enzimol.* **11(A),** 123–133.
Stamm, D. D., Chamberlain, R. W., and Sudia, W. D. (1962). *Am. J. Hyg.* **76,** 61–81.
Sudia, W. D., Stamm, D. D., Chamberlain, R. W., and Kissling, R. E. (1956). *Am. J. Trop. Med. Hyg.* **5,** 802–808.
Sudia, W. D., Chamberlain, R. W., and Coleman, P. H. (1968). *Am. J. Epidemiol.* **87,** 112–126.
Sudia, W. D., Fernandez, L., Newhouse, V. F., Sang, R., and Calisher, C. H. (1975a). *Am. J. Epidemiol.* **101,** 51–58.
Sudia, W. D., McLean, R. G., Newhouse, V. F., Johnston, J. G., Miller, D. L., Trevino, H., Bowen, G. S., and Sather, G. (1975b). *Am. J. Epidemiol.* **101,** 36–50.
Sudia, W. D., Newhouse, V. F., Beadle, L. D., Miller, D. L., Johnston, J. G., Young, R., Calisher, C. H., and Maness, K. (1975c). *Am. J. Epidemiol.* **101,** 17–35.
Sulkin, S. E. (1945). *Science* **101,** 381–383.
Sulkin, S. E., and Allen, R. (1974). *Monogr. Virol.* **8.**
Sulkin, S. E., and Izumi, E. M. (1947). *Proc. Soc. Exp. Biol. Med.* **66,** 249–250.
Sulkin, S. E., Wisseman, C. L., Izumi, E. M., and Zarafonetis, C. (1955). *Am. J. Trop. Med. Hyg.* **4,** 119–135.

Sussman, O., Kerlin, R. E., Carter, W. C., Swineboard, J., and Goldfield, M. (1966). *Bird-Banding* **37,** 183–190.
Syverton, J. T., and Berry, G. P. (1936a). *Proc. Soc. Exp. Biol. Med.* **34,** 822–834.
Syverton, J. T., and Berry, G. P. (1936b). *Science* **84,** 186–187.
Syverton, J. T., and Berry, G. P. (1940). *Am. J. Hyg.* **32,** 19–23.
Syverton, J. T., and Berry, G. P. (1941). *J. Exp. Med.* **73,** 507–530.
Taylor, D. J., Meadows, K. E., and Baughman, I. E. (1966). *Mosq. News* **26,** 502–506.
Tempelis, C. H., and Washino, R. K. (1967). *J. Med. Entomol.* **4,** 315– 318.
Tempelis, C. H., Reeves, W. C., Bellamy, R. E., and Lofey, M. F. (1965). *Am. J. Trop. Med. Hyg.* **14,** 170–177.
Ten Broeck, C., and Merrill, M. H. (1935). *Am. J. Pathol.* **11,** 847.
Thomas, L. A. (1963). *Am. J. Hyg.* **78,** 150–165.
Thomas, L. A., and Eklund, C. M. (1960). *Proc. Soc. Exp. Biol. Med.* **105,** 52–55.
Thomas, L. A., and Eklund, C. M. (1961). *Proc. Soc. Exp. Biol. Med.* **109,** 421–424.
Thomas, L. A., Eklund, C. M., and Rush, W. A. (1958). *Proc. Soc. Exp. Biol. Med.* **99,** 698–700.
Trager, W. (1938). *Am. J. Trop. Med.* **18,** 387–393.
Vigilancia Epidemiologica (1973). Annual Report No. 2.
von Sprockhoff, H., and Ising, E. (1971). *Arch. Gesamte Virusforsch.* **34,** 371–380.
Wallis, R. C. (1953). *Mosq. News* **14,** 33–34.
Wallis, R. C. (1959a). *Mosq. News* **19,** 157–158.
Wallis, R. C. (1959b). *Proc. Entomol. Soc. Wash.* **61,** 219–222.
Wallis, R. C. (1960). *Conn. Agric. Exp. Sta., New Haven, Bull.* **632.**
Wallis, R. C. (1965). *Yale J. Biol. Med.* **37,** 413–421.
Wallis, R. C., and Main, A. J. (1974). *Mem., Conn. Entomol. Soc.* pp. 117–144.
Wallis, R. C., and Whitman, L. (1967). *J. Med. Entomol.* **4,** 273–274.
Wallis, R. C., Taylor, R. M., McCollum, R. W., and Riordan, J. T. (1958). *Mosq. News* **18,** 1–4.
Wellings, F. M., Lewis, A. L., and Pierce, L. V. (1972). *Am. J. Trop. Med. Hyg.* **21,** 201–213.
Wheeler, R. (1966). *Meet. Am. Epidemiol. Soc., San Francisco, Calif.*
Whitney, E. (1963). *Am. J. Trop. Med. Hyg.* **12,** 417–424.
Whitney, E., Jamnback, H., Means, R. G., and Watthews, T. H. (1968). *Am. J. Trop. Med. Hyg.* **17,** 645–650.
W.H.O. Scientific Group (1967). *W.H.O. Tech. Rep. Ser.* **369.**
Williams, J. E., Young, O. P., Watts, D. M., and Reed, T. J. (1971). *J. Wildl. Dis.* **7,** 188–194.
Williams, J. E., Watts, D. M., Young, O. P., and Reed, T. J. (1972). *Mosq. News* **32,** 188–192.
Williams, J. E., Young, O. P., and Watts, D. M. (1974). *J. Med. Entomol.* **11,** 352–354.
Winn, J. F., and Bennington, E. E. (1959). *Proc. Soc. Exp. Biol. Med.* **101,** 135–136.
Work, T. H. (1964). *Science* **145,** 270–272 .
Yuill, T. M., and Hanson, R. P. (1964). *Zoonoses Res.* **3,** 153–164.
Yuill, T. M., Iverson, J. O., and Hanson, R. P. (1969). *Bull. Wildl. Dis.* **5,** 248–253.

THE GENOME OF SIMIAN VIRUS 40*

Thomas J. Kelly, Jr., and Daniel Nathans

Department of Microbiology,
The Johns Hopkins University School of Medicine,
Baltimore, Maryland

* Abbreviations used herein: SV40, simian virus 40; SDS, sodium dodecyl sulfate; DNA form I, covalently closed circular duplex DNA; DNA form II, circular duplex DNA with one or more single strand scissions; DNA–L, full length, linear duplex DNA; endo R, restriction endonuclease; specific endo R's (*Eco*RI, *Hin*dII, etc.) as in Smith and Nathans (1973); DNA–L_{EcoRI}, DNA–L_{HpaII}, linear duplex DNA resulting from cleavage by endo R·*Eco*RI or *Hpa*II, respectively; ts, temperature sensitive; wt, wild type, dl, deletion; *ev*, evolutionary variant; VP, virion protein; MW, molecular weight; AGMK, African green monkey kidney cells; RI, replicative intermediate; EtBr, ethidium bromide.

I. Introduction

The essential feature of a virus is its nucleic acid genome, a cluster of genes that brings new information into the cell it infects. Depending on the circumstances of infection, expression of viral genes can lead to multiplication of the virus and perhaps cell damage, or to a stable virus–host cell relationship in which the cell often acquires new properties. These characteristics of viruses, combining simplicity of genomic structure and easy manipulation with important cellular effects, accounts for their great usefulness in studying gene action and regulation in complex cells.

The advantageous use of viruses is nowhere more evident than in current studies of viral tumorigenesis. Tumor viruses are not fundamentally different from other viruses. In fact, almost every type of DNA-containing animal virus is potentially tumorigenic. Such viruses can express their genetic information for reproduction, or, by partial gene expression, they can cause heritable changes in cell growth, i.e., "transformation" of cells to tumorigenicity, in the living animal and in culture. In transformed cells viral genes persist, often as part of cellular chromosomes, and it is likely that viral gene products are continuously needed to maintain the transformation phenotype. Moreover, transformation to tumorigenicity may be but one type of heritable change in animal cells mediated by latent viruses. Therefore a great deal of attention is now being focused on the molecular biology of tumor viruses, in the expectation that better understanding will be gained of nonlytic virus–cell interactions as well as of gene expression and growth regulation in animal cells, including neoplastic cells.

Among the simplest tumor viruses are the small DNA-containing papovaviruses, including simian virus 40 (SV40), the murine polyoma virus, and the human papovaviruses BK and JC. These viruses exhibit similar physical and biological properties, and in some instances partial homology of DNA sequences. Therefore, although we concentrate in this review on SV40, the general features of genomic organization and function are likely to apply to all the small papovaviruses.

SV40 occurs as an inapparent natural infection of certain species of Asiatic macaques and was discovered in rhesus monkey cell cultures as a vacuolating agent for African green monkey cells (Sweet and Hilleman, 1960). Electron microscopy of purified SV40 virions shows a typical small, icosahedral, nonenveloped particle about 41 nm in diameter (Fig. 1).

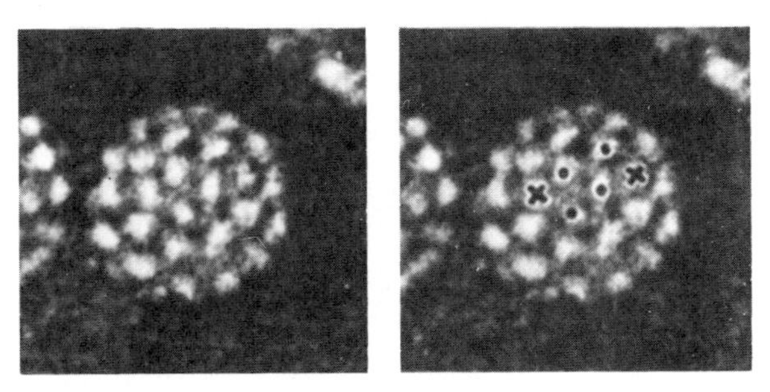

FIG. 1. Electron micrographs of SV40 virions. KEY: ×, 5-coordinated capsomeric unit; ●, 6-coordinated capsomeric unit. [Reproduced from Anderer *et al.* (1967), with permission.]

Each particle has a mass of about 17×10^6 daltons and contains a molecule of duplex DNA in the form of a covalently closed circle of about 3.4×10^6 daltons (see Section II, A, 1). In tissue culture various types of SV40–cell interactions occur, depending on the host cell. In African green monkey kidney cells the virus undergoes a rather typical reproductive cycle resulting in the formation of new infectious virus and death of the cell. These cells are therefore "permissive," and the infectious cycle is called "productive infection." In mouse cells, however, SV40 causes an "abortive infection" in which only part of the viral genome is expressed and no progeny virus is produced. Such cells survive infection, and some of them acquire new, heritable growth properties, i.e., they are "transformed." In some instances transformed cells can grow into tumors when transplanted into appropriate animals. Human and hamster cells are "semipermissive" for SV40: virus production is slower than in fully permissive monkey cell cultures, few progeny viruses are produced, and many cells survive infection. Semipermissive and even permissive cells can also be transformed by SV40.

An outline of the molecular events occurring during productive infection and transformation by SV40 is presented in Table I. In a *productively infected* cell, following attachment, penetration, and uncoating of the virus, RNA with sequence homology to SV40 DNA appears; this is the "early" SV40 RNA, i.e., viral RNA which is formed prior to and independent of the onset of viral DNA replication. At this time also SV40-specific T antigen and U antigen are first detectable, cell DNA synthesis is stimulated (in most permissive cell lines), and the activities or amounts of certain (probably cellular) enzymes related to DNA synthesis are increased. Viral DNA replication then commences in the cell nucleus, and another class of SV40 RNA, "late" RNA, is synthesized in addition to "early" RNA. Now "late" proteins are detectable, which eventually form

TABLE I

OUTLINE OF MOLECULAR EVENTS (SV40–CELL INTERACTIONS) FOLLOWING SV40 INFECTION OF PERMISSIVE MONKEY CELLS AND NONPERMISSIVE MOUSE CELLS

Productive infection[a]	Transformation[b]
Entry and uncoating	Entry and uncoating
"Early" mRNA and protein(s)	Viral mRNA and "early" protein(s)
(Cell DNA synthesis)	(Cell DNA synthesis)
Viral DNA replication	Integration of viral DNA
"Late" mRNA and proteins	"Early" mRNA and protein(s)
Virus assembly	Cell transformation to tumorigenicity
Cell lysis	

[a] Permissive monkey cells.
[b] Nonpermissive mouse cells.

part of the SV40 virion. Following virus assembly in the nucleus, cell lysis occurs with release of infectious progeny. In the case of *transformation* the molecular events are not as well understood, partly because most infected, nonpermissive cells are only "abortively" transformed and relatively few cells become stably transformed except at extremely high multiplicity of infection. In those cells that are stably transformed, SV40 DNA, SV40 RNA, and T and U antigens are detectable in all the progeny cells. In addition an SV40-specific transplantation antigen (TSTA) is detectable by biological tests. Therefore in transformed cells SV40 genes persist, possibly integrated into cellular chromosomes, and at least some of the genes are expressed. The products of these genes may be responsible for initiating and maintaining the transformed state.

Our purpose in this review is to examine in detail some of the molecular events just outlined, and to relate them to the structure and genetic organization of the SV40 genome. Our primary focus will be on the SV40 genome itself, its structure as a DNA molecule and small chromosome, its genetic content, its replication and transcription, and its interaction with cellular chrmosomes. A symposium dealing with these and related topics has recently been published.*

II. STRUCTURE OF THE SV40 GENOME

A. *SV40 DNA*

SV40 DNA can be conveniently isolated from purified virions by treatment with the ionic detergent SDS and/or phenol extraction. In general, two forms of duplex SV40 DNA with identical molecular weights (Fig. 2)

* *Cold Spring Harbor Symposia on Quantitative Biology* **39** (1974).

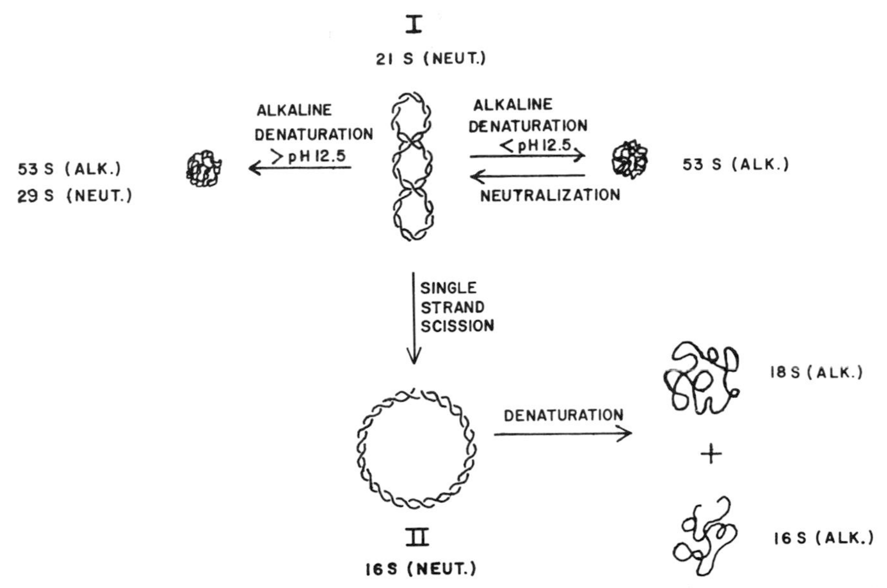

FIG. 2. Forms of SV40 DNA. See text for details. [Adapted from Vinograd and Lebowitz (1966a).]

are recovered by this procedure (Crawford and Black, 1964). The major component, form I, is a covalently closed circular molecule with a sedimentation coefficient in neutral sucrose gradients of about 21 S. Form II is a circular DNA molecule containing one or more single-strand interruptions (nicks). It sediments at about 16 S in neutral sucrose gradients. Depending on the host cell upon which the virus is grown, DNA preparations from purified virions may also contain some linear molecules. These are derived from virus particles which contain host DNA rather than SV40 DNA (pseudovirions). Pseudovirion DNA is double stranded and heterogeneous in molecular weight (11–15 S on neutral sucrose gradients). Its origin and properties will be described in Section II, A, 4.

1. Molecular Weight and Base Composition

The molecular weights of SV40 DNA forms I and II have been estimated by a variety of physical methods including sedimentation velocity (Crawford and Black, 1964; Anderer *et al.*, 1967), sedimentation equilibrium in CsCl (Crawford and Black, 1964), and electron microscopy (Tai *et al.*, 1972; Crawford *et al.*, 1966; Anderer *et al.*, 1967; Gerry *et al.*, 1973). Many of these molecular weight determinations were made before it was generally appreciated that deletions and substitutions accumulate in SV40 stocks passaged at high multiplicity (Uchida *et al.*, 1966, 1968; Yoshiike, 1968; Lavi and Winocour, 1972; Tai *et al.*, 1972; see also Section IV, C). Therefore, only the most recent studies, in which DNA was prepared from virus grown at low multiplicities of infection, are likely to be accurate. Tai *et al.* (1972) measured the relative contour lengths of SV40 DNA and bacteriophage λ DNA present on the same electron microscope grid and obtained a value of 3.6×10^6 for the molecular weight of SV40 DNA. Using a similar method and bacteriophage T7 DNA as an internal length standard, Gerry *et al.* (1973) obtained a value of 3.2×10^6. Thus, SV40 DNA contains roughly 4800–5500 nucleotide pairs and has a coding capacity for 160,000–180,000 daltons of proteins. The GC (guanine–cytosine) content of SV40 DNA as determined by buoyant density measurements (Crawford and Black, 1964) and direct base analysis (Danna and Nathans, 1972) is about 41%, which is very similar to that of mammalian cell DNA. Mulder and Delius (1972) have obtained a denaturation map of linear SV40 DNA produced by cleavage with R • *Eco*RI. The DNA was partially denatured by treatment with alkali in the presence of formaldehyde and mounted for electron microscopy (Inman and Schnös, 1970). The molecules showed a considerable degree of intramolecular heterogeneity in base composition. The earliest melting region (presumably AT rich) mapped between SV40 map positions 0.40 and 0.65, and the latest melting region (presumably GC rich) mapped in the neighborhood of SV40 map position 0.3 (see Section II, C for definition of map positions).

2. *Physical Properties of SV40 Form I DNA*

Covalently closed circular duplex DNA molecules are widely distributed in nature. Besides SV40 form I DNA, this class of molecules includes the DNA of the other papovaviruses, mitochondrial DNA, the intracellular DNA of many bacteriophages, and the DNA of bacterial plasmids. These molecules share a number of interesting physical properties due to the fact that their complementary strands are topologically linked. These properties have been extensively described elsewhere (Vinograd and Lebowitz, 1966a; Crawford, 1969) and will be discussed only briefly here.

When SV40 form I DNA is exposed to denaturing conditions (heat or alkali up to pH 12.5) the hydrogen bonds holding the duplex structure are disrupted, but the two single strands cannot separate (Fig. 2). Upon restoration of normal conditions, the strands rapidly renature to form the original duplex structure (Westphal, 1970; Salzman *et al.*, 1973). This property of rapid renaturation (or "snap-back") is not observed, however, where denaturation is carried out by exposure to a pH above 12.5 (Westphal, 1970; Salzman *et al.*, 1973). Under these conditions the complementary base pairs of the molecules apparently become sufficiently out of register that renaturation cannot occur after neutralization. Such "irreversibly denatured" structures sediment at 26 S in neutral sucrose gradients containing 0.1 *M* NaCl. SV40 form I DNA shows a very high (53 S) sedimentation coefficient in alkaline sucrose gradients (containing 1 *M* NaCl). This is again a consequence of the inability of the two single strands to separate even though all hydrogen bonds are broken. Each strand is constrained in space by the requirement that it be wrapped around the other approximately 500 times. The result is a very compact structure. Under similar conditions singly nicked form II DNA dissociates into a linear single strand which sediments at 16 S and a circular strand which sediments at 18 S.

Many of the physical properties of SV40 form I DNA are a result of the fact that it (and all other naturally occurring closed circular duplexes) contains superhelical turns. Before discussing the physical consequences of superhelicity, it is necessary to define several parameters which are used to describe the topological features of covalently closed circular duplex DNA molecules (Vinograd *et al.*, 1968). The topological winding number (α) is defined as the number of complete revolutions made by one strand about the duplex axis when the axis is constrained to lie in a plane. For a given closed circular duplex molecule, α is a fixed quantity which cannot be changed without interrupting the continuity of one of the two strands. The duplex winding number (β) is the number of complete revolutions made by one strand about the duplex axis in the unconstrained molecule. The value of β is a function of a number of physical parameters such as pH, ionic strength, nature of the cation, temperature, and concentration of intercalating agents. It follows from topological considerations alone than any difference in the values of α and β will be reflected in tertiary (superhelical) turns in the molecule. Thus, the superhelix winding number (τ), defined as the number of revolutions that the duplex makes about the superhelix axis, is given simply by the equation: $\tau = \alpha - \beta$ (Vinograd and Lebowitz, 1966b: Glaubiger and Hearst, 1967). The number of superhelical turns per 10 base pairs is called the superhelix density, σ.

The superhelix density of SV40 DNA form I was estimated by titration with ethidium bromide, a dye which unwinds duplex DNA by intercalating between adjacent base pairs (Bauer and Vinograd, 1968; Gray *et al.*, 1971). The relative superhelicity of the molecule at each dye concentration was monitored by sedimentation velocity and sedimentation equilibrium measurements. It was found that as the number of bound dye molecules increased, the number of superhelical turns in SV40 form I DNA first decreased to 0 (at which point the sedimentation velocity and buoyant density of SV40 form I DNA was the same as that of SV40 form II

DNA) and then increased again. Since ethidium bromide unwinds the duplex (decreases β), this behavior means that in the absence of dye β is greater than α, and SV40 form I DNA contains negative superhelical turns. Based on an estimated unwinding angle for ethidium bromide of 12° (Fuller and Waring, 1964) and the amount of dye required to remove all of the superhelical turns, the value of σ for dye-free SV40 form I DNA was calculated as 0.039 in 2.85 *M* CsCl at 25°C (Bauer and Vinograd, 1968; Gray *et al.*, 1971). Thus, under these conditions, $\tau = -19$, i.e., each molecule contains approximately 19 negative superhelical turns. At lower ionic strength or higher temperature, the number of negative superhelical turns decreases (Upholt *et al.*, 1971). It should be noted that two recent studies (Wang, 1974; Pulleyblank and Morgan, 1975) suggest that the unwinding angle of ethidium bromide is actually in the neighborhood of 26°. Use of this value leads to an estimate of $\tau = -41$ for SV40 form I DNA.

Differential binding of ethidium bromide by form I and form II DNA has been used to separate these forms of SV40 DNA. When the concentration of ethidium bromide is increased beyond that required to remove all of the negative superhelical turns in form I DNA, positive superhelical turns are introduced into the molecule. Since the introduction of these turns raises the free energy of the molecule (to be discussed shortly), its affinity for dye is reduced. Form II DNA molecules which contain at least one nick (or linear DNA molecules) are not subject to this constraint, since the duplex is free to rotate about the phosphodiester bond opposite the nick and superhelical turns are not introduced. It follows that at high concentrations of ethidium bromide form II DNA will bind more dye than will form I DNA. Since binding of ethidium bromide lowers the buoyant density of DNA, this provides a convenient method for the preparative separation of forms I and II (Radloff *et al.*, 1967). For example, in CsCl gradients containing 200 μg/ml ethidium bromide, the buoyant density of form I SV40 DNA is about 50 mg/ml greater than that of form II DNA.

It is not known at present why SV40 form I DNA (and all other naturally occurring closed circular duplex DNA molecules) have negative superhelical turns. There are two general, nonexclusive possibilities (Vinograd and Lebowitz, 1966a; Crawford and Waring, 1967; Eason and Vinograd, 1971). The first possibility is that in the cell an open (nicked) form of the molecule is condensed into a negative superhelical structure by some unknown organizing agent and then sealed. The second and perhaps more likely possiblity is that prior to final ring closure an open form is partially unwound by DNA-binding proteins, intercalators, or other unwinding systems. Both of these possibilities have the effect of reducing

α, the topological winding number, below that expected if ring closure occurred in an undisturbed Watson–Crick duplex structure. Recent evidence (Germond *et al.*, 1975) suggests that the binding of cell histones may play a role in generating the superhelical turns in SV40 DNA (see the following subsection).

The presence of superhelical turns in SV40 form I DNA gives the molecule a more compact conformation than the relaxed form II molecule. This explains form I's higher sedimentation coefficient, lower viscosity, and relative resistance to breakage by hydrodynamic shear. The superhelical turns in SV40 form I DNA are spontaneously lost when a single nick is introduced into the molecule. This occurs by rotation of the duplex about the phosphodiester bond opposite the nick. The implication of this observation is that a molecule with superhelical turns has a higher free energy than an equivalent nicked relaxed molecule or an identical closed molecule having no superhelical turns (Vinograd and Lebowitz, 1966a). Part of the increased free energy results from entropic effects (the superhelical molecule is more ordered), and part results from enthalpic effects (bending and torsional stresses). Since the sign of the superhelix in all naturally occurring closed circles is negative, the free energy of superhelix formation provides a driving force which tends to unwind the duplex. This accounts for the observation that a small fraction of the base pairs in form I DNA melts at a lower temperature or pH value than that required to initate melting of form II DNA (Crawford and Black, 1964; Vinograd *et al.*, 1965, 1968). This early helix–coil transition involves the simultaneous unwinding of duplex and superhelical turns. Further unwinding of the duplex results in the introduction of positive superhelical turns, which again raises the free energy of form I DNA relative to that of form II with an equivalent number of melted bases. The net effect of this is to raise the T_m or pH_m of form I substantially above that of form II (Vinograd *et al.*, 1968). In other words, form I DNA freed of its tertiary turns by the early helix–coil transition is thermodynamically more stable than form II DNA.

Another consequence of the higher free energy of superhelical molecules is the fact that SV40 form I DNA behaves as if it contains unpaired or weakly hydrogen-bonded regions. Gene 32 protein of T4, which has a selective affinity for single-stranded DNA, also binds to SV40 form I DNA but not to form II (Morrow and Berg, 1973). The single-strand–specific nuclease S_1 from *Aspergillus oryzae* (Ando, 1966) cleaves SV40 form I DNA to unit-length linear molecules (Beard *et al.*, 1973). Finally, a water-soluble carbodiimide which reacts preferentially with unpaired bases has been shown to react with form I in preference to form II (Salzman *et al.*, 1974). There is evidence that the unpaired bases in form

I DNA reside in specific regions of the molecule. At moderate salt concentration (75 mM) SV40 form I DNA is cleaved preferentially by S_1 nuclease in one of two regions located at SV40 map positions 0.15 to 0.25 and 0.45 to 0.55. At high salt concentrations (250 mM) cleavage by S_1 nuclease occurs preferentially in the 0.45 to 0.55 region (Beard *et al.*, 1973). The binding site for T4 gene 32 protein was also mapped in this latter region at SV40 map position 0.46 (Morrow and Berg, 1973). It is interesting to note that in the denaturation map of SV40 described earlier (Mulder and Delius, 1972) the earliest melting region lies between SV40 map positions 0.40 and 0.65. The regions of SV40 form I which react with carbodiimide have not yet been mapped precisely, but recent evidence indicates that they all must lie between SV40 map positions 0.11 and 0.66, which is consistent with the preceding data (Salzman *et al.*, 1974).

3. Intracellular Forms of SV40 DNA

When SV40 DNA is extracted free of protein from infected cells four types of molecules are observed: form I, form II, replicative intermediates (RI), and oligomers. Forms I and II have essentially the same properties as the molecules extracted from virions with the exception that "free" intracellular form I DNA is more heterogeneous with regard to the number of superhelical turns per molecule and has a lower average superhelix density than form I DNA from virions (Eason and Vinograd, 1971). Oligomeric forms of SV40 DNA up to hexamers have been isolated from infected cells, and both catenated (interlocked) and concatenated (continuous) oligomers have been observed (Jaenisch and Levine, 1971; Rush *et al.*, 1971). Oligomeric forms represent only a small fraction of the total intracellular SV40 DNA. Catenated dimers may arise by a failure of proper segregation of the two daughter molecules after DNA replication is completed, and concatenated dimers may arise by genetic recombination between two SV40 monomers (Jaenisch and Levine, 1973). The properties of the other major intracellular form of SV40 DNA, replicative intermediate, will be described later (Section V, A).

4. DNA in Pseudovirions

Some of the virions recovered after growth of SV40 contain host DNA rather than SV40 DNA and have been called pseudovirions. The DNA extracted from pseudovirions is linear and double stranded, and heterogeneous in molecular weight. It sediments in neutral sucrose gradients at 11–15 S (Levine and Teresky, 1970; Trilling and Axelrod, 1970). Whether or not SV40 pseudovirions are produced during infection depends upon the particular host cell. Pseudovirions are found in abundance in lysates from primary AGMK or Vero cells, whereas they are not detectable in virus stocks grown on BSC-1 or CV-1 cells. Production of pseudovirions is correlated with the extensive fragmentation of the host genome that occurs during the course of SV40 infection of AGMK or Vero cells (Ritzi and Levine, 1970); they presumably arise by chance encapsidation of cellular DNA fragments of appropriate size.

B. The SV40 Chromosome

Within virions or within infected cells, SV40 DNA is associated with histone in a structure (referred to as the SV40 chromosome) closely resembling chromatin of eukaryotic cells. Recent studies have established that chromatin is a flexible chain composed of globular particles connected by short DNA filaments (Olins and Olins, 1974; Oudet *et al.*, 1975). In the particles, called *nucleosomes* or "nu" (η) bodies, a segment of DNA about 200 base pairs in length is associated with the four histones F2a1, F2a2, F2b, and F3 (Kornberg and Thomas, 1974; Noll, 1974; Burgoyne *et al.*, 1974; Van Holde *et al.*, 1974; Senior *et al.*, 1975; Oudet *et al.*, 1975). The mass ratio of histone to DNA in nucleosomes approximates 1:1, which would accommodate a model in which each nucleosome contains two molecules of each of the four histones (Kornberg, 1974). The DNA in nucleosomes is under constraint, since it is compacted 5- to 7-fold from its length in the absence of histones (Oudet *et al.*, 1975; Griffith, 1975). A number of studies have shown that the chromosome of SV40 has similar properties (Huang *et al.*, 1972; White and Eason, 1971; Hall *et al.*, 1973; Lake *et al.*, 1973; Griffith, 1975; Germond *et al.*, 1975) and therefore may represent a useful model system for studying the details of chromatin structure.

In the case of SV40 virions, a subviral nucleoprotein complex has been isolated by sucrose gradient sedimentation following treatment of virions with alkaline buffers at pH 10.5 (Anderer *et al.*, 1968; Huang *et al.*, 1972; Barban and Goor, 1971; Lake *et al.*, 1973). The complex was shown to contain the viral DNA and the four histones F2a1, F2a2, F2b, and F3 (Huang *et al.*, 1972; Lake *et al.*, 1973). (The complex may also contain small amounts of the virion protein VP3; see Section III, A.) Formation of the complex does not appear to be due to random association of DNA and histones during alkaline dissociation of the virions, since purified SV40 DNA mixed with the virions prior to dissociation does not participate in complex formation (Huang *et al.*, 1972). A nucleoprotein complex with a sedimentation coefficient of 50–60 S has also been isolated from nuclei of SV40-infected cells (White and Eason, 1971; Hall *et al.*, 1973; Griffith, 1975). More than 95% of the protein in this complex was identified as histone, and the mass ratio of histone to SV40 DNA was found to be about 1:1 (Hall *et al.*, 1973; Griffith, 1975). Electron microscopic examination of SV40 DNA–histone complexes isolated either from infected cells (Griffith, 1975) or from virions (Germond *et al.*, 1975) revealed a circular structure very similar in appearance to chromatin fibers (Fig. 3).

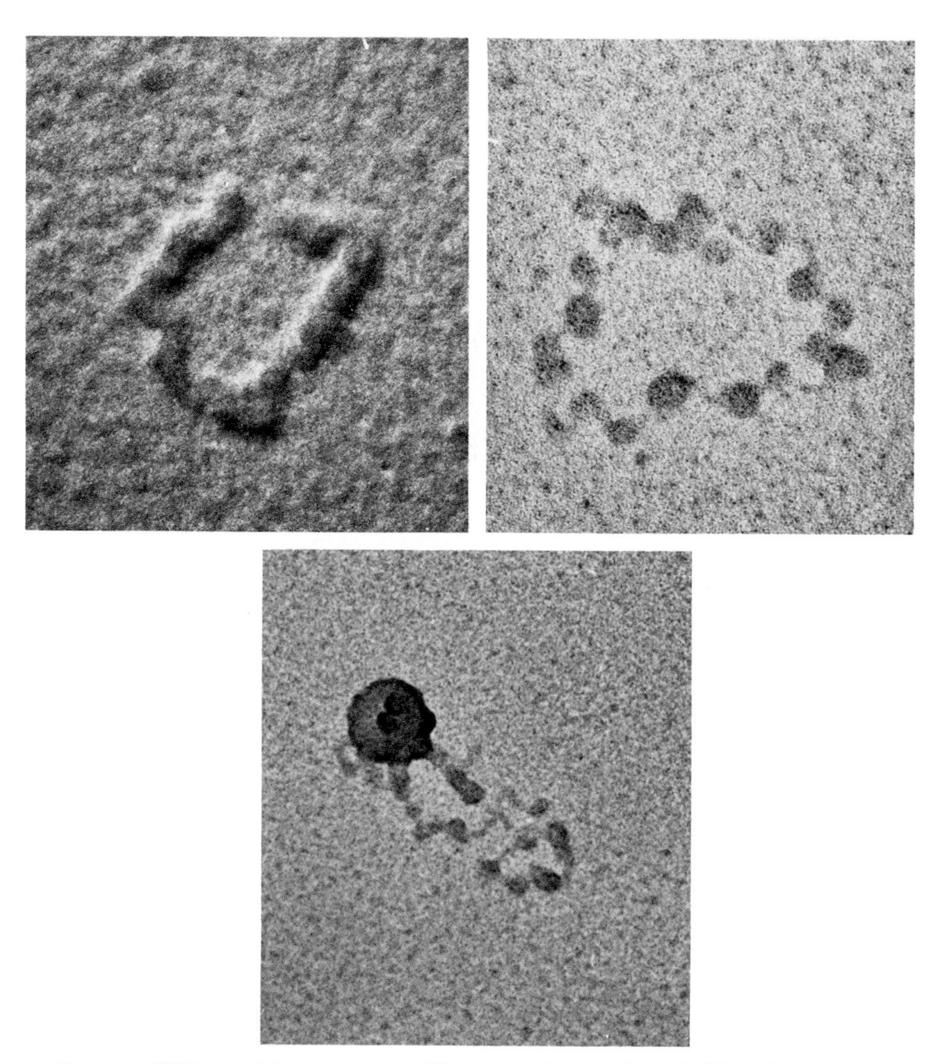

FIG. 3. SV40 minichromosomes: Electron micrographs of SV40 DNA–histone complexes isolated from infected cells and fixed in 0.15 *M* NaCl (*top left*) or 0.01 *M* NaCl (*top right*). Similar complexes can be isolated from SV40 virions. At the *bottom* is a complex "partially extruded" from a virion following very brief treatment at pH 9.8. (Photographs kindly supplied by Jack Griffith.)

When prepared for microscopy in low ionic strength buffers, the complexes were observed to contain about 20 globular particles with dimensions similar to those of the nucleosomes of eukaryotic chromatin (110–130 Å) connected by short filaments of DNA. The contour length of the complexes under these conditions was one-third to one-fifth that of the deproteinized SV40 DNA molecules. At physiological salt concentrations the complexes were even more compact (contour length was one-seventh that of deproteinized SV40 DNA), and the connecting filaments between the nucleosomes were not visible. Most of the protein is removed from such complexes at high ionic strength (1*M* NaCl); however, one to three globular protein "knobs" remain attached to the DNA near SV40 map position 0.7 (Griffith *et al.*, 1975).

Histone-SV40 DNA complexes have been formed *in vitro* using deproteinized SV40 DNA and the purified histones F2a1, F2a2, F2b and F3 (Oudet *et al.*, 1975; Germond *et al.*, 1975). The appearance of these complexes by electron microscopy is similar to that of the complexes prepared from virions or infected cells. *In vitro* prepared histone–SV40 DNA complexes have been employed as a model system for studying the nature of the constraint on DNA in nucleosomes (Germond *et al.*, 1975). Covalently closed SV40 DNA molecules containing no superhelical turns (form Ir) was prepared from form I DNA using a relaxing activity from Krebs ascites cells (see Section V, B). The form Ir molecules were then complexed with histones at various histone to DNA ratios. As the histone-to-DNA ratio increased, the number of nucleosomes per DNA molecule observable by electron microscopy also increased, up to a maximum of about 18. When the form Ir–histone complexes were treated with relaxing activity and deproteinized, the resulting DNA molecules were found to contain negative superhelical turns, and the average number of negative superhelical turns increased as the number of nucleosomes per molecule increased. Although these results do not reveal the exact nature of the constraint on DNA in nucleosomes, they suggest that the net result of that constraint is equivalent to unwinding the duplex. These results also suggest that the binding of histones to SV40 DNA *in vivo* in the presence of an active intracellular relaxing activity may account for most, if not all, of the superhelical turns in SV40 form I DNA.

C. Physical Map of the SV40 Genome

A physical map is one made up of reference points of known actual distances from each other. Physiologically important loci such as regulatory signals, structural genes, template functions, and protein binding sites can then be mapped relative to these points. Several approaches have been employed for defining reference points in SV40 DNA: (1) electron microscopic localization of denaturing regions along the DNA molecules (denaturation mapping); (2) location of deleted segments in cloned deletion mutants; (3) analysis of adeno–SV40 hybrid viruses which contain specific segments of SV40 DNA within their adenovirus genomes; and (4) localization of specific cleavage sites produced by restriction endonucleases. Denaturation mapping of SV40 DNA has been discussed earlier (Section II, A); deletion mutants and adeno–SV40 hybrid viruses will be dealt with later. Here we focus on the most useful and general method, namely, cleavage of DNA by appropriate restriction endonucleases, enzymes which make double-strand scissions in duplex DNA at specific nucleotide sequences.

We should point out that the ultimate physical map of DNA is a nucleotide sequence of the entire genome. This would allow direct comparison of nucleotide sequences of structural genes with amino acid sequences of protein products, as well as correlation of sequence signals between genes with biological tests of function. In the case of SV40, the small size of the genome and the availability of specific DNA fragments have resulted in rapid progress toward these goals and we will include where appropriate that part of the available data related to specific viral functions.

1. Cleavage Maps of the SV40 Genome

The availability of bacterial restriction endonucleases opened a new approach to mapping viral DNA molecules and isolating specific segments of viral genomes [for recent reviews, see Arber (1974); Nathans and Smith (1975)]. The general strategy of this approach is to cleave the molecule of interest with a suitable restriction enzyme, isolate individual fragments by gel electrophoresis, and determine their size and order in the original molecule, thus providing a series of reference points (the enzyme cleavage sites) of defined relative positions. Not all restriction endonucleases are useful for this purpose: so-called Class I enzymes (Meselson and Yuan, 1968; Boyer, 1971) recognize specific nucleotide sequences but cleave at a variable distance from a recognition site. A second group of restriction enzymes (Class II), however, can recognize nucleotide sequences in duplex DNA and cleave specific phosphodiester bonds within recognition sites (Kelly and Smith, 1970; reviews cited above). The cleavage sites for several Class II enzymes useful in the analysis of SV40 are shown in Table II. As seen in the table, each site has 2-fold rotational symmetry, and some enzymes cleave in the center of the symmetrical sequence, resulting in even-ended fragments, whereas others cleave (sym-

TABLE II

RESTRICTION ENDONUCLEASES IN THE ANALYSIS OF SV40 DNA[a]

Enzyme	Restriction site	No. of sites in SV40 DNA	References[b]
*Hind*II	GTPy↓PuAC	7	*a, b, c*
*Hind*III	A↓AGCTT	6	*d, b*
*Hinf*I	G↓ANTC	10	*x*
Hae II	PuGCGC↓Py	1	*e, f*
Hae III	GG↓CC	17	*g, h*
Hha I	GCG↓C	2	*i, j*
Hpa I	GTT↓AAC	5	*k, l, c*
Hpa II	C↓CGG	1	*k, l*
*Eco*RI	G↓AATTC	1	*m, n, o*
*Eco*RII	↓CC(A/T)GG	16	*p, q, r*
*Bam*H–I	G↓GATCC	1	*s*
Alu I	AG↓CT	>32	*i, t, u, v, w*

[a] These data are taken from a table compiled by R. Roberts.

[b] REFERENCES: (*a*) Kelly and Smith (1970). (*b*) Danna *et al.* (1973) (*c*) Fiers *et al.* (1974). (*d*) Old *et al.* (1975). (*e*) R. J. Roberts *et al.* (1975). (*f*) C. A. Hutchison, III and B. Barrell (personal communication). (*g*) K. Murray and A. Morrison (personal communication). (*h*) Subramanian *et al.* (1974). (*i*) R. J. Roberts, P. A. Meyers, A. Morrison, and K. Murray (1976b). (*j*) Dhar *et al.* (1974a). (*k*) Garfin and Goodman (1974). (*l*) Sharp *et al.* (1973). (*m*) Hedgpeth *et al.* (1972). (*n*) Morrow and Berg (1972). (*o*) Mulder and Delius (1972). (*p*) Bigger *et al.* (1973). (*q*) Boyer *et al.* (1973). (*r*) Subramanian *et al.* (1974). (*s*) Wilson and Young (1975). (*t*) K. N. Subramanian and S. M. Weissman (personal communication). (*u*) R. Wu and E. Jay (personal communication). (*v*) Yang *et al.* (1976). (*w*) Roberts *et al.* (1976b). (*x*) K. N. Subramanian, S. M. Weissman, B. S. Zain, and R. J. Roberts, (personal communication).

metrically) near the ends of the sequence in each strand, resulting in fragments with single-stranded, cohesive ends. Fragments of the latter type can cyclize or join end to end by base pairing between single-stranded tails (Mertz and Davis, 1972); as will be discussed later, these properties allow reconstruction of the SV40 genome from endo R cleavage products.

The early use of restriction endonucleases to cleave SV40 DNA at specific sites has resulted in a rather detailed cleavage map of the SV40 genome. To construct the map, enzymes which cleave SV40 DNA at one or more sites were used, and the resulting fragments were separated and sized by gel electrophoresis. Their order was then established by analysis of partial digest products or by sequential digestion with various enzymes. For example, DNA is cleaved into 13 electrophoretically separable fragments by endo R • *Hind*II / III (Danna and Nathans, 1971; Fiers

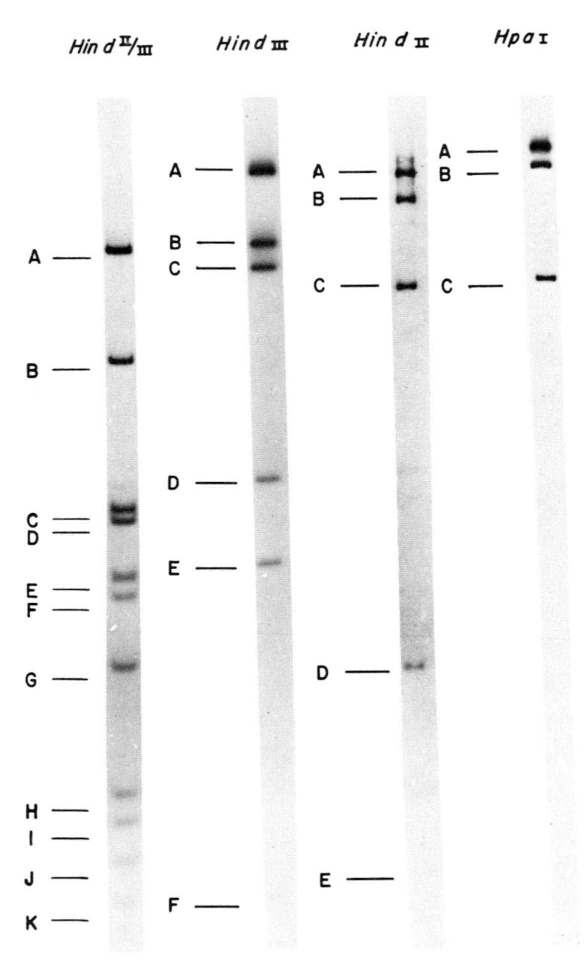

FIG. 4. Fragments of SV40 DNA produced by different restriction endonucleases and separated by electrophoresis in 4% polyacrylamide gels: autoradiogram of ^{32}P-labeled fragments. The origin is at the top. Fragments $<3\%$ of genome length have run off the gel. (Photographs kindly supplied by C.-J. Lai.)

et al., 1974), a mixture of two restriction endonucleases (dII and dIII) present in *H. influenzae* strain d (Fig. 4). *Hind*II alone produces seven fragments; *Hind*III, six fragments; and *Hpa* I, an enzyme related to *Hind*II, produces three large fragments. The size of these fragments has been estimated by their ^{32}P content (starting with uniformly labeled [^{32}P]-SV40 DNA), by electron microscopic length measurement of the larger fragments, or by electrophoretic mobility. The *Hin* fragments range from 22.5% of the length of SV40 DNA (*Hin*-A) to 0.6% of unit length (*Hin*-M). By analysis of partial digest products or of fragments produced by *Hpa* I endonuclease, the order of *Hin* and *Hpa* fragments could be deduced (Danna *et al.*, 1973; Sharp *et al.*, 1973). Based on this order and the size of each fragment, *Hin* and *Hpa* cleavage maps were constructed. The cleavage sites of several other enzymes were localized by similar methods or in relation to the established *Hin* map. Figure 5 summarizes the map positions of cleavage sites and fragments produced by several enzymes (references are noted in the figure legend). Of particular note are the *Eco*RI site, the *Hpa* II site, the *Bam* I site, and the *Hae* II site, each of which represents a single enzyme-susceptible sequence in the molecule and is of special value in electron microscopic mapping procedures (see below). By convention the unique *Eco*RI site (Morrow and Berg, 1972; Mulder and Delius, 1972) is taken as the zero point on the map, and map distances are measured in the *Hin*-F → J → G . . . direction, as shown in Fig. 5; one map unit is the total length of wt SV40 DNA (about 5000 nucleotide pairs).

2. 5′ → 3′ Orientation of DNA Strands

An additional feature of the physical map, important especially for transcriptional analysis, is the 5′ → 3′ orientation of each DNA strand. The two strands of SV40 DNA have been separated by taking advantage of the asymmetric transcription of form I DNA by *E. coli* RNA polymerase (Westphal, 1970). Since this enzyme preferentially transcribes one of the two SV40 DNA strands, in the presence of excess RNA transcript one of the strands (called the "minus" or E strand, which is the template for early SV40 mRNA) forms a DNA–RNA hybrid; the "plus" or L strand does not. Hydroxyapatite fractionation then leads to essentially quantitative separation of the E strand–RNA hybrid from the free L strand (Khoury and Martin, 1972; Sambrook *et al.*, 1972). To establish the 5′ → 3′ orientation of the E and L strands, two conceptually similar experiments have been carried out in which the 3′ or 5′ ends of the E and L strands of SV40 DNA–L_{RI} were localized with reference to the SV40 cleavage map (Fig. 6). In one case, 3′ half strands were removed by

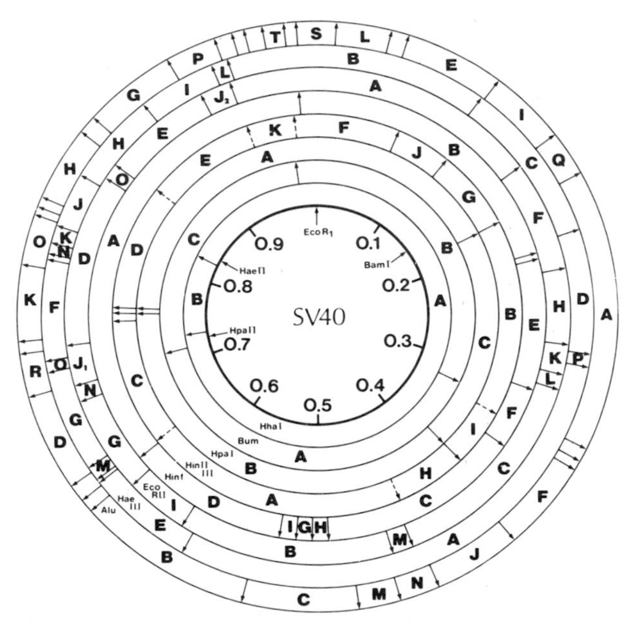

FIG. 5. Cleavage maps of SV40 DNA. (Reproduced, with permission, from the cover of the Cold Spring Harbor Laboratory's Abstracts of Papers at the Tumor Virus Meeting, August, 1975.) Enzymes producing single scissions of SV40 DNA are noted inside the circle; the *Eco*RI site is defined as the zero coordinate. Each concentric ring shows sites of cleavage by individual enzymes. Positions of cleavage sites were taken from the following sources: *Eco*RI, Morrow and Berg (1972); Mulder and Delius (1972). *Bam* I, J. Sambrook and Mathews, M. (unpublished data). *Hpa* II, Sharp *et al.* (1973). *Hae* II, R. J. Roberts *et al.* (1975). *Hha* I, K. N. Subramanian, S. Zain, R. J. Roberts, and S. Weissman (unpublished data). *Bum* I,M. Mathews, (unpublished data). *Hpa* I, Danna *et al.* (1973); Sharp *et al.* (1973). *Hin*dII/III, Danna *et al.* (1973); Yang et al. (1975). *Hin* f, K. N. Subramanian, S. Zain, R. J. Roberts, and S. Weissman (unpublished data). *Eco*RII, Subramanian *et al.* (1974). *Hae* III, Subramanian *et al.* (1974). *Alu* I, Yang *et al.* (1976).

exonuclease III, and the resulting 5′ halves of the separated E and L strands were mapped by annealing to *Hin* fragments (Khoury *et al.*, 1973b). In the other experiment, 3′ ends of SV40 DNA–L_{RI} were enzymatically labeled, and the labeled linear DNA was cleaved with endo R • *Hpa* I (Sambrook *et al.*, 1973). The two resulting 3′-labeled end fragments were then isolated and individual strands separated to determine in which fragment the L or E strand was labeled. Both experiments led to the same conclusion: the *E* strand is $5' \rightarrow 3'$ clockwise and the *L* strand is $5' \rightarrow 3'$ counterclockwise in the conventional cleavage map (Fig. 6). The same conclusion follows also from nucleotide sequence analysis of RNA transcribed from SV40 DNA form I by *E. coli* RNA polymerase in

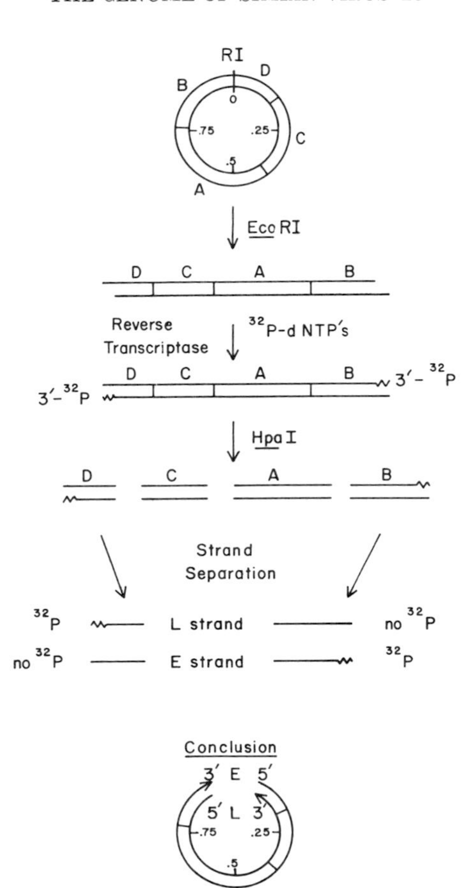

FIG. 6. Method for determining the 5' → 3' orientation of E and L strands of SV40 DNA. Unlabeled SV40 DNA was cleaved by endo R·*Eco*RI, as shown at the top of the figure, and the resulting linear molecules with 5' single-stranded tails were labeled at each 3'-end by incorporation of ^{32}P dNTP's using RNA-dependent DNA polymerase. The 3'-end-labeled DNA was then cleaved into four fragments (A, B, C, D) by *Hpa* I, the strands of each labeled fragment were separated, and the presence of label in the E and L strands was determined. From the results one can deduce that the E strand is 5' → 3' clockwise in the standard cleavage map, and the L strand 5' → 3' counterclockwise. [Adapted from Sambrook *et al.* (1973).]

relation to the nucleotide sequences of specific DNA fragments (Dhar *et al.*, 1974b).

3. Restriction Patterns of DNA from Different Strains of SV40

Restriction endonuclease digest patterns have been used to compare the DNA of various infectious SV40 strains with that of the prototype small plaque virus (strain 776) whose cleavage map is shown in Fig. 5 (Nathans

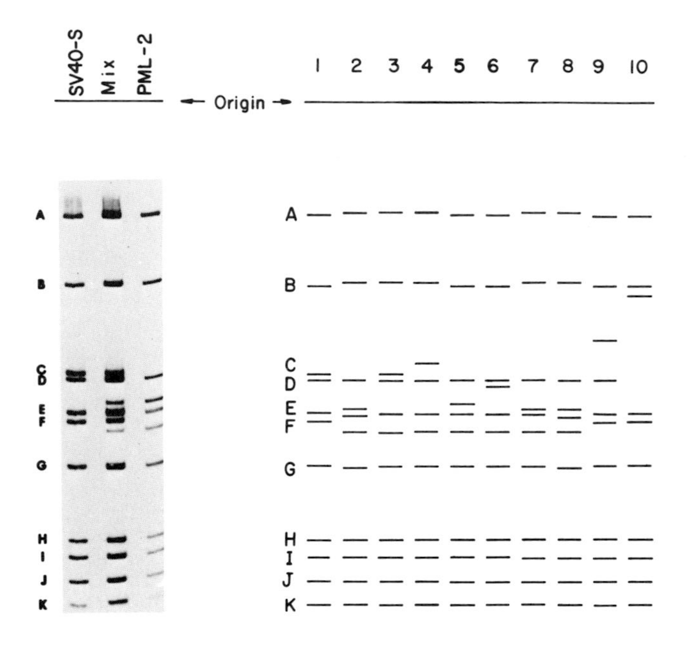

FIG. 7. "Fingerprints" of DNA from various infectious strains of SV40. On the left are autoradiograms of *Hin*dII/III ^{32}P fragments from the small plaque strain (SV40-S), from a human SV40 isolate (PML-2) and a mixture of the two digests (from Sack *et al.*, 1973), in each case following electrophoretic separation. Origin is at the top. On the right are diagrams of *Hin*dII/III digest patterns of DNA derived from other infectious strains and analyzed as indicated for PML-2. Column 1, small plaque strain; 2, minute plaque strain; 3, large plaque strain; 4, strain VA45–54; 5, PML-2 (DAR); 6, PML-1 (EK); 7, a strain detected in primary African green monkey cells; 8, SV68, rescued from 3T3 mouse cells, transformed by the small plaque virus; 9, an infectious evolutionary variant; 10, an infectious constructed variant. [From Nathans (1976).]

and Danna, 1972; Sack *et al.*, 1973; Huang *et al.*, 1972). As summarized in Fig. 7, although the *Hin*dII / III digest patterns of DNA from all infectious strains are similar, each DNA species is distinctive. Particularly variable are *Hin* fragments A, B, C, and F, suggesting that small deletions in these segments of the genome have occurred during the evolution of individual strains (see Section IV, C).

III. SV40 Gene Products

As indicated previously, the maximum coding capacity of the SV40 genome is about 160,000–180,000 daltons of protein. The total capacity is divided roughly equally between the early and late regions of the genome (see Section VI). Although the exact number of polypeptides coded by the SV40 genome is not known, the available genetic data (considered in Section IV) suggests that there may be no more than one early and two late polypeptides which are required for lytic growth. A number of different approaches to the identification of the primary gene products of SV40 have been employed in recent years. In this section we will discuss these approaches and describe the possible candidates for SV40-coded proteins that have been uncovered so far. At the present time only one protein, the major capsid protein VP1, has been unequivocally shown to be virus coded. However, in view of the recent advances in understanding of the organization of genetic information in the SV40 genome and the development of methods for producing SV40 deletion mutants all over the SV40 map (see Section IV), it is expected that future progress in this area will be rapid. Current information is summarized in Table III and presented in detail in the following discussion.

A. *Virion Proteins*

Purified SV40 virions can be dissociated into their constituent polypeptides by treatment with SDS and β-mercaptoethanol at an elevated temperature. When these polypeptides are fractionated by SDS-polyacrylamide gel electrophoresis, seven bands can be observed (Girard

TABLE III
SV40 VIRION PROTEINS AND SV40 SPECIFIC ANTIGENS

Polypeptide or antigen	Physical properties	Virus coded?
VP1	Polypeptide; MW 43,000–45,000	Yes—product of BC cistron
VP2	Polypeptide; MW 32,000–35,000	Probably
VP3	Polypeptide; MW 23,000–31,000	Probably
VP4–7	Histones F2a1, F2a2, F2b, F3; MW 11,000–15,000	No
T antigen	Polypeptide; MW 70–100,000	Probably—product of A cistron
U antigen	?	Probably—part of T antigen?
Tumor specific transplantation antigen (TSTA)	?	?
S antigen	?	No

et al., 1970; Barban and Goor, 1971; Estes *et al.*, 1971; Walter *et al.*, 1972; Ozer, 1972). The major band (VP1) accounts for more than 70% of the virion protein and has a molecular weight of about 43,000. Virion polypeptide 2 (VP2: MW, 32,000) and virion polypeptide 3 (VP3: MW, 23,000) each account for about 10% of the total virion protein. The remaining small polypeptides (VP4–7: MW, 11,000–15,000) are cell histones (see Section II, b).

Stocks of SV40 generally contain DNA-free particles or "empty shells" as well as intact virions. Analysis of the polypeptides in empty shells indicates that they contain VP1–3 in amounts similar to intact virions, but VP4–7 are reduced in amount (Estes *et al.*, 1971; Ozer, 1972; Walter *et al.*, 1972). Intact virions can be dissociated by treatment with alkaline buffers at pH 10.5 (Anderer *et al.*, 1968; Huang *et al.*, 1972; Barban and Goor, 1971; Lake *et al.*, 1973). The procedure yields two fractions separable by sucrose gradient sedimentation. The faster-sedimenting fraction contains the viral DNA complexed to the virion histones, VP4–7. The slower-sedimenting fraction contains VP1 and VP2. This observation, together with the finding that VP1 and VP2 are also present in empty shells, suggests that these two polypeptides are derived from the capsid of the virion. The location of VP3 in the virion is not clear since it can be associated with either the fast- or slow-sedimenting fraction depending on the ionic strength at which dissociation is carried out (Huang *et al.*, 1972).

As will be shown shortly, there is good evidence that VP1 is virus coded. It also seems certain that the virion histones VP4–7 are host-coded. The status of VP2 and VP3 is not clear, although they are good candidates for SV40-coded proteins. Related to the question of whether the virion polypeptides are virus coded is the problem of whether VP1–3 represent unique polypeptides or whether they share amino acid sequences. At the time of this writing, comparative fingerprints of VP1–3 have not been published.

B. SV40-Specific Antigens

Several new antigens can be detected by immunological techniques in cells infected or transformed by SV40. All of these antigens are virus specific, i.e., they can be detected in SV40-infected or -transformed cells from several different species but not in normal cells or in cells infected or transformed by unrelated viruses. In most cases the chemical nature (protein, carbohydrate, etc.) of the antigens is unknown. One or more of the SV40-specific antigens could represent primary SV40 gene products. Alternatively, they could represent the induction, unmasking, or modification of normal cellular constituents as a result of the action of an SV40 gene product. Except for T antigen, there is at present little direct evidence which would allow a decision between these two possibilities.

1. *T Antigen*

T antigen can be operationally defined as an SV40-specific antigen detectable by reaction with sera prepared from hamsters bearing SV40-induced tumors (Black *et al.*, 1963; Pope and Rowe, 1964). T antigen is found in the nuclei of productively infected permissive cells, abortively infected nonpermissive cells, transformed cells, and tumor cells (Black *et al.*, 1963; Hoggan *et al.*, 1965; Rapp *et al.*, 1964a). It does not appear to be a structural component of SV40 virions, since antibody to virions does not react with SV40-transformed cells or tumor cells and anti-T antibody does not react with virions (Black *et al.*, 1963; Rapp *et al.*, 1964; Pope and Rowe, 1964; Sabin and Koch, 1964). Induction of T antigen is an early SV40 function, i.e., it can be detected in productively infected cells before the onset of viral DNA synthesis, and its expression is not blocked by inhibitors of DNA synthesis such as cytosine arabinoside (Rapp *et al.*, 1965). Furthermore, it is induced by the nondefective Ad2–SV40 hybrid Ad2$^+$ND$_4$, which contains all or nearly all of the early region of the SV40 genome (Lewis *et al.*, 1973; Levine *et al.*, 1973; Kelly and Lewis, 1973; Morrow *et al.*, 1973).

SV40 T antigen appears to be a protein(s) at least in part since it is heat labile and sensitive to trypsin (Gilden *et al.*, 1965). Several attempts to purify T antigen using immunological assays (generally complement fixation) have met with only limited success (Lazarus *et al.*, 1967; Kit *et al.*, 1967; Fuks *et al.*, 1967; Potter *et al.*, 1969; Livingston *et al.*, 1974; Osborn and Weber, 1974), so little is known at present about its physical properties, or even whether it is a single molecular species. Osborn and Weber (1975) have examined the sedimentation coefficient in glycerol gradients of T antigen from infected and transformed cells. They have found that T antigen (detected by complement fixation) from productively infected CV–1 or BSC–1 cells or abortively infected 3T3 sediments at 15 S. In contrast, two clones of transformed 3T3 cells yield T antigen with a sedimentation coefficient of 22 S. Transformed cells from several other sources (human, rat, and hamster cells) were found to contain both the 15 S and 22 S species. Two estimates of the MW of the polypeptides in T antigen have been made using SDS–polyacrylamide electrophoresis. Del Villano and Defendi (1973) isolated T antigen on immunoabsorbants prepared by coupling the IgG (immunoglobin G) fraction of hamster anti-T serum to sepharose or agarose. Radioactively labeled extracts from normal cells, SV40-infected cells, or SV40 transformed cells were incubated with the immunoabsorbants and the proteins which bound were eluted and analyzed by SDS–polyacrylamide gel electrophoresis. A significant fraction of the radioactivity in the eluates corresponding to infected and transformed cells but not normal cells was found to reside in a polypeptide with an estimated MW of 70,000. Tegtmeyer (1974) used an indirect immunoprecipitation method and found that a polypeptide with a MW of about 100,000 and several minor species with lower MW's could be precipitated from SV40-infected, but not uninfected, cell extracts following reaction with hamster anti-T serum. The polypeptide was not precipitated following reaction with normal hamster serum.

Several experiments indicate that SV40 mutations of the tsA group (see Section IV, A) can alter the physical properties and kinetics of induction of T antigen. The sedimentation coefficient of T antigen induced in CV–1 cells by tsA mutants (6–7 S) differs from that induced by wt SV40 (15 S) when the infection is carried out at the nonpermissive temperature (Osborn and Weber, 1975). However, a temperature-shift experiment in which cells infected with a tsA mutant were incubated at the permissive temperature for 69 hours and then shifted to the nonpermissive temperature, failed to show a shift in the sedimentation coefficient of T antigen from 15 S to 6–7 S. Pulse-labeling experiments have provided evidence that the 100,000–dalton polypeptide which is specifically precipitable by anti-T serum (discussed earlier) turns over more rapidly in tsA infected cells than in wild-type infected cells (Tegtmeyer, 1974). Finally, recent experiments (P. Tegtmeyer and C.-J. Lai, personal communication) have shown that a mutant of SV40 which has a deletion of the *Hin*dII / III H and I fragments (the region where most tsA mutants map; see Section IV, A) does not induce the 100,000–dalton polypeptide but induces a smaller polypeptide (MW 35,000) which reacts with anti-T serum. These results provide highly suggestive though not conclusive evidence that T antigen (or at least the 100,000-dalton component of T antigen) may be coded by the A cistron of SV40. Since the product of the A cistron has been implicated in the initiation of SV40 DNA replication (see Section IV, E), it is interesting that several studies have indicated that T antigen has some affinity for DNA (Carroll *et al.*, 1974; Osborn and Weber, 1975; Livingson *et al.*, 1974). Indeed, one recent report has provided evidence that T antigen may have greater affinity for the origin of SV40 DNA replication than for other regions of the genome (Reed *et al.*, 1975).

2. *U Antigen*

The identification of SV40 U antigen was a result of the isolation of a nondefective Ad2–SV40 hybrid, Ad2^{+}ND$_1$, which contained about 17% of the SV40 genome (Lewis *et al.*, 1969) (see Section IV, B). It was found that sera from hamsters bearing SV40–induced tumors which contained significant titers of T antibody could be divided into two groups according to whether or not they also reacted with a heat-stable antigen in $Ad2^{+}ND_{1}$ infected cells (Lewis and Rowe, 1971). Sera which reacted with the Ad2 ND_{1}–induced antigen were designated $T^{+}U^{+}$, and those which did not were designated $T^{+}U^{-}$. The existence of these two types of sera provided the operational basis for distinguishing between SV40 T antigen and SV40 U antigen (Lewis and Rowe, 1971), i.e., SV40 T antigen was defined as an antigen which reacts specifically with both $T^{+}U^{+}$ and $T^{+}U^{-}$ sera and is labile to heating to 50°C for 30 minutes. The SV40 U antigen was defined as an antigen which reacts specifically with $T^{+}U^{+}$ sera but not $T^{+}U^{-}$ sera and is stable to heat treatment (Lewis and Rowe, 1971). The U antigen can also be detected with an antiserum (designated $T^{-}U^{+}$) prepared by immunizing monkeys with $Ad2^{+}ND_{1}$–infected cells (Lewis and Rowe, 1971). Using complement-fixation and immunofluorescent methods, Lewis and Rowe (1971) detected SV40 U antigen in both productively infected and transformed cells. In both types of cells, U antigen was found exclusively in nuclei, and the pattern of fluorescent staining was indistinguishable from that of T antigen. The appearance of U antigen following SV40 infection was found to follow roughly the same time course as the appearance of T antigen; its induction was not sensitive to inhibitors of DNA synthesis, i.e., expression of U antigen is an early function of the SV40 genome. Essentially nothing is known about the physical properties of U antigen or whether or not it is virus coded. Indeed, it is not clear at present whether U and T antigens represent different molecular species. On the basis of currently available data it seems entirely possible that U antigen and T antigen represent different immunologic determinants on the same molecule(s) or that U antigen represents an altered structural form of T antigen.

3. *Tumor-Specific Transplantation Antigen*

Animals immunized with SV40 virions acquire the ability to reject subsequent transplants of SV40–induced tumor cells or SV40–transformed cells (Khera *et al.*, 1963; Habel and Eddy, 1963; Koch and Sabin, 1963; Defendi, 1963). This observation forms the operational definition of an antigen, or group of antigens, in tumor (or transformed) cells that has been called SV40 tumor–specific transplantation antigen (TSTA). Rejection is virus specific, since infection with heterologous viruses will not confer immunity to SV40–induced tumor cells, nor will SV40–immune animals reject tumors induced by unrelated viruses. It appears likely that TSTA is associated with the cell surface, as transplantation immunity can be elicited with injections of membrane preparations from SV40 tumor cells (Coggin *et al.*, 1969), and newborn animals injected with membranes from SV40 tumor cells become immunologically tolerant to SV40 TSTA. Girardi and Defendi (1970) were able to demonstrate that TSTA is also induced during SV40 productive infection in AGMK cells. Induction is prevented by inhibitors of protein synthesis but not by inhibitors of DNA synthesis. By this criterion, induction of TSTA is an early SV40 function. It is not known at present whether TSTA represents a primary gene product of SV40 or a cellular membrane component which is induced, unmasked, or modified following SV40 infection.

4. Surface Antigen(s)

Additional antigens associated with the surface of SV40–transformed or productively infected cells have been detected using several assays including immunofluorescence, mixed hemagglutination tests (MHA) or cytotoxicity tests (Tevethia *et al.*, 1965; Kluchareva *et al.*, 1967; Metzgar and Oleinick, 1968; Häyry and Defendi, 1968). The sera used in these assays were derived from animals injected with SV40 virions or SV40–transformed cells. Although it is not clear that the antigens detected by these various tests are identical, for convenience we will refer to them all as S antigen. The induction of S antigen is virus specific, and the S antigens induced by SV40 in a variety of mammalian cells across react. However, there is compelling evidence that S antigen is not virus coded. Diamandopoulos *et al.* (1968) discovered that some hamster cells transformed following exposure to SV40 did not express SV40 T antigen but did express SV40 S antigen as measured by immunofluorescence. These T^-S^+ cells were examined by A. S. Levine *et al.* (1970) for the presence of SV40–specific DNA using a very sensitive hybridization assay and were found to be completely negative. Furthermore, Häyry and Defendi (1970) found (using the MHA test) that SV40–specific S antigen could be exposed on the surface of normal cells or cells transformed by other viruses by treatment with proteolytic enzymes. Thus, it appears that S antigen is a host antigen which is specifically "unmasked" by SV40 but not by other viruses. Furthermore, maintenance of S antigen in its "unmasked" state does not require the continuous expression of SV40 genetic material. The relationship between one or more S antigens and SV40 TSTA is not clear at present. However, Tevethia *et al.* (1968) have shown that the T^-S^+ hamster lines of Diamandopoulos do not contain detectable levels of SV40 TSTA. This observation argues strongly that S antigen and TSTA need not be identical.

C. Protein Synthesis in SV40–Infected Cells

Several laboratories (Fischer and Sauer, 1972; Anderson and Gesteland, 1972; Walter *et al.*, 1972; Tegtmeyer, 1975) have examined the pattern of protein synthesis in SV40–infected monkey cells. Infected cells were exposed to medium containing radioactive amino acids for short periods of time, and the polypeptides which incorporated the labeled precursor were fractionated by SDS–polyacrylamide gel electrophoresis. Possible SV40-coded proteins were then identified by comparing the labeled polypeptides from infected cells with those from mock-infected cells. There are two major problems associated with this approach. First, SV40 does not repress the synthesis of host proteins, and unfortunately no inhibitor has yet been found which selectively blocks host protein synthesis while leaving viral protein synthesis unimpaired. Thus, at most only a few percent of the total radioactivity incorporated by SV40-infected cells resides in viral-specific polypeptides. Second, infection with SV40 brings about a large number of changes in cellular metabolism, including the induction of host enzyme activities associated with DNA synthesis (Kit, 1968). Hence, the identification of a labeled polypeptide in extracts of infected cells which is absent from uninfected cells does not necessarily imply that it is virus coded. All of the groups which have taken this approach have succeeded in identifying new polypeptides in infected cells that correspond in mobility to the larger viral structural proteins VP1, VP2, and VP3. Identity of the intracellular VP1 with virion VP1 has been established by fingerprint comparisons (Anderson and Gesteland, 1972). Intracellular VP1 is present in much larger amounts than any of the other viral-specific polypeptides, and Anderson and Gesteland (1972) have estimated that it accounts for 5–10% of the total cellular protein late in infection. As might be expected, polypeptides with mobilities identical to VP4–6 are present in both infected and uninfected cells. In addition to the major virion structural proteins, several other polypeptides specific for virus-infected cells have been indentified. Anderson and Gesteland (1972) reported one such polypeptide (IVP3.5) which had a molecular weight of 15,000 and was interpreted as a possible degradation product of VP1. They also observed that a band corresponding to a cellular polypeptide of MW 80,000 often increased in intensity relative to other cellular components following SV40 infection. Walter *et al.* (1972) reported three polypeptides of MW 70,000, 60,000, and 80,000 that were specific for virus-infected cells, and Fischer and Sauer (1972) observed two such polypeptides both of which were smaller than VP1 (molecular weight not reported). The significance of all of these nonstructural polypeptides is not clear at present.

Tegtmeyer (1974) approached the problem of identifying nonstructural SV40-induced proteins in labeled cell extracts by characterizing the polypeptides which are specifically precipitated by treatment with hamster anti-SV40 T serum followed by rabbit anti-hamster globulin. As described in the section on T antigen, this procedure resulted in the identification of a major polypeptide of MW 100,000 and several smaller polypeptides which were specific for SV40-infected cells.

D. In Vitro Synthesis of SV40 Proteins

One decisive way to determine whether a particular polypetide is virus coded is to synthesize it *in vitro* using a cell-free translation system programmed by purified viral DNA or RNA. Prives *et al.* (1974a,b) and Lodish *et al.* (1974) showed that whole-cell messenger RNA isolated from SV40-infected cells could direct the synthesis of VP1 by cell-free extracts derived from wheat germ, Chinese hamster ovary cells, or rabbit reticulocytes. The identity of VP1 synthesized *in vitro* with VP1 from purified virions was established by SDS–polyacrylamide gel electrophoresis, fingerprint analysis, and immunoprecipitation. Messenger RNA from mock-infected cells did not direct the synthesis of VP1. Prives *et al.* (1974a,b) took this approach one step further and showed that SV40-specific mRNA purified from infected cells by selective hybridization to SV40 DNA also programmed the synthesis of VP1. The active RNA species corresponded to the 16 S class of SV40-specific polysomal message which had previously been identified by Weinberg *et al.* (1972). These data provide definitive evidence that VP1 is virus coded, and it is very likely that further exploitation of this general approach will lead to the identification of other SV40-coded proteins.

Several attempts have been made to synthesize SV40 proteins *in vitro* using transcription and translation systems programmed by SV40 DNA (Gelfand and Hayashi, 1969; Bryan *et al.*, 1969; Crawford *et al.*, 1971; B. E. Roberts *et al.*, 1975). Roberts *et al.* (1975) have demonstrated synthesis of VP1 in a system in which transcription of purified SV40 DNA is first carried out using *E. coli* RNA polymerase, and the resulting RNA is then used to program a translation system prepared from wheat germ. A large number of other polypeptides with molecular weights up to 85,000 are synthesized in this system, including some (MW: 60,000, 50,000, 25,000) which appear to react specifically with antibodies against SV40 T antigen. It seems quite likely that this type of system will be very useful in the future for identifying SV40-coded proteins. Given the present availability of specific restriction fragments of the SV40 genome, this general method should also prove useful in mapping the structural genes of SV40-coded proteins on the viral chromosome.

IV. SV40 Mutants

As with other viruses, mutants of SV40 have proven critical in investigations of genomic organization and function. Several types of mutants are now available: (1) conditional mutants which show temperature-sensitive steps in virus development (ts mutants); (2) a series of adeno–SV40 hybrid viruses that contain substitutions of segments of SV40 DNA of varying length and in some cases express SV40 functions in infected cells; (3) evolutionary variants, generated during serial passage of SV40 at high multiplicity of infection, containing deletions, duplications, and/or substitutions within their genomes; (4) deletion mutants generated by enzymatic modification of SV40 DNA; (5) insertion mutants, also constructed enzymatically; (6) reiteration mutants, generated by intracellular amplification of DNA segments.

A. ts Mutants

SV40 ts mutants have been isolated by mutagenizing developing virus with nitrosoguanidine or treating virions with hydroxylamine or ultraviolet light, selecting plaques at 32°C (permissive temperature), and testing for plaque formation at 40°C or 41°C (nonpermissive temperature). In this way more than 200 mutants have been isolated whose plating efficiencies at 32°C relative to 41°C vary from about 10^4 to 10^6.

1. *Complementation Classes*

Complementation tests between pairs of mutants have been carried out by various procedures, some less quantitative than others. In one method, yields of virus at nonpermissive temperature were measured (Tegtmeyer and Ozer, 1971; Kimura and Dulbecco, 1972); since there is a variable degree of leakiness of ts mutants, complementation was quantitated by comparing yields from cells coinfected by two mutants with the sum of yields from cells infected by each mutant alone. In another laboratory, a qualitative test was developed in which cytopathic effect on cell monolayers was assessed at nonpermissive temperature by use of agar slants containing virus (Chou and Martin, 1974); occasionally results were equivocal. A third procedure employed a plaque assay in which the number of infectious centers produced from cells infected by two different mutants was compared to the number produced from singly infected cells (Lai and Nathans, 1974a, 1975a; Mertz and Berg, 1974a); testing of plaques for recombinants was not required. Results of complementation tests by one or another of these procedures are summarized in Table IV. As seen in the table, the ts mutants so far isolated fall into five complementation classes: A, B, C, BC, and D. Using *virions* at nonpermissive temperature, A mutants complement B, C, and BC mutants; B mutants complement A and C mutants; C mutants complement A and B mutants; and BC mutants complement only A mutants. D mutants do not complement or show delayed complementation with other mutants under these assay conditions (Robb and Martin, 1972; Chou and Martin, 1975). From their relative occurrence and reversion frequency (and map position—see

TABLE IV
COMPLEMENTATION GROUPS OF TS MUTANTS

Mutant class	Complemented by				
	tsA	tsB	tsC	tsBC	tsD[a]
tsA	−	+	+	+	+
tsB	+	−	+	−	+
tsC	+	+	−	−	+
tsBC	+	−	−	−	+
tsD[a]	+	+	+	+	−

[a] Note that *tsD virions* do not complement any other mutant class when cells are infected at nonpermissive temperature. When cells are infected at permissive temperature and later shifted to nonpermissive temperature, or when tsD DNA is used, the above results are obtained (see text).

the next subsection), BC mutants do not appear to be double mutants. One interpretation of the complementation pattern of B, C, and BC mutants is that they are all in one cistron, B and C mutants complementing by protein–protein interaction. Thus complementation tests with mutant virions suggest the presence of three genes (A, B/C, and D), although on the basis of these tests alone D mutants could be a special class of A or B/C mutants.

The distinctness of tsD mutants as a separate complementation group has recently been inferred from results of complementation tests in which DNA from D mutants was used instead of virions or infection with virions was initiated at the permissive temperature (Mertz, 1975; C.-J. Lai and D. Nathans, unpublished data). As will be discussed later, D mutant virions appear to be defective in uncoating of viral DNA at nonpermissive temperature and therefore express no viral functions and cannot complement other mutants (Robb and Martin, 1972; R. Martin *et al.*, 1974). However, with tsD DNA, or by allowing uncoating of D mutants at permissive temperature, complementation (as measured by plaque formation) occurs between tsD and tsA mutants and between tsD and tsB mutants (Table IV).

2. A Map of ts Mutants

Temperature-sensitive mutants of SV40 are being mapped by recombination (Dubbs *et al.*, 1974), by marker rescue using endo R fragments of SV40 DNA (Lai and Nathans, 1974b, 1975a; Mantei *et al.*, 1975), by nuclease cleavage at the site of mismatched bases in ts/wt heteroduplex DNA molecules (Shenk *et al.*, 1975), and by complementation with deletion mutants (Lai and Nathans, 1974c). At present most information comes from marker rescue experiments.

Marker rescue experiments with SV40 mutants were based on the methodology developed with coliphage ϕX174 (Hutchison and Edgell, 1971) and adapted to SV40 and animal cells (Lai and Nathans, 1974b). In this procedure (outlined in Fig. 8) single-stranded circles of mutant DNA are hybridized to specific, denatured wt SV40 DNA fragments to form partial heteroduplex molecules, which are then used to infect a monkey cell monolayer at nonpermissive temperature; in the infected cell a complete heteroduplex is formed, followed by replication and/or correction of mismatched base pairs. Where the fragment contains the sequence homologous to the mutant site, the mutant is "rescued" and plaques are formed at high temperature. Since the position of fragments in the genome is known from the cleavage map (see Fig. 5), the mutant site can be localized. In practice this method requires highly purified fragments, since nonspecific fragments enhance the activity of any contaminating fragments corresponding to the mutant site.

In the procedure for mapping point mutants by nuclease cleavage at the site of mismatched bases in wt/mutant heteroduplex DNA molecules (Shenk *et al.*, 1975), mutant DNA and wild-type (or revertant) DNA are separately linearized by cleavage with a single cut restriction enzyme. Heteroduplexes are then formed by melting and reannealing a mixture of the two linear DNA's, followed by cleavage of the heteroduplex with the single-strand specific nuclease S_1, purified from *Aspergillus oryzae* (Ando,

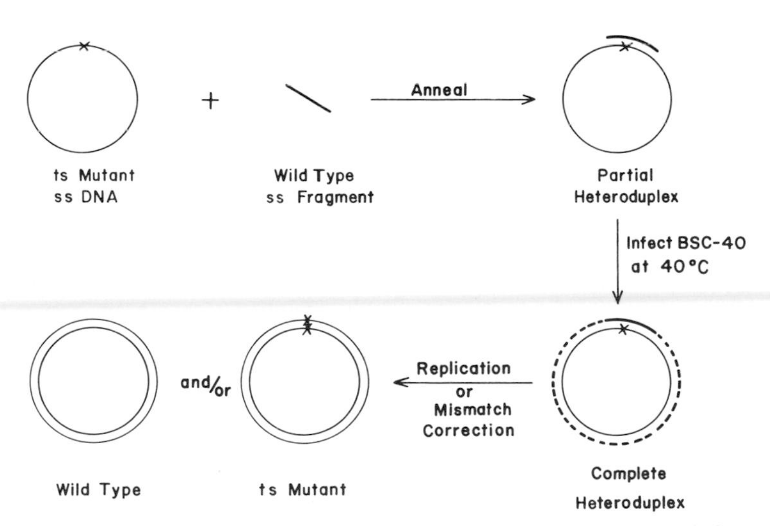

FIG. 8. Scheme for mapping SV40 ts mutants by marker rescue. × indicates the mutational site. For details, see the text. (Figure supplied by C.-J. Lai.)

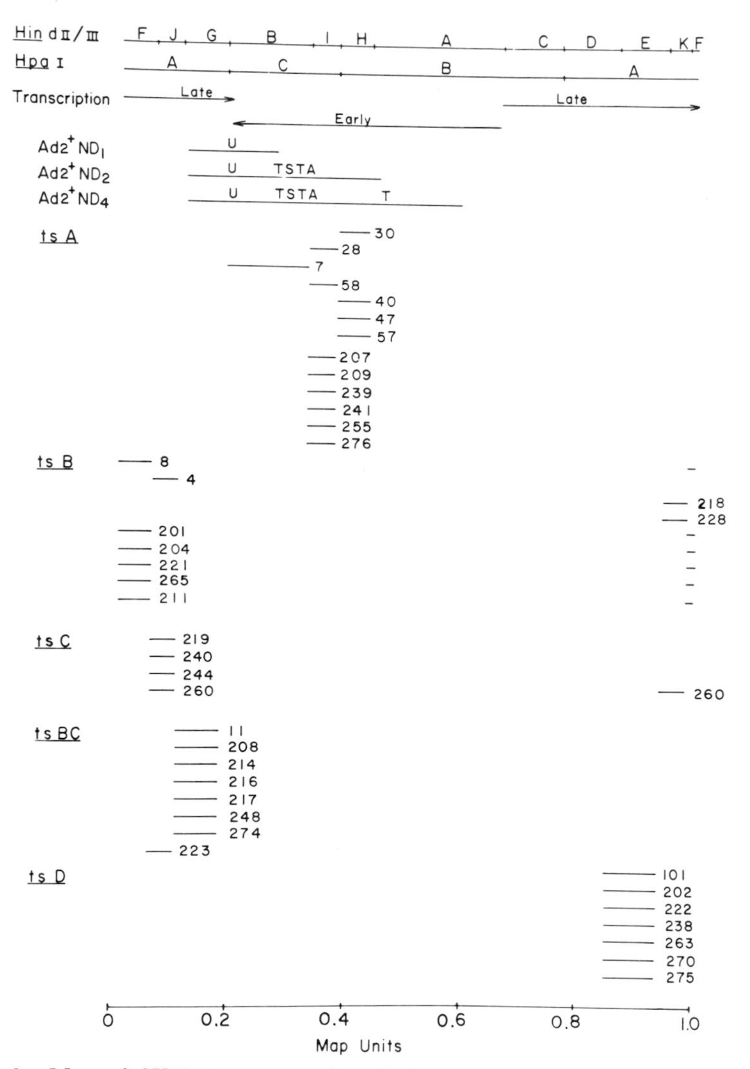

FIG. 9. Map of SV40 ts mutants in relation to the *Hind* cleavage map. The genome is shown as a linear map starting at the *Eco*RI site. Also shown are the *Hind*II/III and *Hpa* I cleavage maps, early and late transcription regions (see text), and map positions of SV40 DNA segments in adeno-SV40 hybrid viruses (see text). For each mutant the bar indicates the fragment which corrects the mutational defect. [From Lai and Nathans (1975a).]

1966). The resulting DNA fragments are then separated and sized by agarose gel electrophoresis. From the size of the resulting dominant fragments, one can deduce the position of the preferred site of S_1 scission relative to the site of cleavage by the restriction enzyme used to linearize the DNA. Repetition of the procedure with DNA cut by a second one-site restriction enzyme localizes the S_1 scission unambiguously. The preferred S_1 cleavage site is assumed to correspond to a mismatched base pair involving the mutated base. Compared to marker rescue, the heteroduplex-S_1 procedure has the potential of localizing a mutant site in DNA more precisely. However, at the present stage of development of this promising method, it is not clear that a single mismatched base pair is sufficient for S_1 cleavage of duplex DNA; in fact, some heteroduplex molecules made up of genetically distinct SV40 DNA strands are not cleaved, suggesting that a mismatched pair is insufficient or that certain mismatched base pairs are not enzyme targets. Also, in practice, different results may be obtained for the same ts mutant with wt/ts heteroduplex and with revertant/ts heteroduplex, making localization of the mutational site uncertain. In contrast to these limitations for mapping point mutants, the interpretation of results of mapping deletions (Shenk *et al.*, 1975) or substitutions (Bartok *et al.*, 1974) by the heteroduplex–nuclease method is straightforward (see Section IV, D).

The map positions of mutants rescued by specific *Hin* fragments are shown in Fig. 9 (Lai and Nathans, 1975a). As seen in the figure, the 41 mutants mapped cover about half of the genome; there is a notable absence of mutants mapping between 0.43 and 0.85 map units, and only one mutant that maps between 0.17 and 0.35 map units. Group A mutants all map in the "early" region of the genome (see Section VI); 12 of 13 such mutants are clustered in about one-fourth of the "early" region between 0.32 and 0.43 map units. Mutants in complementation classes defective in a late function (discussed next) map in the "late" region and also show clustering: 8 of 9 B mutants map between 0.94 and 0.06 map units; all C mutants map between 0.06 and 0.11 map units; and 7 of 8 BC mutants map between 0.11 and 0.17 map units. D mutants all map between 0.85 and 0.94 map units, i.e., in a small segment of the "late" region. One C mutant proved to be a double mutant with one mutation in the B region and another in the C region, both mutations being required for the tsC phenotype. Thus the mapping data supports prior complementation tests (described earlier) suggesting that B, C, and BC mutations are in one cistron, complementation between B and C mutants occurring by protein subunit interaction.

3. Mutant Phenotypes

The phenotypes of ts mutants in each complementation class are summarized in Table V.

a. Mutants. As seen in Table V, there is only one class of true DNA synthesis-defective mutants, the A group. At nonpermissive temperature, little or no viral DNA is synthesized in tsA-infected cells (Tegtmeyer and Ozer, 1971; Kimura and Dulbecco, 1972; Chou *et al.*, 1974). Tempera-

TABLE V
PHENOTYPES OF SV40 TS MUTANTS

Complementation class	DNA synthesis		ts virions	DNA infection	Antigens				Transformation
	Virus	Host			T	U	C	V	
tsA	−	±	No	ts	+	−	−	−	−[a]
tsB	+	+	Yes	ts	+	+	+	+	+
tsC	+	+	Yes	ts	?	?	?	?	+
tsBC	+	+	Yes	ts	+	−	+	−	+
tsD	−	−	No	wt	−	−	−	−	−

[a] Initiation and maintenance of transformation are temperature sensitive (ts).

ture-shift experiments indicate that "A protein" function in infected permissive cells is required relatively early in the infectious cycle (compared to "B protein") (Tegtmeyer, 1972; Chou and Martin, 1975), consistent with that protein's involvement in viral DNA replication and the map position of A mutants. Temperature shift experiments also have demonstrated that, once viral DNA synthesis has begun (at the permissive temperature) in tsA-infected cells, partially replicated viral DNA molecules can be converted to completed form I molecules at nonpermissive temperature, but formation of new replicating molecules does not occur (Tegtmeyer, 1972). These observations have led to the important conclusions that the A protein is involved in initiating SV40 DNA replication but not in the elongation or termination steps of viral DNA replication (see Section V, B). What the initiation defect of A mutants is at the molecular level remains unknown.

A second interesting property of A mutants is their effect on cellular DNA synthesis. As pointed out earlier (see Table I), wt SV40 stimulates cellular DNA synthesis in most permissive and nonpermissive cells (Gershon *et al.*, 1966; Hatanaka and Dulbecco, 1966; Henry *et al.*, 1966; Werchau *et al.*, 1966). This activity appears to be dependent on viral gene expression, since ultraviolet irradiation of virions inactivates this activity at about one-third the rate of inactivation of plaque-forming ability (Chou and Martin, 1975); moreover, certain deletion mutants of SV40 (see Section IV, E) fail to stimulate cell DNA synthesis (Brockman and Scott, 1975). Based on the time course of virus-stimulated cellular DNA synthesis (Hatanaka and Dulbecco, 1966; Ritzi and Levine, 1970), this activity of SV40 appears to be an "early" function. In tsA-infected cells, cellular DNA synthesis is still stimulated at temperatures where viral DNA replication is barely detectable (Tegtmeyer, 1972; Chou and Martin, 1975). However, at somewhat higher temperatures, the ratio of high to low temperature stimulation is much reduced compared to wt-infected cells (Chou and Martin, 1975). Therefore, it does appear that tsA mutants are defective in stimulating host DNA synthesis, but the consistently greater resistance to high temperature of this activity compared to viral DNA replication remains puzzling.

Measurements of viral RNA in cells infected with a tsA mutant indicates that at nonpermissive temperature early viral RNA is synthesized but not late RNA (Cowan *et al.*, 1973). After viral DNA synthesis has occurred at permissive temperature, late viral RNA continues to be made after shift to high temperature. These observations are similar to those made with many DNA phages, indicating that transcription of late genes is related to phage DNA replication. In the SV40 case, the observations suggest that A protein function is not continuously required for late transcription, but the mechanism for early to late gene transcription remains obscure.

Among the most interesting properties of tsA mutants is their inability or reduced ability to transform cells at high temperature (Tegtmeyer, 1972; Kimura and Dulbecco, 1973) coupled with the temperature-dependent phenotype of cells transformed by A mutants at permissive temperature (Butel *et al.*, 1974; Kimura and Itagaki, 1975; Martin and Chou, 1975; Osborn and Weber, 1974, 1975; Tegtmeyer, 1975; Brugge and Butel, 1975). Mutants of no other class have such characteristics. All tsA mutants adequately tested have shown these properties, including mutants which map in genome segments *Hin*-B, -I, and -H, respectively (Fig. 9). Moreover, temperature-dependent growth properties of A mutant–transformed cells have been observed in cells from a variety of sources: rat embryo (Osborn and Weber, 1974, 1975); mouse 3T3 line (Swiss), rabbit kidney, Syrian hamster (Tegtmeyer, 1975); Chinese hamster lung cells (Martin and Chou, 1975); human Fanconi skin cells, Syrian hamster embryo fibroblasts, mouse 3T3 line (Balb), hamster tumor cells (Butel *et al.*, 1974); and rat embryo fibroblast line (Kimura and Itagaki, 1975). Several parameters of transformation have been assessed at permissive and nonpermissive temperatures in mutant-transformed cells versus wt-transformed cells: growth rate; saturation density on a plastic surface; growth in media with reduced serum concentration; colony formation on plastic, on nontransformed cell monolayers, or in soft agar; cell morphology; presence of actin "cables"; presence of T antigen; uptake of 2-deoxyglucose. In some cases clones of transformed cells were tested; in other cases uncloned populations were used (all cells of which were T antigen positive), obtained by repeated passage of virus-infected cells.

Results have been generally consistent: most or all of the parameters of transformation measured showed greater temperature sensitivity in tsA-transformed cells than in wt-transformed cells. One of the properties of tsA-transformed cells most sensitive to high temperature was their ability to form colonies on nontransformed cell monolayers. In some instances, the ability of mutant-transformed cells to produce transformed colonies was measured at permissive temperature after maintenance of the cells for many generations at nonpermissive temperature. The fact that the transformed growth pattern was restored by the temperature shift indicates that at least this transformation property was not irreversibly inactivated at nonpermissive temperature and that failure to observe transformed colonies at high temperature was not due to cell death. Although rare A mutant-transformed cells or clones failed to show temperature dependence of one or another parameter of transformation, in none of these instances was the presence of a resident wt SV40 genome excluded; moreover, in all but one case the same mutants were shown to cause temperature-dependent growth in other types of cells. Therefore, the overall conclusion of these important studies is that functional A gene product is required to maintain transformation, including all the measured parameters of the SV40-transformed phenotype.

Although A mutants are uniformly defective in "initiation" of transformation, as noted earlier, it has not been possible to distinguish experimentally between "initiation" and maintenance of transformation when maintenance is temperature dependent. Therefore it is not known whether the A protein is also involved in "initiating" transformation in a specific way, for example, by promoting integration of viral genomes into cellular DNA.

The demonstrated involvement of SV40-A protein (which as discussed earlier appears to be T antigen) in maintenance of transformation, stimulation of cell DNA synthesis, and initiation of SV40 DNA replication has provoked much speculation on a possible common molecular basis for these various activities. In one type of hypothesis, A protein is thought to recognize a specific nucleotide sequence signal in SV40 DNA and in cell DNA, binding at these sites and somehow causing cellular enzymes to initiate DNA replication there (Levine, 1972; Martin *et al.*, 1974; Butel *et al.*, 1974). (As noted in Section III, B, there is recent evidence that SV40 T antigen binds preferentially to SV40 DNA at the origin of DNA replication.) In an alternate hypothesis, A protein is thought to modulate more directly the expression of a series of cellular genes, including those coding for DNA replication enzymes (Hatanaka and Dulbecco, 1966; Kit, 1968). Neither hypothesis specifies the precise molecular events connecting effects on cellular DNA and altered cell growth.

b. B, C, and BC Mutants. Mutants in the B, C, and BC complementation classes share certain properties (Table V): (1) some mutants in each of these classes have temperature-sensitive virions, indicating that a virion protein is altered (Tegtmeyer, 1972; Ishikawa and Aizawa, 1973; Chou and Martin, 1974); (2) in mutant-infected cells, viral DNA replication occurs and host cell DNA synthesis is stimulated; (3) temperature-shift experiments indicate that the temperature-dependent function of these mutants occurs later in the infectious cycle than does that of tsA mutants (Tegtmeyer, 1972; Chou and Martin, 1975). Hence B, C, and BC mutants show normal early function but are defective in late function, due to mutant structural protein. As indicated in Table V, two distinct phenotypes have been described in this group of mutants, related to the extent of virus development at restrictive temperature, although few mutants have been studied in detail (Tegtmeyer and Ozer, 1971). In one type (B complementation group, e.g., B8) V antigen as well as C (capsid) antigen is detectable in infected cells, and noninfectious virus particles are formed. In the other type [BC complementation group, e.g., BC11, formerly called B11 (Mertz and Berg, 1974a; Lai and Nathans, 1975a)] C antigen, but not V antigen, is found in cells infected at restrictive temperature, and virus particles are not formed. One interpretation of these results, consistent with the notion that B and BC mutants are in one cistron (see Sections IV, A, 1 and 2), is that the B/C gene codes for the major virion protein VP1, which in monomeric form is C antigen and in capsomers is V antigen, and B-mutant VP1 can form capsids of virus particles whereas BC-mutant VP1 cannot. Temperature-sensitive C mutants have not been as well characterized, but a mutant with C complementation properties does not yield V antigen in infected cells (Kimura and Dulbecco, 1972).

Another interesting but unexplained difference between a B and a BC mutant has been reported: cells infected by ts BC11 at restrictive temperature show little or no SV40 U antigen by indirect fluorescent antibody staining, whereas B8-infected cells appear similar to wt-infected cells (Robb *et al.*, 1974). Since U antigen results from an early SV40 function and the BC mutation is in the late region of the genome, the explanation for this observation is likely to be complex.

Transformation by B, C, and BC mutants is entirely comparable to transformation by wt SV40 (Martin and Chou, 1975). Neither efficiency of transformation nor maintenance of transformation is more temperature dependent than wt virus.

c. D Mutants. The phenotype of tsD mutants is quite distinctive (Robb and Martin, 1972; Chou and Martin, 1975). Infection by D mutant virions at restrictive temperature results in viral adsorption and penetration, but no viral functions are detectable in the cells, including viral RNA, early or late antigens, viral DNA replication, or cellular DNA synthesis. The step which is temperature sensitive occurs very early in the infectious cycle, as judged by temperature-shift experiments. Also, unlike other ts mutants, the infectivity of D mutants is dependent on factors other than temperature, e.g., serum concentration in the medium and whether cells are stationary or growing. Most strikingly, infection by D mutant DNA is entirely normal through one round of infection at high temperature, though of course the progeny are D mutant virions. This collection of properties has led to the conclusion that D mutants have a mutant virion protein from which viral DNA cannot be freed within infected cells under restrictive conditions. Since this uncoating is probably a cellular function, the dependence of the infectivity of D mutants on the state of the cells may be related to variable cellular uncoating activity. Although D mutant virions are not unusually heat sensitive, the map position of D mutants in the late region of the genome (see Fig. 9) supports the notion that a structural protein is defective. An attractive possibility is that VP3 (or VP2), a minor virion protein (see Section III, A), is the D gene product.

D mutants are often less active than wt virus in transformation at nonpermissive temperature (Martin and Chou, 1975), a fact readily explainable by an uncoating defect. In contrast to A mutant–transformed cells, when cells are transformed by D mutants at permissive temperature and shifted to restrictive temperature, they retain the transformation phenotype. However, it has been reported that some D mutant transformants appear to have much reduced T antigen at elevated temperature as judged by immunofluorescence (Robb, 1973, 1974).

B. Adenovirus–SV40 Hybrids

Genetic recombinants between adenoviruses and SV40 (Ad–SV40 hybrids) have proven useful in analyzing the organization of genetic information in the SV40 genome. As will be described in detail shortly, the genomes of Ad-SV40 hybrids are enclosed in apparently normal adenovirus capsids and contain specific segments of the SV40 genome covalently inserted into the adenovirus genome. In the past several years the SV40 DNA segments carried by a number of different hybrids have been mapped on the SV40 chromosome. Since the SV40 genetic information in Ad–SV40 hybrids is often expressed in infected cells, it has proved possible to assign certain specific SV40 functions to specific regions of the SV40 genome.

The first hybrid to be isolated and characterized was obtained from a stock of human adenovirus 7 which had been "adapted" to growth in monkey kidney cells for vaccine production (Hartley *et al.*, 1956). After passage for several years in monkey cells, this stock was found to be contaminated with SV40. In order to eliminate the contaminant the stock was passaged twice in the presence of neutralizing antibody against SV40. At the end of this procedure the stock (E46$^+$) contained no detectable free SV40 virions but did contain an Ad7–SV40 hybrid. The hybrid particles could be recognized because they induced SV40 T antigen in infected cells and caused tumors in hamsters that resembled SV40-induced tumors morphologically and contained SV40 T antigen (Huebner *et al.*, 1964; Rowe and Baum, 1964; Rapp *et al.*, 1964b). The ability of the hybrid particles to express these SV40 functions could be completely abolished by treatment of the inoculum with Ad 7 neutralizing antibody but was unaffected by treatment with SV40 neutralizing antibody. In subsequent studies (cited below) the Ad7–SV40 hybrids in the E46$^+$ stock were shown to be true genetic recombinants between Ad7 and SV40. The question arises as to why passage of a mixed stock of Ad7 and SV40 in the presence of antiserum against SV40 should have led to the selection of such recombinants. The answer seems to lie in the fact that SV40 greatly enhances the yield of adenovirus from infected monkey cells (Rabson *et al.*, 1964). Normally, infection of monkey cells with the human adenoviruses at low multiplicities is abortive. Early functions are expressed and DNA synthesis occurs, but little or no capsid protein is produced. If the cells are simultaneously infected with SV40, however, abundant capsid protein is synthesized, and the yield of infectious adenovirus virions is increased by a factor of 10^3. The exact biochemical basis for the "helper" effect of SV40 is not known. [However, recent studies (Baum *et al.*, 1968, 1972; Friedman *et al.*, 1970; Fox and Baum, 1974; Hashimoto *et al.*, 1973; Nakajima *et al.*, 1974; Baum and Fox, 1974; Eron *et al.*, 1975) indicate that the block to synthesis of adenovirus capsid protein in monkey cells may be at the level of translation and that "helper" function requires the expression of an early SV40 gene.] It follows from these facts that when the progeny of a mixed (Ad and SV40) infection are passaged in the presence of SV40 antiserum there will be strong selection pressures against both wt parents. However, recombinant progeny (hybrids) which contain adenovirus genetic information and at least that portion of the SV40 genome required for "helper" function will have a selective advantage. Since the isolation of the E46$^+$ hybrid, a number of other Ad–SV40 hybrids have been obtained by similar procedures (Lewis *et al.*, 1966, 1973; Easton and Hiatt, 1965; Beardmore *et al.*, 1965; Schell *et al.*, 1966; Lewis and Rowe, 1970).

The adenovirus and SV40 DNA sequences in the genomes of Ad–SV40 hybrids are covalently linked as shown by the fact that they do not separate in alkaline CsCl equilibrium gradients or alkaline sucrose gradients (Baum *et al.*, 1966; Crumpacker *et al.*, 1970; Henry *et al.*, 1973). The detailed structure of the DNA molecules from a number of hybrids have been examined by electron microscopic heteroduplex methods (Kelly and Rose, 1971; Kelly and Lewis, 1973; Morrow *et al.*, 1973; Kelly *et al.*, 1974a,b). The hybrid DNA molecules, like those of wt adenovirus, are linear. Each hybrid genome contains a single substitution, i.e., a segment of adenovirus DNA is deleted and is replaced by a segment of SV40 DNA (Fig. 10). The sizes of the Ad DNA deletions and the SV40 DNA insertions are different for the various hybrids which have been studied (see Fig. 10), and in a given hybrid the deletion and insertion are not generally equal in size. The E46$^+$ hybrid just discussed was found by Kelly and Rose (1971) to contain an insertion equivalent to roughly 75% of the SV40 genome and a deletion of about 10% of the Ad7 molecule. The location of this substitution was found to be 0.05 fractional lengths from one end of the hybrid DNA molecule. Subsequent studies (Kelly, 1975; Lebowitz and Khoury, 1975) have demonstrated that the detailed structure of the inserted SV40 DNA in E46$^+$ is rather complex. The segment contains the entire early region of the SV40 genome but only a small portion of the late region, and part of the early region is repeated in tandem.

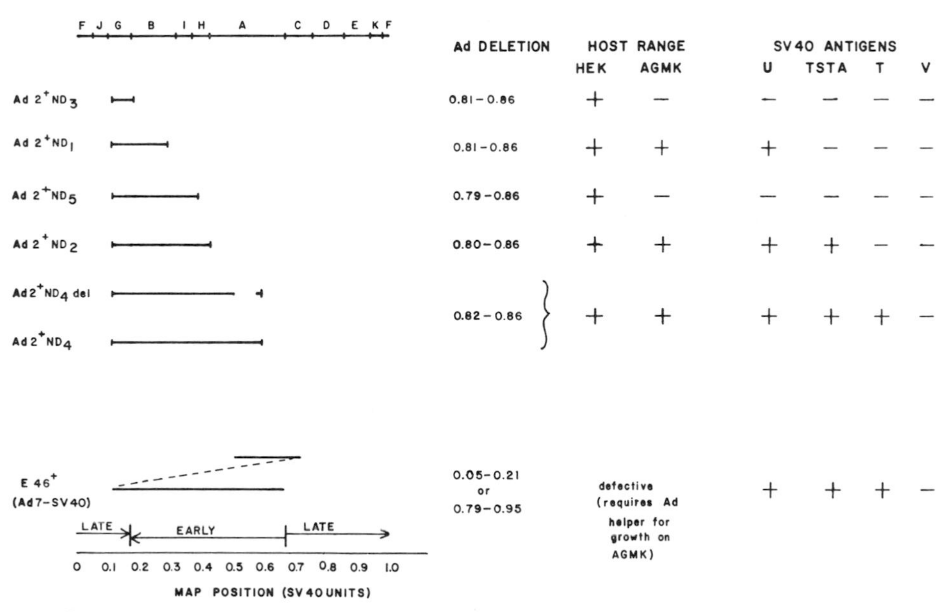

FIG. 10. Structure and properties of Ad-SV40 hybrid viruses. For details, see text.

These data correlate well with findings (1) that $E46^+$ induces only early SV40 antigens, U, TSTA, and T (Huebner *et al.*, 1964; Rowe and Baum, 1964; Rapp *et al.*, 1964b; Lebowitz and Khoury, 1975), and (2) that the predominant SV40-specific RNA sequences in $E46^+$ infected cells are similar, if not identical, to those found in the cytoplasm during the early phase of SV40 infection (Lebowitz and Khoury, 1975).

Ad–SV40 hybrids can be either defective or nondefective. In the case of defective hybrids, the substitution is presumably located in a region of the adenovirus genome that is essential for viral replication. Such hybrids cannot replicate in the absence of a "helper" wild-type adenovirus. The $E46^+$ hybrid is a good example of a defective Ad–SV40 hybrid. All stocks of $E46^+$ are actually mixed virus populations containing both hybrid particles and wild type particles, with the latter usually in excess. On AGMK cells such stocks plaque with two-hit kinetics, indicating that coinfection with both types of particles is required for viral replication (Rowe and Baum, 1964). The nature of the essential Ad7 function missing from the hybrid genomes is not known. Other defective Ad–SV40 hybrids include $Ad2^{++}HEY$ and $Ad2^{++}LEY$, both of which have large deletions of adenovirus DNA and contain more than one genome equivalent of SV40 DNA (Lewis and Rowe, 1970). The structure and properties of these hybrids will be described in Section VII.

A series of nondefective Ad2–SV40 hybrids have been isolated by Lewis *et al.* (1969, 1973, 1974). Cloned stocks of these hybrids are free of wild-type adenovirus and plaque with one-hit kinetics on human cells. All of the nondefective Ad2–SV40 hybrids which have been studied to date were derived from the same virus pool, and the structures of their genomes are closely related (Kelly and Lewis, 1973; Morrow *et al.*, 1973; Lebowitz *et al.*, 1974). As in the case of other Ad–SV40 hybrids, the genome of each nondefective Ad2–SV40 hybrid contains a single substitution (Fig. 10). Furthermore, one of the substitution end points (located at Ad2 map position 0.86) is the same for all of them. The SV40 DNA segments of the hybrids extend various distances to the left of the common end point, forming a completely overlapping set (Fig. 10). The SV40 DNA segments range in size from 0.07 SV40 units ($Ad2^+ND_3$) to 0.59 SV40 units ($Ad2^+ND_4$), and the Ad2 deletions from 0.045 Ad2 units ($Ad2^+ND_4$) to 0.071 Ad2 units ($Ad2^+ND_5$). Since these hybrids are nondefective, these data suggest that the region of the Ad2 genome extending from Ad2 map position 0.79 to map position 0.86 (deleted in $Ad2^+ND_5$) is not essential for Ad2 replication.

Using electron microscopic heteroduplex methods (Kelly and Lewis, 1973; Morrow *et al.*, 1973; Lebowitz *et al.*, 1974) or other hybridization procedures (Lebowitz *et al.*, 1974), the SV40 DNA segments in the nondefective Ad2–SV40 hybrids were mapped within the SV40 genome. The SV40 DNA segments all begin at 0.11 in the SV40 physical map and extend various distances in a clockwise direction (see Fig. 10). Levine *et al.* (1973) and Khoury *et al.* (1973a) have analyzed the SV40-specific RNA molecules made in Vero cells infected with the nondefective Ad2–SV40 hybrids. Their results indicate that: (1) most, if not all, of each SV40 DNA segment is transcribed into RNA; (2) the template for the SV40-specific RNA molecules made *in vivo* is the E strand of SV40; and (3) the RNA sequences induced by $Ad2^{+}ND_{4}$ which contains the largest SV40 segment, include most, if not all, of the sequences found in the cytoplasm during the early phase of SV40 productive infection. The third of these observations locates the early region of the SV40 genome as roughly between the boundaries of SV40 map positions 0.11 and 0.59, in good general agreement with the results obtained using restriction enzymes (see Section VI). Various early SV40-specific antigens (U, TSTA, and T) are induced by the different nondefective hybrids. When the pattern of antigen induction is compared with the map positions of the SV40 DNA segments of the hybrids, a self-consistent map can be constructed which assigns induction of a given antigen to a particular region of the SV40 genome. The order of the antigen-inducing regions is U–TSTA–T proceeding clockwise around the SV40 map from the common end point at map position 0.11 (Fig. 10). It is necessary to be cautious in interpreting this map of the SV40 antigen inducing regions for several reasons. First, it is not known at present whether all of the early SV40 antigens are virus coded. Second, even if the early antigens are virus coded, they would not necessarily represent three independent gene products as opposed to three antigenic determinants on the same molecule. [In this regard it is important to note that only one early SV40 gene required for lytic growth has been detected by genetic methods (see Section IV, A).] Finally, one hybrid, $Ad2^{+}ND_{5}$, behaves anomalously with regard to antigen induction. The SV40 DNA segment of this hybrid contains all of the sequences which $Ad2^{+}ND_{1}$ contains; however, the latter induces U antigen, and the former does not. The reason for this is not known; however, since $Ad2^{+}ND_{5}$-infected cells contain all of the SV40-specific RNA sequences which are present in $Ad2^{+}ND_{1}$-infected cells, the inability of $Ad2^{+}ND_{5}$ to induce U antigen may be related to some defect at the translational level. In summary, although it is possible to associate specific parts of the early region of the SV40 genome to the induction of specific antigens, the detailed nature of the association is not clear. One other SV40 function, namely, the adenovirus "helper" function, can be mapped using the nondefective hybrids. When the host range of the hybrids was examined (Lewis *et al.*, 1973), it was found that all those hybrids which expressed U antigen plaqued on monkey cells with about the same efficiency as on human cells. The smallest SV40 DNA segment which induces U antigen is contained in $Ad2^{+}ND_1$. This places the region necessary for "helper" function between SV40 map positions 0.11 and 0.29.

C. Evolutionary Variants

As with many other viruses, during serial passage of SV40 at high multiplicity of infection, defective viruses appear whose genomes contain deletions, reiterations, and/or rearrangements of viral DNA segments (Uchida *et al.*, 1968; Yoshiike, 1968; Uchida and Watanabe, 1969; Tai *et al.*, 1972). Most of the defective viruses contain sizable deletions and can therefore be separated from wt virus by centrifugation to equilibrium in CsCl gradients (Yoshiike, 1968). The DNA of these light virions is present as covalently closed (form I) DNA, and varies from about 65% of the length of wt SV40 DNA to slightly longer than unit length, probably reflecting the size limits for encapsidation. In addition, particularly after several successive passages at high multiplicity, genomes evolve that contain segments of cellular DNA covalently linked to SV40 DNA (Lavi and Winocour, 1972; Martin *et al.*, 1973); eventually, the encapsidated DNA contains primarily cellular DNA sequences (Brockman *et al.*, 1973). On the basis of the structure of variant genomes (discussed next), it appears that variants are generated by extensive recombination within and between SV40 DNA molecules and between SV40 DNA and cellular DNA. When wt SV40 is present to supply *trans* functions, defective genomes that retain at least the origin of viral DNA replication will multiply and become encapsidated.

Examination of populations of variant DNA molecules by electron microscopic heteroduplex mapping (Tai *et al.*, 1972; Risser and Mulder, 1974; Mertz *et al.*, 1974; Yoshiike *et al.*, 1974a) or by restriction endonuclease cleavage (Brockman *et al.*, 1973; Rozenblatt *et al.*, 1973) indicates that deletions in the viral genome can occur at many different sites, although in the large plaque strain of SV40 a more restricted distribution of deletions has been found than in the small plaque or minute plaque strains (Yoshiike *et al.*, 1974a). In regard to the substitutions resulting from recombination with cellular DNA, as noted in Section VII, D, there is some indication that substitutions occur at preferred sites in the SV40 genome.

1. Isolation of Cloned SV40 Variants

In order to analyze evolutionary variants more precisely and to use them in genetic and physiological studies, it was necessary to clone individual variants. Methods have been developed for cloning two general types of defective variants of SV40 present in the evolving population of viruses. Those which retain enough SV40 DNA to express at least one viral gene (complementing variants) have been cloned by selective plaque formation with a ts mutant of SV40 as helper (Brockman and Nathans, 1974; Mertz and Berg, 1974a); see Fig. 11. The starting material can be a preparation of light virions purified by CsCl centrifugation, short variant genomes isolated by agarose gel electrophoresis, or viral DNA resistant to a restriction enzyme that makes one break in wt SV40 DNA. Using as helper an (early) tsA mutant, variants that produce functional A protein form plaques at nonpermissive temperature; with a (late) ts B, C, or BC mutant, variants that produce functional B/C protein form plaques. In this way mutants with defects in any *trans* complementible gene can be cloned. To clone those variants which retain the origin of replication but express no SV40 genes, a nonselective plaquing technique has been used in which wt SV40 served as helper, thus supplying all *trans* viral functions (Brockman and Nathans, 1974). Plaques have been surveyed for the presence of a variant by electrophoretic analysis of viral DNA (or endo R digests thereof) prepared from cells infected with plaque suspensions. Variant genomes have greater electrophoretic mobility than helper virus genomes and/or show altered endo R digestion patterns. Stocks of cloned variants have been prepared by coinfecting cells with both the variant and helper virus, i.e., by adding the appropriate ts helper virus (Mertz *et al.*, 1974) or by infecting the edge of a susceptible monolayer with a few drops of plaque suspension to initiate infection

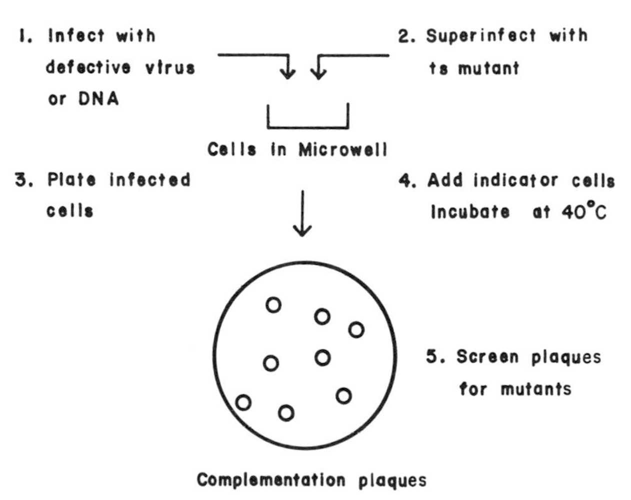

FIG. 11. Method of cloning SV40 variants by complementation with ts mutants. See text for details.

(Brockman and Nathans, 1974). The titer of variant and helper virus in the stock was then assessed by complementation plaquing for those variants selected by complementation. Noncomplementing variants were quantitated by physical methods after separation of variant particles or DNA from the helper virus.

In addition to the foregoing types of nonviable variants, viable deletion mutants which form small plaques have been cloned from serially passaged virus by infecting cells with viral DNA resistant to a restriction enzyme that makes one break in SV40 DNA (Mertz and Berg, 1974b). Specific SV40 variants have also been isolated from clones of cells transformed by defective SV40 by triple fusion of the transformed cells with wt SV40–transformed cells and permissive monkey cells (Yoshiike *et al.*, 1974a,b). Under these conditions, defective genomes rescued from the defective virus-transformed cell propagate in the triple heterokaryons together with virus rescued from wt SV40–transformed cells.

2. Examples of Cloned Variants

The genomes of several complementing variants cloned from serially passaged SV40 stocks have been mapped in relation to wt SV40 by restriction endonuclease analysis and by electron microscopic heteroduplex mapping (Mertz *et al.*, 1974; Brockman *et al.*, 1974, 1975). A few examples are shown in Fig. 12. As seen in the figure, some variants have deletions in the early region of the genome, whereas others have deletions in the late region. Moreover, two of the variant genomes shown, and most others studied, have reiterations of a DNA segment which includes the origin of SV40 DNA replication. Such reiteration presumably confers selective replicative advantage on these variants, thus making them more abundant in the evolving virus population (see also Section V, B).

Also shown in Fig. 12 is an example of a viable deletion mutant isolated from a serially passaged virus as already described (Mertz and Berg, 1974b). A series of such variants have been reported in which the size of the deletion varies from about 80 to 190 nucleotide pairs around map position 0.74, thus indicating that this region of the genome is not essential for virus development, although these mutants produce small plaques. It is interesting that the minute plaque strain of SV40 isolated from the small plaque strain (Takemoto *et al.*, 1966) appears to have a deletion in this region of the genome, in addition to alterations elsewhere (Nathans and Danna, 1972).

Two examples of noncomplementing variants cloned in the presence of wt SV40 are shown in Fig. 13. Also shown is the genome structure of a third variant detected by its supersensitivity to endo R • *Eco*RI

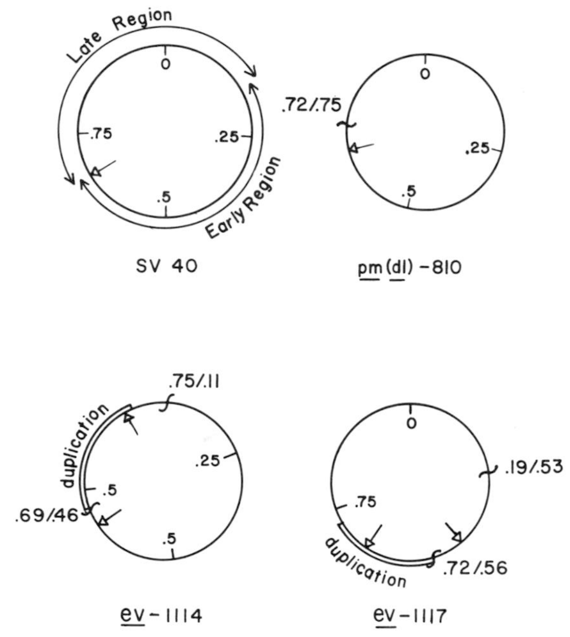

FIG. 12. Examples of cloned complementing variants of SV40. Each variant genome is represented as a circular map with standard map coordinates as indicated by the numbers, in relation to the standard SV40 map. KEY: ⌙⌝, site of recombination between map coordinates indicated in the generation of a deletion or duplication; *short arrows,* origin of DNA replication; *double areas,* duplication as noted; *pm(dl)*-810 is a viable deletion variant (Mertz and Berg, 1974b); *ev*-1114 and *ev*-1117 are nonviable variants (Brockman *et al.,* 1975).

(Fareed *et al.,* 1974; Khoury *et al.,* 1974). [This variant has also been cloned by coinfecting cells with the *Eco*RI digest product and wt SV40, resulting in reiteration of the infecting segment (Davoli and Fareed, 1974).] As seen in Fig. 13, each of the variant genomes contains tandem repetitions of a basic DNA segment, and in all cases the SV40 DNA present includes the origin of viral DNA replication. In addition, the DAR variant genome has an inverted segment, and the other variants contain substitutions of cellular DNA.

D. Constructed Mutants

The ability to cleave SV40 DNA at specific sites or randomly with endonucleases and to clone defective viruses by complementation with ts mutants has provided a simple way to construct and subsequently to clone an almost unlimited variety of mutants.

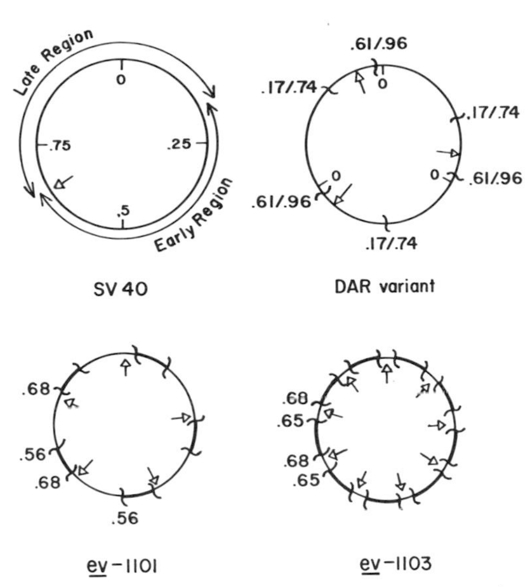

FIG. 13. Examples of noncomplementing evolutionary variants. Each variant genome consists of tandem repeats of a smaller segment of DNA. DAR variant (Fareed *et al.*, 1974) has three repeating units and also contains an inverted segment of SV40 DNA (0.74 to 0.61 map units) within each repeating unit; *ev*-1101 and *ev*-1103 (Lee *et al.*, 1975) contain five and nine repeating units, respectively, each consisting of cellular DNA linked to SV40 DNA. KEY: ⌙, site of recombination during generation of the variant genome; *short arrows*, origin of SV40 DNA replication; ▬, cellular DNA; —, SV40 DNA. Numbers outside each circle indicate standard SV40 map units.

1. Deletion Mutants

Simple deletion mutants have been generated by enzymatic excision of DNA segments from SV40 DNA (Fig. 14) using a multi-cut restriction enzyme or two different enzymes successively (Lai and Nathans, 1974c; Mertz *et al.*, 1974). From the resulting large linear fragments, individual deletion mutants were cloned selectively by complementation plaque formation with a ts mutant, as described in Section IV, C for complementing evolutionary variants. Cyclization of excised linear molecules of SV40 DNA occurred intracellularly by either of two mechanisms: (1) by cohesive termini produced by an appropriate restriction endonuclease (Mertz and Davis, 1972; Lai and Nathans, 1974c; Mertz *et al.*, 1974; see also Section II, C); or (2) by intramolecular recombination near the ends of a linear fragment (Lai and Nathans, 1974c). The recombinational mechanism generated a large variety of deletion mutants from a given linear molecule and therefore does not appear to require extensive base

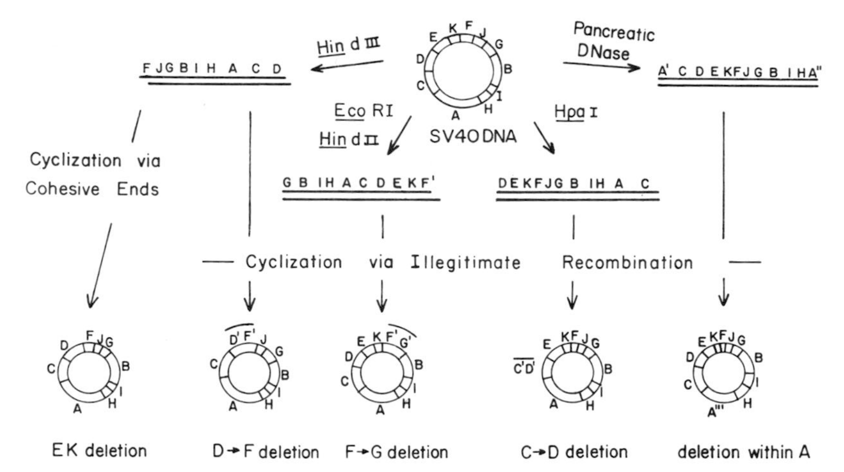

FIG. 14. Generation of SV40 deletion mutants by the use of specific or nonspecific endonucleases to generate linear molecules followed by complementation cloning.

pairing ("illegitimate" recombination). When full-length linear molecules with cohesive termini were used to generate deletions, 5′ ends were first trimmed back with 5′ exonuclease in order to inhibit cyclization via cohesive termini, which would generate a wt genome (Carbon *et al.*, 1975a). Once cloned, constructed deletion mutants can be mapped by restriction endonuclease digestion, by electron microscopic heteroduplex mapping, or by S_1 nuclease digestion of heteroduplex molecules, as described in Section II, C. The S_1 nuclease method is especially valuable for mapping very small deletions (Shenk *et al.*, 1975). It should be noted that the methodology developed for constructing and cloning deletion mutants is also applicable to the isolation of substitution mutants, opening up the possibility of constructing transducing SV40 particles.

2. *Insertion Mutants*

Insertion mutants of SV40 have been constructed by enzymatic addition of oligo dT or oligo dA to the 3′ termini of linear DNA produced by single-cut restriction endonucleases followed by construction of heteroduplex molecules, one strand of which contains a 3′ oligo dA end and the other a 3′ oligo dT end (Carbon *et al.*, 1975b). Depending on the site of insertion, viable or nonviable mutants may be produced. It should be possible to extend this technique to construct mutants with larger segments of inserted DNA (Jackson *et al.*, 1972; Lobban and Kaiser, 1973).

3. Reiteration Mutants

As noted in Section IV, C, 2, certain DNA fragments containing cohesive termini and the origin of viral DNA replication have been propagated in monkey cells coinfected with wt SV40 (Davoli and Fareed, 1974). Some of the virions that result contain genomes with reiterations of the original fragment (see Fig. 13, DAR variant). The mechanism of reiteration is not known but could involve recombination between identical minicircles of DNA. This procedure has subsequently been extended to the isolation of a number of other reiteration mutants of this type (Ganem *et al.*, 1975). Since cohesive termini are not required for intracellular cyclization of a linear DNA fragment (as already discussed), this procedure should be generally applicable to fragments containing the initiation site for DNA replication.

E. Biological Properties of Deletion Mutants and Variants

Studies with heterogeneous populations of defective SV40 virions indicated that some particles induced T antigen and transformed mouse cells, whereas other particles lacked these properties (Uchida *et al.*, 1968). Physiological studies with recently cloned and characterized deletion mutants and variants are still in an early stage. As indicated earlier, deletions of DNA near map position 0.73 are nonlethal, although mutant plaques are smaller than wt plaques, suggesting that this genome segment, while not essential, is required for optimal yield of virus (Mertz and Berg, 1974b). An attractive possibility is that the deletions occur in the D gene and that the D protein (possibly VP3) is nonessential for virus infection and growth but increases the efficiency of infection.

Deletions mutants are being used to identify viral gene products by relating the production of short polypeptides in infected cells to the position of the deletion. A deletion mutant missing a segment of DNA (*Hin*-H and -I) between 0.32 and 0.43 map units (where tsA mutants map) induces the formation in infected cells of a new, short polypeptide that reacts with anti-T serum, suggesting that T antigen is the A gene product (P. Tegtmeyer and C.-J.Lai, unpublished observations). If the new polypeptide is shown to be part of wild-type T antigen by chemical criteria, this would be strong evidence that T antigen is the A gene product. Similarly, a deletion mutant missing the genome segment between 0.99 and 0.10 map units (where tsB, C, and BC mutants map) produces in infected cells a new, short polypeptide precipitable by anticapsid serum (C.-J. Lai and D. Nathans, unpublished observations). Again, peptide analysis should indicate whether the new polypeptide is part of VP1, the major capsid protein.

The physiological activities of a few cloned complementing variants have been reported briefly (Brockman and Scott, 1975). In these experiments mutants were separated from helper virus by serial equilibrium centrifugation in CsCl solutions and were quantitated by complementation plaquing. A variant that has the entire early region of the genome and adjacent segments but lacks SV40 DNA between 0.75 and 0.11 map units (*ev*-1114 of Fig. 12) was able to induce T antigen, replicate its DNA in the absence of helper virus, stimulate host cell DNA synthesis, and transform mouse and hamster cells. Variants with deletions in the early region [0.19 to 0.53 map units (*ev*-1117)] and between 0.24 and 0.54 map units (*ev*-1119) showed none of these SV40 functions. The authors concluded that the early region plus adjacent segments is sufficient for all of the aforementioned viral functions. Extension of these studies to mutants with more restricted deletions may allow finer dissection of early functions.

F. Activity of SV40 DNA Fragments

Related to the studies just described is the prior finding that segments of SV40 DNA can transform cells (Graham *et al.*, 1974). Activity depended on the use of calcium phosphate to enhance the cellular uptake of DNA (Graham and van der Eb, 1973). A linear fragment of SV40 DNA comprising 74% of the genome between the *Eco*RI site (0 map units) and the *Hpa* II site (0.74 map units) had about the same specific transforming activity as complete circular molecules of SV40 DNA. The other 26% fragment was inactive. Therefore a segment which retains the entire early region of the genome, but is missing much of the region where tsB and tsD mutants map, is sufficient for transformation. Extension of these important studies should help define the minimal continuous stretch of SV40 DNA required for transformation.

V. Replication of SV40 DNA

A. Structure of SV40 Replicative Intermediates

Analysis of the replication of SV40 DNA was greatly accelerated by the finding of Hirt (1967) that cellular DNA can be selectively precipitated at 4°C in 1 *M* NaCl–0.6% SDS leaving viral DNA in the supernatant. A. J. Levine *et al.* (1970) used this procedure to extract replicating SV40 DNA molecules from infected monkey cells. After a short (5–10 minute) pulse of [^{3}H]-thymidine, most of the labeled viral DNA was found to sediment in neutral sucrose gradients in a rather broad zone with a peak at about 25 S. This rapidly labeled 25 S species was shown to be a precursor of SV40 form I (21 S) by pulse-chase experiments. The basic structural features of SV40 replicative intermediate (RI) are diagrammed in Fig. 15 (A. J. Levine *et al.*, 1970; Sebring *et al.*, 1971; Jaenisch *et al.*, 1971). Electron microscopic examination revealed that the majority of the SV40 RI's had two branch points, three branches, and no visible ends. Two of the branches were equal in length and were assumed to represent the daughter duplexes. The third branch (presumed unreplicated portion of the RI) was superhelical [Fig. 15; see Sebring *et al.*

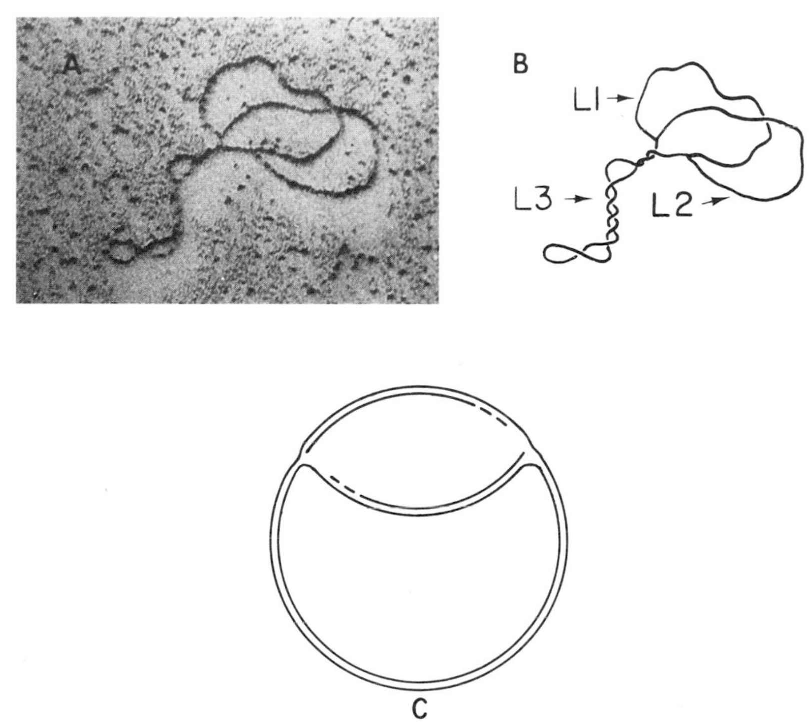

FIG. 15. SV40 DNA replicating intermediate: (A) electron micrograph of SV40 replicating intermediate; (B) interpretive drawing. Replicative intermediates contain two forks, three branches, and no visible ends. Two of the branches (L1 and L2) are of equal length and represent the replicated segments of the molecule. The third branch (L3) contains superhelical twists and represents the unreplicated portion of the molecule. (C) Summary of the basic structural features of SV40 replicating intermediate. Both parental strands are covalently closed. The daughter strands are held to the parental strands by hydrogen bonds alone. Both daughter strands grow by a discontinuous mechanism, and there may be single-stranded regions (in the *trans* configuration) at each fork. [A and B from Sebring *et al.* (1971).]

(1971); also Jaenisch *et al.* (1971)]. When pulse-labeled RI was sedimented in alkaline sucrose gradients, the newly synthesized strands separated from the parental strands. The lengths of the newly synthesized strands were less than or equal to that of a complete viral DNA strand (Sebring *et al.*, 1971; Jaenisch *et al.*, 1971). The parental strands of SV40 RI were shown to be covalently closed circles. Labeled parental DNA strands sedimented in alkaline sucrose gradients as cyclic coil forms with sedimentation coefficients in the range of 37–47 S (Sebring *et al.*, 1971).

The fact that the parental strands of SV40 RI are covalently closed accounts for the observation that these molecules are partially superhelical. The sense of the superhelical turns was determined by assaying the effect of EtBr on the sedimentation velocity or electron microscopic appearance of SV40 RI (Mayer and Levine, 1972; Sebring *et al.*, 1974). Both methods indicated that the number of superhelical turns in RI first decreased and then increased with increasing concentrations of EtBr. This shows that the sense of the superhelical turns in RI is negative (i.e., the same as SV40 form I). Superhelical RI can be converted to relaxed RI by a brief treatment with pancreatic DNAse that results in the introduction of one or more nicks in the parental strands (Jaenisch *et al.*, 1971).

It is possible to fractionate SV40 RI molecules according to their degree of replication by equilibrium centrifugation in CsCl gradients containing high concentrations of EtBr (200 μg/ml) (Sebring *et al.*, 1971). Under these conditions the "youngest" replicating molecules have a buoyant density which is close to that of SV40 form I DNA. Molecules with greater extents of replication band at progressively lower densities, with the "oldest" molecules banding near the position of SV40 form II DNA. This phenomenon is a consequence of the fact that the parental strands of SV40 RI are covalently closed. Because of this constraint, binding of the EtBr to the unreplicated branch of the molecule is limited as in the case of SV40 form I DNA. The daughter duplexes, on the other hand, are free to swivel at both ends and, therefore, would be expected to bind as much EtBr per unit length as would SV40 form II DNA. As replication proceeds, the size of the daughter duplexes increases at the expense of the unreplicated branch, and the buoyant density decreases from that characteristic of form I DNA to that characteristic of form II DNA.

SV40 RI appears to contain small single-stranded regions. A. J. Levine *et al.* (1970) showed that pulse-labeled SV40 DNA binds to benzoylated–naphthoylated DEAE cellulose under conditions where fully double-stranded DNA fails to bind. Jaenisch *et al.* (1971) obtained electron micrographs which showed possible single-stranded regions at the replicating forks of SV40 RI. Each fork contained one such region, and the two regions in each molecule were in a *trans* configuration (see Fig. 15).

B. Events during SV40 DNA Replication

Two independent lines of evidence indicate that SV40 DNA replication is initiated at a unique site on the SV40 genome and proceeds bidirectionally to a terminus about 180° around the circle from the origin. Danna and Nathans (1972) labeled SV40-infected BSC-1 cells with a brief pulse of [^{3}H]-thymidine and then isolated the newly completed form I DNA. It was expected that this population of molecules would be most highly labeled near the terminus of replication and least highly labeled near the origin. The molecules were cleaved with R • *Hin*dII/III, and the relative specific activity of the various fragments was determined. From the observed labeling pattern it was concluded that SV40 DNA replication begins near SV40 map position 0.67 (in *Hin*dII/III fragment C) and proceeds bidirectionally, ending near SV40 map position 0.17 (in *Hin*dII/III fragment G). Fareed *et al.* (1972) isolated replicating SV40 DNA molecules and cleaved them with endonuclease R • *Eco*RI. Since this enzyme breaks SV40 DNA only once (at map position 0), a unique reference point was created in the replicating molecules. Electron microscopic examination indicated that the origin of replication was 33% of an SV40 genome length from the *Eco*RI cleavage site (i.e., at either 0.67 or 0.33 on the SV40 physical map) and that replication was bidirectional.

It seems likely that initiation of SV40 DNA replication requires the recognition of the origin site (*ori*) by one or more SV40 or host-coded proteins. A possible candidate for such a protein is the product of the SV40 A gene. This protein is continuously required for SV40 DNA replication (Tegtmeyer and Ozer, 1971) and appears to act at the level of initiation (Tegtmeyer, 1972; see also Section IV, A). The nature of *ori* is not known. However, recent sequencing studies by Dhar *et al.* (1974a) have identified sequences of nucleotides in the neighborhood of *ori* that display 2-fold axes of rotational symmetry. Since SV40 replication is bidirectional, it is an appealing possibility that *ori* would have this kind of symmetry. However, further studies are required to determine whether one of these sequences is the initiation signal.

The origin of replication is probably the only *cis*-specific function required for SV40 DNA replication. The origin is uniquely conserved in SV40 variants produced by high multiplicity passage that are highly substituted with host DNA and retain only a small fraction of the SV40 genome (Lee *et al.*, 1975; see also Fig. 13). In fact such variants often contain several origin sites. Termination of SV40 DNA replication appears to occur wherever two growing forks meet and therefore does not appear to require a specific recognition signal. Brockman *et al.* (1975) have isolated a variant of SV40 *ev*-1114 in which a segment of DNA containing *ori* is duplicated (Fig. 12). Replication of the DNA of this variant can begin at either origin but always terminates approximately 180° around the circle from where it began. Furthermore, neither of the regions where termination of replication occurs in this variant corresponds to the region where termination occurs in the wt SV40 genome. Similar results were obtained by Lai and Nathans (1975b) with deletion mutants of SV40.

After initiation, DNA synthesis proceeds in both directions at approximately equal rates (Fareed *et al.*, 1972; Danna and Nathans, 1972). To date no SV40 gene products have been implicated in elongation of the nascent polynucleotide chains, so it may be that this process is mediated entirely by host enzymes. The daughter strands appear to be synthesized by a discontinuous mechanism similar to that of *E. coli* (Okazaki *et al.*, 1968), in which short DNA chains are formed at the replication forks and subsequently joined together. Fareed and Salzman (1972) exposed SV40-infected monkey cells to very short (15–45 second) pulses of [^{3}H]-thymidine and extracted SV40 RI. A significant fraction of the labeled precursor was found to be incorporated into DNA chains which sedimented at 4 S in alkaline sucrose gradients. The 4 S DNA was associated with RI at all stages of replication and was shown by pulse-chase experiments to be a precursor of the growing daughter strands. Under renaturing conditions the 4 S chains show extensive self-annealing (Fareed *et al.*, 1973b; Laipis and Levine, 1973), indicating that both daughter strands at each fork are replicated discontinuously.

During SV40 DNA replication, the elongation of the daughter strands requires unwinding of the parental strands. The topological winding number of the parental strands (see Section II, A) must decrease from a value of about 500 to a final value of 0 [Sebring *et al.* (1971)]. Salzman *et al.* (1973) have obtained direct physical evidence that this decrease in the topological winding number of the parental strands occurs more or less continuously during the replication process. This raises a problem, since (as indicated earlier) the parental strands of SV40 RI as isolated from the cell are covalently closed. To obviate this difficulty, Sebring *et al.* (1971) proposed that nicking, unwinding, and sealing of the parental strands occur intermittently during the replication cycle. If the nick is present only transiently, then most of the RI isolated from a cell at any given moment would contain covalently closed parental strands. An activity which can carry out the intermittent nicking and sealing of DNA molecules has recently been found in extracts of mouse embryo cell nuclei (Champoux and Dulbecco, 1972) and partially purified from KB cells (Keller and Wendel, 1974). A similar activity has previously been found in *E. coli* (Wang, 1971). The activity is assayed by its ability to remove superhelical turns from (i.e., "relax") covalently closed circular DNA. The removal of superhelical turns clearly requires the introduction of a break in one of the two strands of the closed circular DNA; however, the break is only temporary, as after the reaction with the relaxing activity the DNA is still covalently closed. The reaction proceeds without any external energy donor (Wang, 1971; Champoux and Dulbecco, 1972; Keller and Wendel, 1974). The partially purified relaxing activity from KB cells is a protein with an estimated molecular weight of 70,000–90,000 (Keller and Wendel, 1974). The idea that the relaxing activity may act as a "swivel" during DNA replication (Wang, 1971; Champoux and Dulbecco, 1972) is supported by the finding that nucleoprotein complexes containing replicating SV40 DNA also contain relaxing activity (Sen and Levine, 1974). It is interesting that when the relaxing activity is added to SV40 form I DNA at 0°C, 19 discrete intermediates between form I and relaxed form I can be detected by agarose gel electrophoresis. It has been suggested that these intermediates result from the removal of one superhelical turn at a time (Keller and Wendel, 1974). If this is the case, SV40 form I must contain a minimum of 20 superhelical turns.

Little is known at present about the mechanism of segregation of the two daughter molecules from SV40 RI. The immediate postsegregational product may be SV40 form II DNA. Fareed *et al.* (1973a) showed that after a short pulse of [^{3}H]-thymidine the intracellular pool of SV40 DNA contains a small amount of labeled form II DNA (1–3% of incorporated isotope). Analysis of this form II DNA in alkaline sucrose gradients revealed that the newly synthesized DNA was present in a linear strand with a sedimentation coefficient of 16 S. It was further shown that the interruption in this linear strand was located in the region of the SV40 genome where replication terminates. It seems likely that these form II molecules represent an intermediate between RI and progeny form I; however, for technical reasons it was not possible to show this directly.

The length of time required to complete one round of SV40 DNA replication has been estimated by various groups as between 5 and 25 minutes (A. J. Levine *et al.*, 1970; Danna and Nathans, 1972; Fareed *et al.*, 1973a). Levine (1974) has pointed out that if the rate of chain elongation were the same for SV40 DNA as for mammalian DNA, one round of SV40 replication should require only about 25 seconds. This discrepancy could be related to the unwinding problem discussed earlier. Since replication of SV40 DNA requires the intermittent nicking and sealing of the parental strands, the availability of the enzyme(s) which carry out this function could be rate limiting.

C. SV40 DNA Synthesis in Vitro

Detailed characterization of SV40 DNA replication at the biochemical level will ultimately require the development of an *in vitro* system which is capable of carrying out the entire replication process from initiation of new chains to segregation of the daughter molecules. A promising start in this direction has been made in recent studies of SV40 DNA synthesis in purified nuclei (Qasba, 1974) or whole-cell lysates containing intact nuclei (DePamphilis *et al.*, 1975) from SV40-infected cells. It has been shown that SV40 RI prelabeled *in vivo* can be converted into covalently closed superhelical SV40 form I DNA by using the whole-cell lysate system (DePamphilis *et al.*, 1975). DNA synthesis *in vitro* occurred at about one-third the *in vivo* rate for 30 minutes at 30°C. After a 1 hour incubation, about half the prelabeled RI molecules were converted to form I molecules, about 5% to form II, and about 5% to covalently closed dimers. The remaining RI did not complete replication, although 75% of the prelabeled daughter strands were elongated to one genome length. Thus, the system appears to carry out the elongation, termination, and segregation steps in SV40 DNA replication with good fidelity. However, pulse-chase and density-shift experiments indicated that no new rounds of replication were initiated *in vitro*. The reason for this defect is not clear. In contrast with the whole-cell lysate system, purified nuclei show a greatly reduced ability to convert RI to form I (Qasba, 1974; DePamphilis and Berg, 1975). This appears to be due to a failure of joining of the 4 S DNA chains into longer strands. The ability of nuclei to convert RI to form I can be restored by adding back a cytoplasmic fraction from either infected or uninfected cells. This observation should allow the purification of at least one component of the *in vitro* replication system.

VI. Transcription of the SV40 Genome

RNA with sequence homology to SV40 DNA (i.e., SV40 RNA) is readily detected in SV40-infected or transformed cells by various assays involving DNA–RNA hybridization (Reich *et al.*, 1966; Oda and Dulbecco, 1968; Aloni *et al.*, 1968; Sauer and Kidwai, 1968). Probably the most nearly quantitative assay involves the reaction in solution of denatured SV40 DNA (or individual strands) with cellular RNA, or subfractions of cellular RNA, under annealing conditions (Khoury and

Martin, 1972; Sambrook *et al.*, 1972). DNA–RNA hybrids are then quantitated by selective adsorption to hydroxyapatite or, alternatively, by treating the reaction mixture with single-strand–specific nuclease (e.g., S_1 nuclease from *Aspergillus oryzae*) and measuring residual acid-precipitable DNA or RNA. To determine the percentage of an RNA preparation that is SV40 RNA, one can use excess denatured DNA and limiting concentrations of radioactive RNA. To determine the percentage of SV40 DNA sequences represented in a population of RNA molecules, one reacts limiting amounts of radioactive, single strands of DNA or DNA fragments with excess RNA; S_1 nuclease–resistant DNA is then a measure of SV40 sequences present in the RNA.

The results of hybridization assays for viral RNA present in SV40–infected cells depend on the time of sampling relative to the replicative cycle, on whether the RNA is derived from whole cells, nuclei, or cytoplasm, and whether one examines newly synthesized RNA (pulse-labeled RNA) or accumulated ("stable") SV40 RNA.

A. Stable SV40 RNA from Productively Infected Cells

Stable SV40 RNA falls into two general temporal classes, as noted in Table I: "early" RNA found prior to the onset of viral DNA replication, constituting about 0.01% of newly synthesized cellular RNA; and "late" RNA found after viral DNA replication begins, constituting about 4% of newly synthesized cellular RNA (Oda and Dulbecco, 1968; Aloni *et al.*, 1968; Sauer and Kidwai, 1968). By competitive hybridization of stable SV40 RNA to viral DNA it was shown that early RNA is a subset of late SV40 RNA, i.e., late RNA contains early RNA plus specifically late species. Taken together, early and specifically late stable RNA contain nucleotide sequences equivalent to about half of the sequences in duplex viral DNA (Martin and Axelrod, 1969).

Two procedural advances allowed more precise mapping of stable SV40 RNA found in productively infected cells. First, the separation of the strands of SV40 DNA, as described in Section II, C, permitted the assignment of early and late RNA to individual template strands (Lindstrom and Dulbecco, 1972; Khoury and Martin, 1972; Sambrook *et al.*, 1972). By hybridization of excess unlabeled SV40 RNA (isolated from whole cells) with labeled individual DNA strands, it was found that early RNA was derived entirely from about 40% of one strand (E or minus strand, the same template strand transcribed by *E. coli* RNA polymerase), and specifically late RNA was derived from about 60% of the opposite strand (L or plus strand). Secondly, the availability of DNA fragments from specific regions of the SV40 genome permitted the mapping of early and late stable RNA on the cleavage map (Khoury *et al.*, 1973b; Sambrook *et al.*, 1973). For this purpose, individual strands of ^{32}P fragments prepared from restriction enzyme digests were reacted with excess early or late infected cell RNA and the percentage of DNA in hybrid form assessed. When total cellular RNA was used with separated strands of *Hind* fragments (see Fig. 5 for the *Hin* cleavage map), early RNA isolated from infected cells treated with cytosine arabinoside to prevent viral DNA replication reacted with the E strand of contiguous *Hin* fragments B, I, H, A. Late RNA, on the other hand, reacted with the L Strand of *Hin*-A, -C, -D, -E, -K, -F, -J, -G, -B, and to a minor extent with *Hin*-H and -I. On the basis of these results it was concluded that *Hin*-A and -B contain transitions between early and late transcription. This conclusion appeared to be confirmed by S_1 nuclease experiments demonstrating that part of each strand of *Hin*-A and *Hin*-B is represented in SV40 RNA. On the assumption that early and late transcription regions were each continuous in the genome, the early region was placed at 0.24 to 0.56 on the map, and the late region was placed at 0.56 to 0.24 map units. Subsequent studies, however, have indicated that the early region of the genome is more extensive and the late region less extensive than these results indicate.

Among the studies pointing to a more extensive early region were those using DNA from Ad–SV40 hybrid viruses, which contain different segments of SV40 DNA (discussed later). These studies suggested that early stable SV40 RNA includes essentially all the sequences present in the E strand of *Hin*-B (Patch *et al.*, 1974). More recently, it has been found that in contrast to the results obtained when *whole-cell* RNA was used, *cytoplasmic* RNA extracted from cells 48 hours after infection (which presumably contains processed early and late SV40 mRNA) hybridizes with the E strands of *Hin*-B, -I, -H, and -A, with the L strands of *Hin*-C, -D, -E, -K, -F, -J, and -G (Khoury *et al.*, 1975a). Essentially no *cytoplasmic* RNA reacts with the L strands of *Hin*-B, -I, -H, or -A, nor with the E strands of the other *Hin* fragments. [In another study of this type, however, cytoplasmic SV40 RNA was thought to react with part of the L strands of *Hin*-A and -B (Sambrook *et al.*, 1973).] Thus the early region comprises about half the genome (0.17 to 0.65 map units), and the late region the other half.

More precise mapping of SV40 RNA in infected cells has been carried out by comparing nucleotide sequences in isolated mRNA with those in selected fragments of viral DNA (Dhar *et al.*, 1974b, 1975). In these experiments ^{32}P-labeled cytoplasmic SV40 RNA was fractionated on oligodeoxythymidylate cellulose columns to isolate poly A–containing RNA, and the latter was then hybridized to and eluted from filters containing SV40 DNA fragment *Hin*-G or DNA from Ad–SV40 hybrid virus Ad^{+}ND$_{1}$ (see Section IV, B), which contains parts of SV40 *Hin*-G and -B segments. Nucleotide sequence analysis of these portions of SV40 mRNA indicated that the 3′ ends of early and late SV40 mRNA were derived from a segment of SV40 DNA near map position 0.17, as detailed in Fig. 16. Similar analyses were carried out on the 5′ ends of SV40 mRNA's which hybridized to *Eco*RII fragment G (0.635 to 0.70 map units). The 5′ ends of early and late SV40 mRNA were found to overlap for a stretch of 70–80 nucleotides at or near the *Hin*-A • C junction, i.e., the site of origin of DNA replication (Dhar *et al.*, 1975). These results are thus in agreement with the hybridization results already described and more precisely localize the limits of the early and late transcription regions. From the combined data and the strand orientation noted earlier, one can draw a transcription map as shown in Fig. 17. Early transcription is counterclockwise on the map, and late transcription is clockwise.

The size of SV40 mRNA's found in productively infected cells has been estimated by sedimentation in sucrose gradients and by electrophoresis in acrylamide–SDS gels with or without formamide (Tonegawa *et al.*, 1970; Weinberg *et al.*, 1972). Early SV40 mRNA has one size class, estimated as 19 S and about 8 × 10^{5} daltons. This size class (isolated from RNA made late in infection) has been found to hybridize to the *Hae*-A frag-

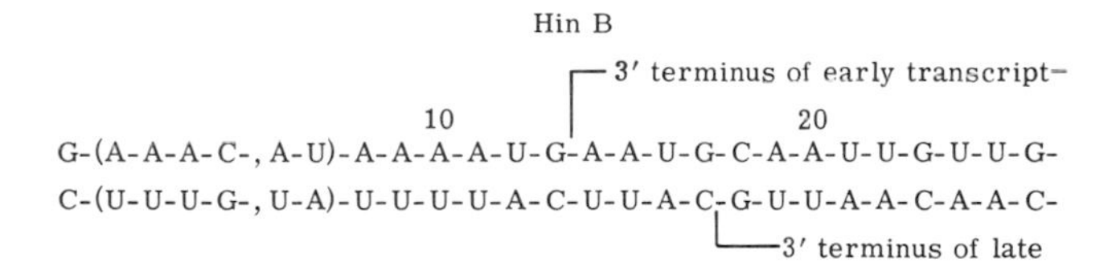

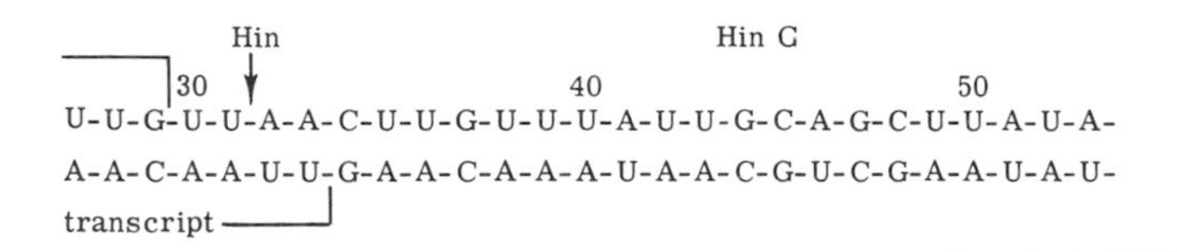

60 70
A-U-G-G-U-U-A-C-A-A-A-U-A-A-A-G-C-A-A-U-A-G-C-A
U-A-C-C-A-A-U-G-U-U-U-A-U-U-U-C-G-U-U-A-U-C-G-U

5′ terminus of
E. coli RNA polymerase transcript

FIG. 16. Nucleotide sequences at the 3′ ends of SV40 mRNA in relation to sequences within DNA fragments *Hin*-G and B. [From Dhar *et al.* (1974b).]

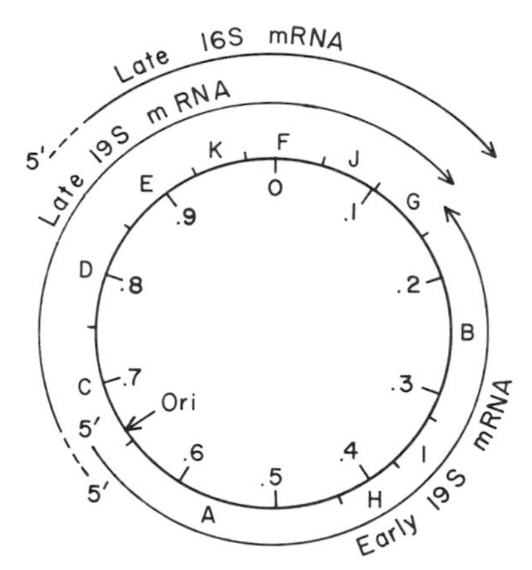

FIG. 17. Transcriptional map of SV40, in relation to the *Hind* cleavage map. For details, see the text.

ment of SV40 DNA, which is derived from the early region of the genome (Weinberg *et al.*, 1974). However, the 19 S RNA extracted prior to DNA replication has not been directly mapped. Late stable mRNA has two size classes of SV40 RNA: 16 S (about 6×10^5 daltons) and 19 S (about 8×10^5 daltons). The 16 S class hybridizes with *Hae*-B, -C, -E, and -F, derived from the late region of the genome, but not with *Hae*-A or -D, derived from the early region. The specifically late 19 S species of SV40 mRNA appears to contain all the sequences present in the 16 S RNA. On the basis of these data, taken together with the results of transcription mapping already described, it appears likely that the early (19 S) transcript is derived from the entire early region of the genome, that the specifically late 16 S transcript is derived from the genome between about 0.9 and 0.2 map units, and that the specifically late 19 S transcript is derived from the genome between about 0.7 and 0.2 map units (Fig. 17). Clearly more detailed experiments, especially comparison of nucleotide sequences of mRNA and DNA, will be needed to establish the map coordinates more firmly.

It should be noted in Fig. 17 that the 5′ ends of early and late SV40 mRNA are derived from the same genome region that contains the initiation site for DNA replication. A similar topography has been found for polyoma as well (Kamen *et al.*, 1974). Whether this reflects a relationship between control of transcription and replication is not known; it could be a clue to the still elusive mechanism for switching from early to late mRNA production. Also unknown is whether SV40 transcription begins at specific promotors near the *Hin*-A • C junction or whether the 5′ ends of cytoplasmic mRNA arise by processing (discussed in the next subsection).

B. Nuclear SV40 RNA from Productively Infected Cells

As noted earlier, the results of SV40 RNA analysis depend on whether one examines pulse-labeled RNA or stable RNA and are also dependent on whether the RNA is nuclear or cytoplasmic in origin. Pulse-labeled RNA present in infected cell nuclei contains species considerably larger than one single-strand length of SV40 DNA (Tonegawa *et al.*, 1970). Moreover, pulse-labeled SV40 RNA readily forms RNA-RNA duplex molecules of unit length or greater on annealing (Aloni, 1972, 1974), indicating that RNA is transcribed from complementary regions of each DNA strand. This inference has been confirmed by the observation that nuclear SV40 RNA (in contrast to cytoplasmic RNA) hybridizes to both strands of several *Hin* fragments (Khoury *et al.*, 1975a). Thus there is evidence that nuclear SV40 RNA includes molecules equivalent to entire strands of SV40 DNA. Possibly related to the nuclear RNA species which contain anti–late and anti–early sequences are a class of high molecular weight nuclear RNA molecules which contain both SV40 and host-cell DNA sequences (Rozenblatt and Winocour, 1972). These molecules have been found late in productive infection after a labeling period of 6 hours, and they range in size from 2 to 10 million daltons, as estimated by sedimentation in dimethylsulfoxide-containing sucrose gradients. Since they hybridize serially to both SV40 DNA and to cellular DNA after isolation in a denaturing gradient, the viral and cell sequences appear to be covalently linked. This class of SV40 RNA is thought to arise by transcription of integrated SV40 DNA (cf. Section VII, C).

The significance of nuclear SV40 RNA molecules which contain sense and antisense sequences is not clear at present. One possibility is that primary transcription products are complete or multiple-length E or L strand transcripts derived from free or integrated viral DNA; SV40 mRNA found in the cytoplasm would then result from posttranscriptional removal of part of the primary transcript. According to this hypothesis, the sites in the viral genome corresponding to the 5′ and 3′ ends of cytoplasmic mRNA (see Fig. 17) determine processing signals for cleavage, 5′ capping, or 3′ poly A addition. Alternatively, "symmetrical transcription" could arise by aberrant initiation or termination of transcription, i.e., as a physiologically insignificant side reaction, mRNA being derived from a subset of transcripts starting and ending at specific promotor and termination sites near the junctions between the early and late regions of the genome. A clear distinction between these hypotheses will depend on demonstrating or disproving a precursor–product relationship between the nuclear high molecular weight SV40 RNA and cytoplasmic mRNA. The results could have general import for transcription from cellular DNA.

C. SV40 RNA in Transformed Cells

Viral RNA was detected in SV40-transformed or tumor cells by Reich *et al.* (1966), who estimated that about 0.01% of labeled RNA was virus-specific. Similar experiments with SV40-transformed 3T3 mouse cells established that all transformed clones examined contained RNA that hybridized to SV40 DNA (Oda and Dulbecco, 1968; Aloni *et al.*, 1968). Nuclear SV40 RNA was found to be of variable size, estimated as up to three times the length of a strand of SV40 DNA, whereas cytoplasmic SV40 RNA was about half the length of an SV40 DNA strand (Lindbergh and Darnell, 1970; Tonegawa *et al.*, 1970). Moreover, some of the high molecular weight SV40 RNA appeared to be covalently linked to cellular RNA (Wall and Darnell, 1971; see also Section VII, C). By competitive hybridization with infected cell RNA, the SV40 RNA from transformed cells was shown to be similar to early lytic RNA. Consistent with this was the observation that when individual strands of ^{32}P-labeled SV40 DNA were hybridized to excess cell RNA from a series of transformed lines, reaction occurred predominantly with the E strand—the early template strand (Khoury *et al.*, 1973c; Ozanne *et al.*, 1973). However, the percentage of E strand that hybridized, as determined by resistance of the DNA to S_1 nuclease, varied from about 30% to about 60% in different lines. Moreover, in a few instances, a very small fraction (2–9%) of the L (late) template strand also reacted with cell RNA. These results thus suggested that in some transformed cells segments of the viral genome outside the early region were transcribed from each strand, in addition to at least part of the early region.

More detailed mapping of viral RNA from transformed cells has been carried out by using separated strands of endo R fragments of SV40 DNA in the hybridization reactions, as had been done for mapping SV40 RNA in productively infected cells (Khoury *et al.*, 1975b; Ozanne *et al.*, 1973). The overall findings can be summarized as follows: the E strands of *Hin*-A, -H, -I, and -B (see Fig. 17) reacted completely, or nearly so, with all the transformed cell RNA's tested; the E strand of *Hin*-C reacted partially in all cases, but was never saturated; with four of seven lines the E strand of *Hin*-D also reacted partially; none of the other *Hin* fragments reacted significantly, in particular no significant reaction of L strand fragments was found. These results indicate that in all lines the entire early region of the SV40 genome is transcribed, and in addition a variable portion of the genome "upstream" from the early region is also transcribed (see Fig. 17), but stable RNA from this segment is less abundant than that from the early region. It should be noted that these experiments were performed with total cellular RNA, and therefore the RNA includes processed and unprocessed transcripts. The significance of the small amount of RNA apparently derived from the L strand is not clear.

Two models have been proposed to explain the foregoing mapping results (Khoury *et al.*, 1975b), both based on transcription of an integrated viral genome (see Section VII): one postulates specific promotor and termination sites; the other, specific RNA processing signals (Fig. 18). In the first model, SV40 DNA is assumed to have recombined with cell DNA within the *Hin*-C or -D segment. Transcription then occurs predominantly via "normal" SV40 early transcription signals and, to a lesser extent, via upstream cellular promotors, yielding viral RNA species with or without *Hin*-C or *Hin*-C and -D sequences. In the second model, a recombination between SV40 and cell DNA may occur anywhere outside the early genome segment, and transcription begins and ends in adjacent cell DNA, the final RNA resulting from processing. In this case the presence of Hin-C and -D sequences in the RNA is not explained. Again, detailed study of the structure and relationship between high molecular weight SV40 nuclear RNA and SV40 cytoplasmic RNA is needed.

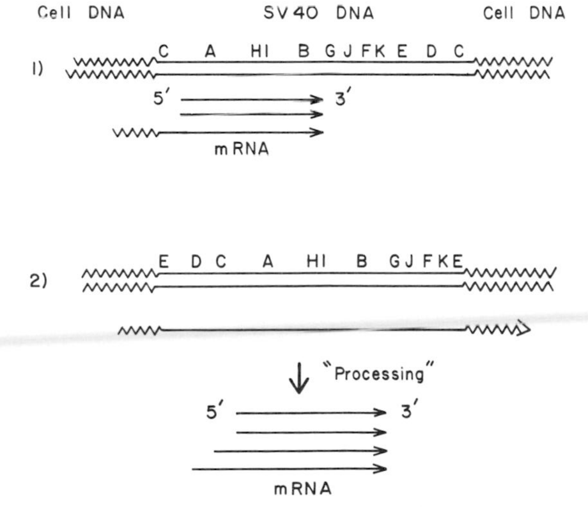

FIG. 18. Models for transcription of integrated SV40 genes in transformed cells. For details, see the text. [Adapted from Khoury *et al.* (1975b).]

VII. Integration of SV40 DNA into Cellular Chromosomes

Since clones of SV40-transformed cells contain SV40 RNA and virus-specific antigens, it was logical to assume that at least part of the viral genome persists in an inheritable form. By analogy with lysogenic or plasmid-containing bacteria, viral DNA could persist as an integrated provirus or as a separate replicon. In the experiments to be reviewed next, transformed cells have been shown to contain SV40 genes by biological and biochemical tests, the amount of SV40 DNA has been quantitated, and the relationship to cellular chromosomes has been explored.

A. *Quantitation of SV40 DNA in Transformed Cells*

The direct detection and quantitation of SV40 DNA in transformed cells has required the development of very sensitive assays, since each host cell has about 10^6 times as much DNA as an SV40 virus particle. Two general approaches have been employed. Westphal and Dulbecco (1968) prepared SV40-specific RNA of high specific radioactivity using *E. coli* RNA polymerase and purified SV40 form I DNA as template. This RNA (in excess) was then hybridized to cell DNA immobilized on nitrocellulose filters according to the method of Gillespie and Spiegelman (1965). Approximately 0.2% of the input radioactive RNA hybridized to 100 μg of untransformed cell DNA. A significant increase (up to 3-fold) in hybridization was observed with DNA extracted from four different SV40 transformed cells. Using reconstruction experiments to obtain a calibration curve, Westphal and Dulbecco (1968) estimated that the SV40 DNA content of these four lines ranged from a low of 7 to a high of 58 genome equivalents per diploid quantity of cell DNA. Subsequent to these experiments it was discovered that the reconstruction experiments used to calibrate the method suffered from a rather large systematic error due to the selective loss of DNA–RNA hybrids from the nitrocellulose filters (Haas *et al.*, 1972). This error led to an overestimation of the amount of SV40 DNA per transformed cell. This problem has recently been solved by using cRNA whose specific activity is accurately known (Botchan and McKenna, 1973), thus obviating the necessity for reconstruction experiments. With this improvement, the filter hybridization method gives results which are in excellent agreement with the hybridization kinetic method described below.

The second method for quantitating SV40 DNA in transformed cells (Gelb *et al.*, 1971) is based upon the fact that renaturation of a given sequence of denatured DNA is a second-order reaction whose time course is given by the equation:

$$1/f_{ss} = 1 + KC_0t \quad (1)$$

Here f_{ss} is the fraction of the DNA remaining single stranded at time t; K is the rate constant; and C_0 is the initial concentration of DNA (Britton and Kohne, 1968; Wetmur and Davidson, 1968). Application of the method involves the preparation of a highly radioactive SV40 DNA probe which is mechanically sheared into small fragments. Following denaturation, the probe is allowed to renature in the presence of a known concentration of cell DNA fragments sheared to the same size as the probe. The extent of renaturation of the probe is measured as a function of time using hydroxyapatite chromatography. It follows from Eq. (1) that the presence of sequences complementary to SV40 DNA in the cell DNA will increase the rate of renaturation of the probe according to the relation:

$$1/f_{ss} = 1 + KC_pt\ (1 + r) \qquad (2)$$

Here C_p is the initial concentration of the probe DNA, and r is the molar ratio of SV40 sequences in the cell DNA to those in the probe DNA. From the value of r determined experimentally, it is a simple matter to compute the number of SV40 genome equivalents per diploid quantity of cell DNA. Reconstruction experiments (Gelb *et al.*, 1971) have established that the sensitivity of the method is sufficient to detect considerably less than one genome equivalent per cell. Gelb *et al.* (1971) measured the SV40 DNA content of five different SV40-transformed lines using this approach. The baseline rate of renaturation of the probe was measured in the presence of salmon sperm DNA, which was assumed to contain no sequences homologous to SV40. Significant increases in the rate of renaturation of the probe were observed when SV40-transformed cell DNA was substituted for the salmon sperm DNA. From the magnitude of these increases, it was calculated that the SV40 DNA content of the five SV40-transformed lines ranged from about one to three genome equivalents per diploid quantity of cell DNA (Table VI). Ozanne *et al.* (1973) examined a number of other SV40-transformed cell lines using this method and observed up to 9 genome equivalents per diploid quantity of cell DNA (Table VI). It should be noted that transformed cells generally contain significantly more DNA than one diploid quantity (about 4×10^{12} daltons). In the few transformed lines where the cellular DNA content has been measured, the number of SV40 genome equivalents per cell has ranged from 2 to 15.

Gelb *et al.* (1971) reported that DNA from untransformed 3T3 or primary AGMK cells contained sequences complementary to SV40 DNA at a level of about 0.5 genome equivalents per diploid quantity of cell DNA measured relative to the control salmon sperm DNA. They specu-

TABLE VI
AMOUNT OF SV40 DNA IN VARIOUS TRANSFORMED AND REVERTANT CELL LINES

Cell line	Parental line	Number of SV40 genome equivalents per diploid quantity of cell DNA	References
SVT2	Balb C/3T3	1.56, 2.2	*a, b*
SVPy11	3T3	1.42, 1.3	*a, b*
SV–UV–15 clone 7	3T3	1.04	*a*
SV–UV–15 clone 5	3T3	3.86	*a*
SV3T3 clone 9	3T3	8–9	*b*
SV101	3T3	8–9	*b*
SVB30	Balb C/3T3	6.1	*b*
F1SV101	Revertant of SV101	8–9	*b, c*
CA^r 30.4	Revertant of SV3T3	8–9	*b, c*
CA^r 32.6	Revertant of SV3T3	8–9	*b, c*
CA^r 41.6	Revertant of SV3T3	8–9	*b, a*

[a] REFERENCES: (*a*) Gelb *et al.* (1971). (*b*) Ozanne *et al.* (1973). (*c*) Pollack *et al.* (1968). (*d*) Ozanne (1973).

lated that if these sequences were clustered in one region, they might represent an "integration site" for transforming SV40 DNA (see Section VII, D). However these data were not confirmed by Ozanne *et al.* (1973), who found that the renaturation kinetics of labeled SV40 DNA were identical in the presence of DNA from salmon sperm, calf thymus, 3T3 cells, HeLa cells, and adenovirus-transformed rat cells. Ozanne *et al.* (1973) concluded that if mammalian cells contain DNA sequences homologous to SV40, such sequences are not species specific and, in fact, are shared with at least one nonmammalian species. The reason for the discrepancy between these two experiments is not clear at present; however, it seems possible that the labeled SV40 DNA employed as a probe by Gelb *et al.* may have contained a low level of molecules with host substitutions (see Section IV, C).

When the SV40 DNA content of transformed cells is assayed by the renaturation kinetics method, it is conventional to assume that all sequences in the viral genome are equally represented in the genome of the cells. In theory, unequal representation of different segments of the viral genome in transformed cell DNA should be detectable as an alteration in the shape of the renaturation curve of the probe from that predicted by Eq. (2). However, it has recently been shown (Sharp *et al.*, 1974; Sambrook *et al.*, 1974) that under the conditions generally employed in this type of experiment an examination of the shape of the renaturation curve is not sensitive enough to detect rather substantial departures from the assumption of equal representation of all viral sequences. This problem was overcome by using specific restriction fragments of SV40 DNA as probes in the renaturation kinetic experiments (Sambrook *et al.*, 1974; Botchan *et al.*, 1974). In the case of the one SV40-transformed line (SVT2) which has so far been studied by this method, the results clearly indicate that different regions of the SV40 genome are represented at different frequences (Table VII). The implications of this observation are not clear at present. It is possible that multiple integration events (perhaps involving defective SV40 genomes) occurred during the generation of the SVT2 line (discussed later). Alternatively, the biased representation of SV40 sequences in SVT2 could be a consequence of events which occurred during the many passages since the original integration event.

Cells may contain SV40 DNA sequences without expressing all of the phenotypic characteristics of transformation. For example, revertants of SV40-transformed cell lines selected for growth to a low saturation density (Pollack *et al.*, 1968) or for resistance to Concanavalin A

TABLE VII

AMOUNTS OF DIFFERENT REGIONS OF THE SV40 GENOME PRESENT IN THE SVT2 LINE OF TRANSFORMED MOUSE CELLS[a]

Fragment	Number of copies per diploid quantity of SVT2 DNA
A	6.52
B	0.81
C	5.52
D	2.04

[a] From Sambrook *et al.* (1974). SV40 DNA was digested with the *Hpa* I and *Eco*RI restriction enzymes giving four fragments (see map below). The fragments were purified and used as probes in hybridization kinetic experiments, as described in the text.

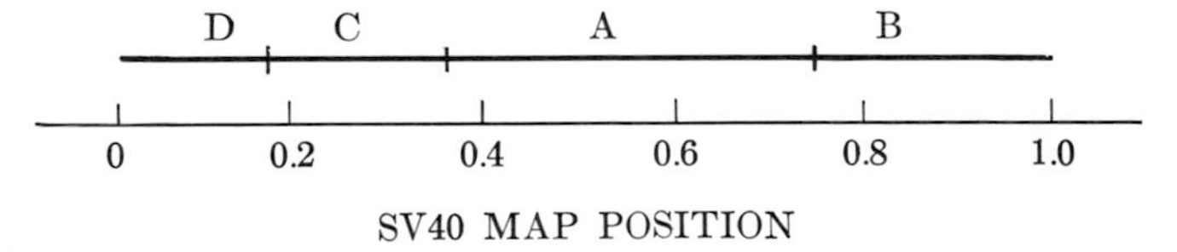

(Ozanne and Sambrook, 1971) were found to contain the same number of SV40 genome equivalents per diploid quantity of cell DNA as did their respective transformed parental lines (Ozanne *et al.,* 1973). These revertants also contained the SV40 T antigen. These data indicate that although the presence of SV40 DNA is probably necessary for the maintenance of the transformed phenotype (discussed next), it is by no means sufficient.

B. Integrated State of SV40 DNA in Transformed Cells

The exact physical state of SV40 DNA in transformed cells is not known at present, but the results of a number of experiments strongly suggest that it is attached to host DNA by covalent linkages (i.e., "integrated"). Sambrook *et al.* (1968) fractionated DNA from SV40-transformed cells (SV 3T3) by a number of methods designed to separate high molecular weight cell DNA from viral DNA. SV40 DNA sequences were assayed using the RNA–DNA filter hybridization technique previously employed by Westphal and Dulbecco (1968). It was found that the SV40 sequences remained associated with cell DNA after equilibrium centrifugation in CsCl–EtBr gradients, velocity sedimentation in alkaline sucrose gradients, and Hirt fractionation. In reconstruction experiments, all of these procedures were shown to be capable of separating cell DNA from SV40 DNA extracted from purified virions. Similar results were obtained by Hirai and Defendi (1971) for a line of SV40-transformed Chinese hamster cells. Further evidence for the association of SV40 sequences with high molecular weight cell DNA was obtained by Boyd and Butel (1972), who carried out transfection experiments with DNA from SV40-transformed cells. When permissive cells were inoculated with transformed-cell DNA in the presence of DEAE dextran to promote uptake, a small but significant fraction of the cells subsequently liberated infectious SV40 virions. Boyd and Butel (1972) showed that the viral DNA sequences "rescued" by this procedure resided exclusively in the high molecular weight cell DNA fraction. Finally, the experiments of Wall and Darnell (1971) provide some indirect evidence for the integrated state of SV40 DNA in transformed cells. Using RNA–DNA hybridization methods, these workers obtained data suggesting that at least some of the SV40-specific RNA molecules isolated from transformed cell nuclei contain covalently linked host RNA. These molecules may arise from transcriptional "read-through" from host DNA into adjacent integrated SV40 DNA, or vice versa (see Fig. 18).

The detailed organization of integrated SV40 sequences in transformed cells is unknown. However, there is evidence that in a given transformed clone the viral DNA is inserted at a limited number of chromosomal sites (Botchan and McKenna, 1973). DNA from the transformed line SVT2 was digested with *Eco*RI restriction endonuclease and fractionated by gel electrophoresis. While the bulk cell DNA yielded a very large number of restriction fragments of different sizes, fragments containing viral DNA sequences fell into two more or less discrete size classes. This result indicates that, at least for the SVT2 line, the location of the integrated SV40 sequences is not completely randomized by continual excision and reintegration during growth of the transformed cells.

C. Integration of SV40 DNA during Productive Infection

Several recent experiments suggest that integration of SV40 DNA can occur during productive infection of permissive cells. Hirai and Defendi (1972) infected the CV-1 line of AGMK cell with SV40 at a multiplicity of 0.5 pfu per cell in the presence of cytosine arabinoside. At various times after infection DNA was extracted from isolated nuclei and fractionated on alkaline sucrose gradients. At 6 hours after infection most of the SV40 DNA sequences, detectable by hybridization with labeled SV40–cRNA, sedimented at 53 S or slower. However, by 30 hours after infection the bulk of the SV40 sequences were found in fractions sedimenting at 100 S or faster in the region where host DNA sedimented. When the infection was carried out in the absence of cytosine arabinoside, a large quantity of newly synthesized SV40 DNA was found in fractions at 53 S or slower at 30 hours after infection; however, significant quantities of SV40 sequences continued to be associated with the faster-sedimenting fractions. Hölzel and Sokol (1974) carried out a careful quantitative study of the association of SV40 DNA with cellular DNA during lytic infection. These workers used a four-step fractionation procedure designed to efficiently separate cellular DNA from DNA with the size and physical properties of viral DNA. Reconstruction experiments indicated that cellular DNA purified by this procedure from SV40-infected CV-1 cells at 50–60 hours after infection could contain a maximum of 0.006% contaminating free viral DNA; however, the amount of SV40 sequences found associated with high molecular weight DNA, as measured both by hybridization to SV40–cRNA and by reassociation kinetics approximated 2%. This corresponds to more than 20,000 SV40 genome equivalents per cell. When the time course of the putative association of viral sequences with cell DNA was determined, it was found that association was first detectable at 24 hours after infection and increased throughout the period of viral DNA synthesis.

It should be pointed out in connection with these experiments that the separation of cell DNA from viral DNA basically relies upon two differences in the physical properties of these molecules: (1) cell DNA has a higher average molecular weight than viral DNA; (2) viral DNA is covalently closed, and cell DNA is not. A possibility which has not yet been rigorously excluded is that during lytic infection a small fraction of intracellular SV40 DNA exists as oligomers of very high molecular weight. If such oligomers contained one or more single-strand interruptions, their physical properties would be indistinguishable from those of cell DNA. While this possibility also holds for the experiments concerned with integration of SV40 DNA in transformed cells just described, the fact that transformed cells contain only a few genome equivalents per cell renders it extremely unlikely that the high molecular weight DNA contains oligomers of SV40 DNA.

The biological significance of integration during lytic infection is a matter for speculation. There are at present no data which suggest that integration plays any obligatory role in the intracellular life cycle of SV40, and it may simply be a reflection of the fact that infected cells contain an active recombination system. Integration during lytic infection does, however, provide a convenient explanation for the origin of SV40 variants containing host DNA substitutions which accumulate during high multiplicity passage (see Section IV, C). It seems likely that the genomes of these variants arise by integration of an SV40 genome into host DNA, followed by aberrant excision similar to the mechanism proposed to account for the production of λ transducing phage (Campbell, 1962).

D. Mechanism of Integration

Integration of SV40 DNA during transformation presumably occurs via some sort of recombination between the host and viral genomes. Many transformed cell lines contain the entire SV40 genome, since wt SV40 virions can be recovered following fusion of the transformed cells with permissive cells (discussed later). At least in these cases, therefore, the recombination event(s) which results in integration must preserve all of the viral genetic information. A plausible working hypothesis is that integration occurs by a single reciprocal recombination between a circular viral genome and some region in the host genome, as in the well-studied case of bacteriophage λ (Campbell, 1962). This recombination event, which results in the insertion of an intact SV40 genome into the linear continuity of the host DNA, might or might not proceed via a site-specific mechanism. Site specificity could be mediated by a special recombination system which operates on specific nucleotide sequences in viral and host

DNA (analogous to the *int* system of λ). Alternatively, integration could occur in regions of the host DNA that are homologous to portions of the SV40 genome. In this case integration would be site-specific only to the extent that such regions of homology were limited in number and extent. Finally, integration may involve neither of these mechanisms and may belong to a class of genetic events included under the term "illegitimate" recombination. This term has been used to describe a wide variety of rare recombination events which are not obviously dependent on homology, such as insertions, deletions, translocations, and inversions. The mechanisms involved in these events are totally obscure at present. If SV40 integration belongs in this category, there is no expectation that it would be site specific.

At present there is very little experimental evidence regarding the site specificity of SV40 integration. Chow *et al.* (1974) have approached the problem of site specificity of integration during lytic infection by studying the structure of substituted variants of SV40 which arise during high-multiplicity passage. By applying electron microscopic heteroduplex methods to substituted variants which had been cleaved with the *Eco*RI or *Hpa*II restriction enzymes, Chow *et al.* (1974) were able to map the end points of the substitutions of a large number of substituted variants derived by passage of SV40 in several different cell lines. It was found that the substitution end points were not distributed at random throughout the SV40 physical map; in fact, the majority of variants (>75%) contained a substitution end point in one of three small regions in the SV40 genome at map positions 0.312 ± 0.009, 0.248 ± 0.008, and 0.205 ± 0.007. In many cases, especially those with one substitution end point at map position 0.312, the other end point of the substitution was at 0.587 ± 0.016. Two cloned substituted variants (*ev*-1101 and *ev*-1103) containing reiterated host and SV40 DNA sequences have also been studied [see Lee *et al.* (1975); also Fig. 13]. One of the junctions between host DNA and cell DNA in both of these variants is near SV40 map position 0.68. However, the host DNA's in the two variants do not appear to have sequence homology. Since it seems likely, as mentioned earlier, that substituted variants arise as a result of integration followed by aberrant excision, these data suggest that there is some specificity to integration, at least during lytic infection. However, it should be remembered that substituted variants may be subject to selection pressures which we are unaware of and which might lead to a nonrandom distribution of substitution end points even if the primary integration event were random.

Croce *et al.* (1973, 1974a,b) and Croce and Koprowski (1974a,b) have examined the question of the specificity of SV40 integration at the chromosomal level. Two different SV40 transformed human cell lines deficient in hypoxanthine–guanine phosphoribosyl transferase (HGPRT) were fused with mouse L cells deficient in thymidine kinase (TK). After selection in medium containing hypoxanthine, aminopterin, and thymidine [HAT (Littlefield, 1964)], hybrid colonies were picked and cloned. Such human-mouse hybrids preferentially lose human chromosomes during propagation (Weiss and Green, 1967), while retaining the complete mouse chromosome complement. A large number of hybrid clones derived from each of the two SV40-transformants were analyzed for the presence of various human chromosomes and for the expression of SV40 T antigen. In both cases expression of T antigen was strongly correlated with the presence of human chromosome 7. One hundred percent of T antigen–positive clones contained chromosome 7, whereas only about 20% of the T antigen–negative clones contained chromosome 7. The other human chromosomes were apparently uncorrelated with the presence of T antigen. When T antigen–positive clones containing chromosome 7 were cloned a second time, they segregated into T antigen–positive and T antigen–negative subclones, and only the former retained human chromosome 7. A similar correlation was observed between the presence of chromosome 7 and the ability of a given hybrid clone to express SV40 tumor-specific transplantation antigen (TSTA) or to produce SV40 capsid (V) antigen following fusion with permissive monkey cells. While these experiments raise the possibility that the SV40 genetic information expressed in transformed human cells is located on a specific chromosome, they cannot be considered conclusive until more transformed human cell lines have been examined.

E. Rescue of SV40 from Transformed Cells

Once cells have been transformed by SV40 and integration has occurred, the transformed phenotype is quite stable. Clones of transformed cells do not segregate untransformed cells with an appreciable frequency. However, under certain circumstances it appears that integration is reversible, albeit at very low frequency. When the cells of most SV40 transformed lines are fused with permissive monkey cells (using inactivated Sendai virus), a small fraction of the cells liberate infectious SV40 virions (Gerber, 1966; Koprowski *et al.*, 1967; Watkins and Dulbecco, 1967; Tournier *et al.*, 1967). The fraction of cells which yield virus under these circumstances depends upon the efficiency of fusion but generally does not exceed a few percent of the transformed cells. The mechanism of virus rescue is not understood but presumably requires some kind of excision event. The permissive cell supplies some factor or factors necessary for SV40 replication. It is not known whether the permissive cell also supplies some factor required for excision or whether excision continually occurs at a low frequency in the transformed cell population. Some SV40-transformed lines do not liberate infectious SV40 following fusion with permissive cells. Virus can, however, be recovered from some of these lines using the DNA transfection technique of Boyd and Butel (1972) described earlier. Some transformed lines do not yield virus by either method. This is especially likely in the case of permissive or semipermissive cells transformed by SV40. These transformed lines may contain a defective SV40 genome (see Section IV, C). Evidence supporting this view has been obtained by Knowles *et al.* (1968). These workers were able to rescue infectious SV40 by forming triple heterokaryons containing permissive monkey cells and cells of two independent transformed lines, neither of which yielded virus when fused alone to permissive cells. It seems likely that the mechanism of rescue in this case involves complementation of the defective SV40 genomes rescued from the two transformed lines, followed by recombination to yield wt virus.

An interesting model system for excision of integrated SV40 genomes is provided by certain Ad2–SV40 hybrids. Lewis and Rowe (1970) isolated two Ad2–SV40 hybrids ($Ad2^{++}HEY$ and $Ad^{++}LEY$) which appeared to contain the entire SV40 genome since infectious SV40 could be rescued following infection of monkey cells. The two hybrids differed markedly in the efficiency with which they yielded SV40: in the case of $Ad2^{++}HEY$ (high-efficiency yielder) essentially every cell infected with a hybrid particle gave rise to SV40 progeny, whereas in the case of $Ad2^{++}LEY$ (low-efficiency yielder) only about one in 10^4 cells infected with a hybrid particle yielded infectious virions. Kelly *et al.* (1974a,b) studied the structure of the hybrid DNA molecules of $Ad2^{++}HEY$ and $Ad2^{++}LEY$ using electron microscopic heteroduplex methods. Both hybrid populations were found to be heterogeneous (Fig. 19). $Ad2^{++}HEY$ contained three hybrids whose SV40 DNA content differed by roughly integral multiples of one SV40 genome (about 0.4, 1.4, and 2.4 SV40 genome equivalents, respectively). In each case the SV40 DNA was integrated as a single continuous segment at the same site in the Ad2 DNA molecules. Furthermore, in those hybrids which contained more than one SV40 genome equivalent, the extra SV40 DNA was arranged as a tandem repetition. $Ad^{++}LEY$ contained two hybrids whose SV40 content differed by about one SV40 genome (about 0.05 and 1.05 genome equivalents, respectively); the SV40 DNA of the two hybrids was integrated at the same site in the Ad2 genome (but different from the integration site of $Ad2^{++}HEY$), and the hybrid which contained more than one SV40 genome had a tandem repetition. These data suggested that the hybrids within each population were interconvertible by recombination events that integrate or excise an intact SV40 genome. It was proposed that the mechanism of excision involved intramolecular recombination

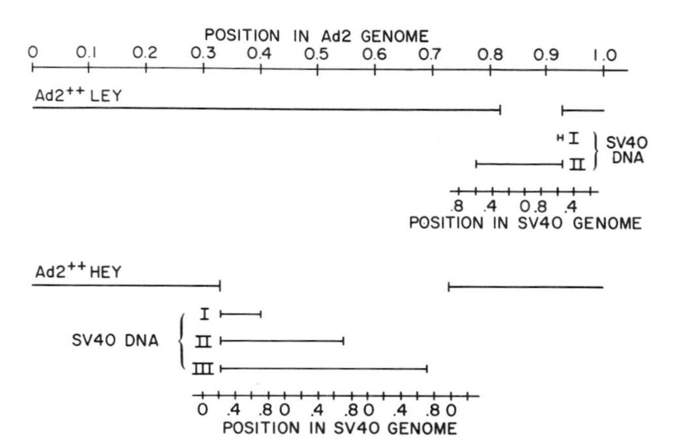

FIG. 19. The structure of AD2++HEY and AD2++LEY genomes. [From Kelly *et al.* (1974b).]

between a given region of the integrated SV40 DNA and the homologous region in the tandem repetition. The fact that the Ad2++HEY hybrids contained much longer tandem repetitions than the Ad2++LEY hybrid provided a reasonable explanation for the difference in efficiency with which the two hybrids yielded SV40 in permissive cells. It is not clear at present whether data obtained in this model system is relevant to virus rescue from transformed cells. However, it should be pointed out that if during the transformation process SV40 DNA inserts into a region of homology in the host genome, the resulting integrated genome would be expected to have a tandem repetition similar to Ad2++HEY and Ad2++LEY.

VIII. Concluding Remarks

During the past 5 years major advances have been made in understanding the structure, organization, and function of the SV40 genome, as summarized in this review. On the basis of these studies a physiological map, albeit incomplete, can now be drawn, as shown in Fig. 20. In addition to summing up what is known about the organization of the SV40 genome, the map also points up deficiencies in current knowledge: the incomplete genetic analysis, particularly in the early region and the beginning of the late region; the nature and exact locus of signals related to replication and transcription and their possible interrelatedness; the relationship between primary transcripts and viral mRNA and between the two different classes of late mRNA; the regulation of early and late gene expression; the enzymology of viral DNA replication; whether specific sites exist in the viral and cellular genomes for recombination leading to

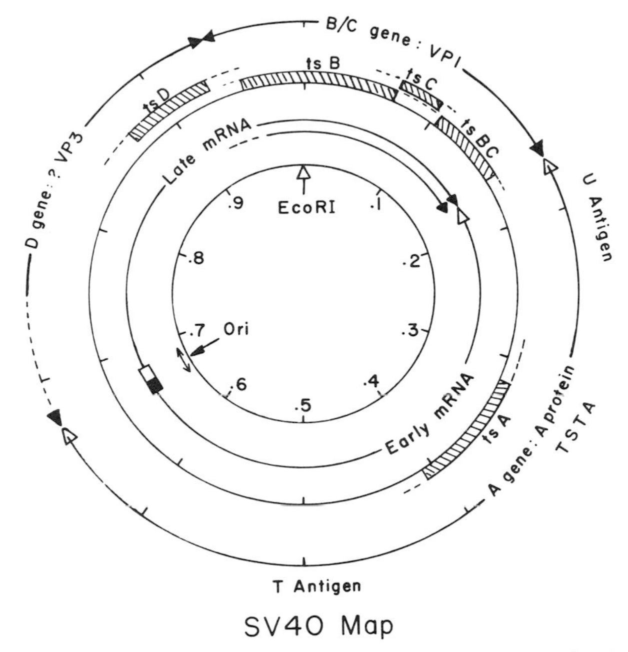

FIG. 20. Summary map of the SV40 genome. See the text for details.

integration; the function of the minor virion protein VP3; the nature of the A gene product(s) and its function in viral and cellular DNA replication and transformation; and the role of cellular gene expression in productive and transforming infection. All of these questions are under active study, and at least some of them are likely to be answered prior to publication of this review.

ADDENDUM

There have been several noteworthy developments since this review was submitted, some of which are summarized below.

1. *T Antigen*

The evidence that T antigen is coded by the early region of the SV40 genome is now quite strong. As described in the body of the review, a polypeptide with a molecular weight of approximately 100,000 which reacts specifically with anti-T serum has been detected in SV40-infected and transformed cells using immunoprecipitation methods (Tegtmeyer, 1974; Tegtmeyer *et al.*, 1975). A mutant of SV40 which has a deletion of the DNA segment between SV40 map positions .32 and .43 does not induce this polypeptide, but does induce a smaller polypeptide (MW 35,000) which reacts with anti-T serum and yields tryptic peptides that are a subset of those of the 100,000 dalton polypeptide (Rundell *et al.*, 1976). Additional evidence that T antigen is SV40 coded has been obtained using the wheat germ cell-free translation system (Prives *et al.*, 1976). In this system early 19 S mRNA purified from SV40-infected cells directs the synthesis of two prominent polypeptides whose sizes are estimated at 93,000 and 17,000 daltons. These polypeptides react with anti-SV40 T serum, and comigrate electrophoretically with two polypeptides which can be specifically immunoprecipitated from SV40-infected cell extracts with anti-T serum. The relationship between these two polypeptides and the 100,000 dalton polypeptide identified by Tegtmeyer and co-workers is not clear at present.

2. Viable Deletion Mutants

A set of viable deletion mutants of SV40 has been isolated by Shenk *et al.* (1976) by infecting permissive cells with SV40 DNA linearized with pancreatic DNase in the presence of Mn^{++}. The deletions range from $<$ 15 base pairs to about 200 base pairs and are clustered at the following sites (see Fig. 20): (1) between 0.17 and 0.18 map units; (2) between 0.54 and 0.59 map units; and (3) between 0.68 and 0.74 map units. Mutants in the last two classes generally grow to lower titer than wild type virus, and the ratio of transforming units to plaque-forming units is about the same as that of wild type SV40. Thus, studies with these mutants define two rather extensive, nonessential segments of the SV40 genome that may regulate virus development in an as yet unknown way. Still to be excluded, however, is the possibility that these regions are part of structural genes coding for nonessential amino acid sequences in virus proteins.

3. Nucleotide Sequence around the Origin of SV40 DNA Replication

Subramanian *et al.* (1976) have recently reported the complete nucleotide sequence of the *Eco*RII-G fragment of SV40, which extends from 0.635 to 0.70 map units in the conventional SV40 map. As seen in Fig. 20, this segment includes the origin of SV40 DNA replication and the templates for the 5′ ends of early and late viral mRNA's. The sequence data of Subramanian *et al.* are reproduced in Fig. 21. Of note are the numerous symmetrical sequences, and in particular, the extensive, GC-rich inverted repeat sequence extending from nucleotide 139 to 165; since SV40 DNA replication is bidirectional, one might expect the signal to have two-

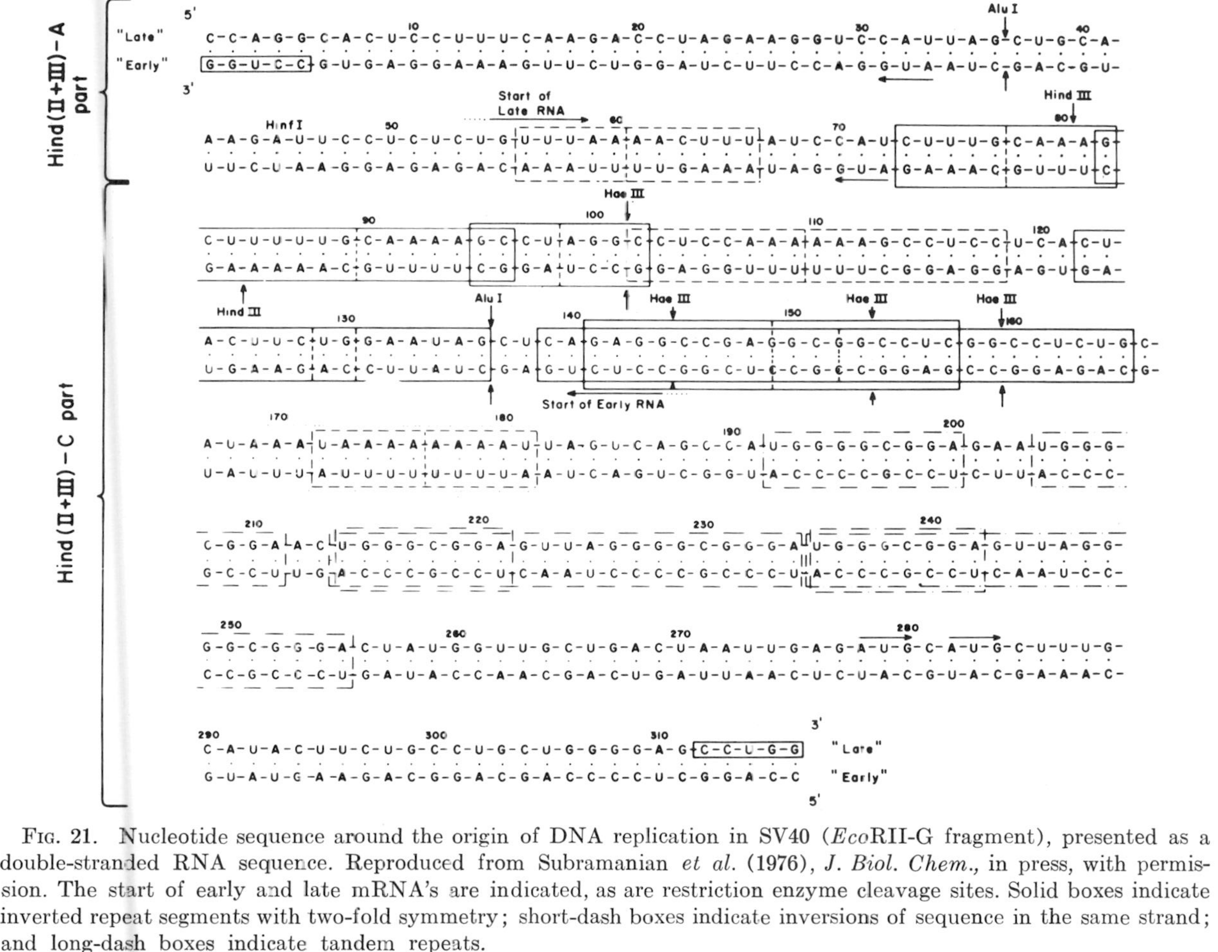

FIG. 21. Nucleotide sequence around the origin of DNA replication in SV40 (*Eco*RII-G fragment), presented as a double-stranded RNA sequence. Reproduced from Subramanian *et al.* (1976), *J. Biol. Chem.,* in press, with permission. The start of early and late mRNA's are indicated, as are restriction enzyme cleavage sites. Solid boxes indicate inverted repeat segments with two-fold symmetry; short-dash boxes indicate inversions of sequence in the same strand; and long-dash boxes indicate tandem repeats.

fold rotational symmetry of this type. More precise localization of *ori* will probably require mutational changes in this region. At present, studies with deletion mutants and evolutionary variants indicate that *ori* must lie between 0.655 (about nucleotide 73) and 0.68 map units (about nucleotide 196). The sequence shown in Fig. 21 is also rich in intrastrand inversions and tandem repeat sequences. One can speculate that some of these sequences play a role in regulating transcription or in coupling transcription and replication. Investigation of these possibilities is awaited with great interest.

4. Nucleotide Sequence at the Start of the B/C (VP1) Gene

Fiers *et al.* (1975) have sequenced the *Hin*-K fragment of SV40 DNA (between .945 and .985 map units, see Fig. 20), and located codons corresponding to the first seven amino terminal amino acids of VP1 as well as the sequence proximal to it (van de Voorde *et al.,* 1976). These data place the start of VP1 (AUG initiator codon) 10 nucleotides from the *Hin*-E • K junction. Also of interest is the sequence preceding the AUG codon (UAU • AA), which is the same sequence as that found in the ribosome-binding site of another eukaryotic mRNA, brome mosaic virus RNA (Das Gupta *et al.,* 1975).

5. Transcription

Late 16 S and 19 S SV40 mRNA species have been mapped more precisely than previously by making use of *Hind*II + III fragments (May *et al.,* 1975; Khoury *et al.,* 1976). Late 16 S RNA has sequences corresponding to *Hin*-K, F, J, and G, suggesting it is derived from DNA between .95 and .17 map units (see Fig. 20). This corresponds to the B/C (VP1) gene. Late 19 S mRNA has sequences corresponding to *Hin*-D, E, K, F, J, G and possibly C, suggesting it is derived from DNA starting between .65 and .75 map units and ending at about .17 map units.

A transcriptionally active SV40 nucleoprotein complex has been isolated from the nuclei of infected permissive cells by Gariglio and Mousset (1975) using sarkosyl lysis of nuclei and differential centrifugation of cellular chromatin. Identification of proteins in the complex, and particularly characterization of nascent RNA transcripts, should help determine whether specific promotor sites are present in SV40 DNA.

6. SV40 Integration

Information about the structure of integrated SV40 DNA has recently been obtained by analysis of restriction enzyme digests of transformed

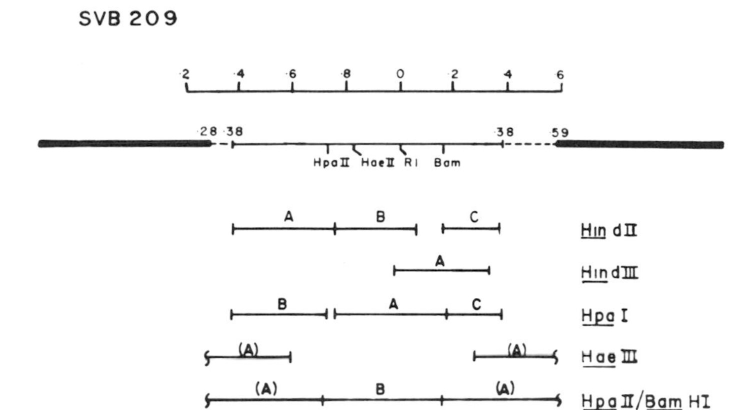

SVB 213

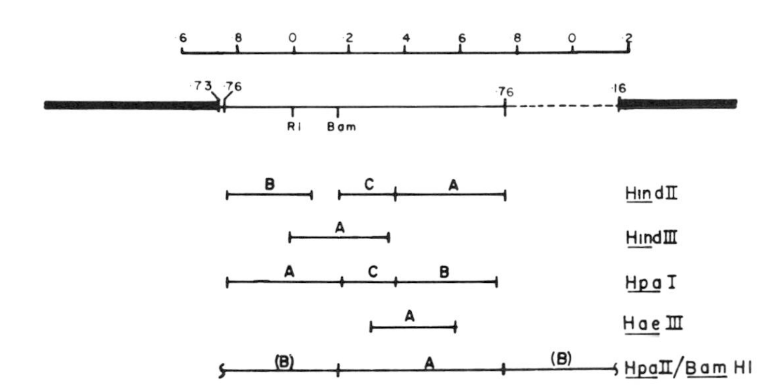

FIG. 22. The structure of the integrated viral DNA in two SV40-transformed BALB/c 3T3 cell lines. Heavy lines denote host sequences; light lines denote viral sequences. The scale above each map is marked in SV40 map units. The endpoints of the integrated viral segment in line SVB209 (upper) lie in the early region of the SV40 genome. (The left endpoint lies between SV40 map positions 0.28 and 0.38 while the right endpoint lies between SV40 map position 0.38 and 0.59). The endpoints of the integrated viral segment in line SVB213 (lower) lie in the late region of the SV40 genome (left endpoint: 0.73–0.76 and right endpoint: 0.76–0.16). Restriction fragments detectable following digestion of transformed cell DNA with various enzymes are indicated below each map. Fragments whose labels are enclosed in parenthesis are missing from the digests, although they are readily detectable in control digests of similar amounts of SV40 DNA (Ketner and Kelly, unpublished).

cell DNA (Ketner and Kelly, 1976). DNA was isolated from a series of independent SV40 transformants of the BALB/c 3T3 cell line and digested with restriction enzymes which are known to cleave the SV40 genome at single sites. The digest was fractionated by agarose gel electrophoresis and restriction fragments containing SV40 DNA sequences were localized by the transfer-hybridization method of Southern (1976). Each of the transformed lines examined by this method yielded a different pattern of restriction fragments containing viral DNA. This finding indicates that the structure of the integrated SV40 DNA and/or its location in the cell genome is different in each independent transformant and suggests that integration of the SV40 genome during transformation is not absolutely site-specific. Subsequent studies in which DNA from these transformants was digested with restriction enzymes that cleave SV40 at multiple sites have led to the construction of physical maps of the integrated SV40 DNA (Ketner and Kelly, unpublished). Figure 22 shows maps of the integrated viral DNA in two different SV40-transformed BALB/c 3T3 lines. The structures of the integrated viral DNA segments are different for the two lines. In particular the joints between viral and cell DNA occur in different regions of the SV40 genome. These results provide further support for the hypothesis that integration of SV40 DNA during transformation can occur at more than one site. A recent study in which the transfer-hybridization method was applied to SV40-transformed rat cells has also led to the conclusion that there are multiple sites where recombination can occur between SV40 and cell DNAs (Botchan, Topp, and Sambrook, 1976).

Acknowledgment

We thank our many colleagues who sent us preprints of unpublished work. The authors' research included in this review was supported by grants from the National Cancer Institute (1–P01–CA16519), the American Cancer Society, Inc. (VC–132A), and the Whitehall Foundation.

References

Aloni, Y. (1972). *Proc. Natl. Acad. Sci. U.S.A.* **69,** 2404.
Aloni, Y. (1974). *Cold Spring Harbor Symp. Quant. Biol.* **39,** 165.
Aloni, Y. E., Winocour, E., and Sachs, L. (1968). *J. Mol. Biol.* **31,** 415.
Anderer, F. A., Schlumberger, H. D., Koch, M. A., Frank, H., and Eggers, H. J. (1967). *Virology* **32,** 511.
Anderer, F. A., Koch, M. A., and Schlumberger, H. D. (1968). *Virology* **34,** 452.
Anderson, C. W., and Gesteland, R. F. (1972). *J. Virol.* **9,** 758.
Ando, T. (1966). *Biochim. Biophys. Acta* **114,** 158.

Arber, W. (1974). *Prog. Nucl. Acid Res. Mol. Biol.* **14,** 1.
Barban, S., and Goor, R. S. (1971). *J. Virol.* **7,** 198.
Bartok, K., Garon, C. F., Berry, K. W., Fraser, M. J., and Rose, J. A. (1974). *J. Mol. Biol.* **87,** 437.
Bauer, W., and Vinograd, J. (1968). *J. Mol. Biol.* **33,** 141.
Baum, S. G., and Fox, R. I. (1974). *Cold Spring Harbor Symp. Quant. Biol.* **39,** 567.
Baum, S. G., Reich, P. R., Huebner, R. J., Rowe, W. P., and Weissman, S. M. (1966). *Proc. Natl. Acad. Sci. U.S.A.* **56,** 1509.
Baum, S. G., Wiese, W. H., and Reich, P. R. (1968). *Virology* **34,** 374.
Baum, S. G., Horwitz, M. S., and Maizel, J. V., Jr. (1972). *J. Virol.* **10,** 211.
Beard, P., Morrow, J. F., and Berg, P. (1973). *J. Virol.* **12,** 1303.
Beardmore, W. B., Havlick, M. J., Serafini, A., and McLean, I. W. (1965). *J. Immunol.* **95,** 422.
Bigger, C. H., Murray, K., and Murray, N. E. (1973). *Nature (London), New Biol.* **244,** 7.
Black, P. H., Rowe, W. P., Turner, H. C., and Huebner, R. J. (1963). *Proc. Natl. Acad. Sci. U.S.A.* **50,** 1148.
Botchan, M., and McKenna, G. (1973). *Cold Spring Harbor Symp. Quant. Biol.* **38,** 391.
Botchan, M., Ozanne, B., Sugden, B., Sharp, P. A., and Sambrook, J. (1974). *Proc. Natl. Acad. Sci. U.S.A.* **71,** 4183.
Botchan, M., Topp, W., and Sambrook, J. (1976). *Cell* (in press).
Boyd, V. A. L., and Butel, J. S. (1972). *J. Virol.* **10,** 399.
Boyer, H. W. (1971) *Annu. Rev. Microbiol.* **25,** 153.
Boyer, H. W., Chow, L. T., Dugaiczyk, A., Hedgpeth, J., and Goodman, H. M. (1973). *Nature (London), New Biol.* **244,** 40.
Britton, R. J., and Kohne, D. E. (1968). *Science* **161,** 529.
Brockman, W. W., and Nathans, D. (1974). *Proc. Natl. Acad. Sci. U.S.A.* **71,** 942.
Brockman, W. W., and Scott, W. (1975). *Fed. Proc., Fed. Am. Soc. Exp. Biol.* **34,** No. 1724. (Abstr.)
Brockman, W. W., Lee, T. N. H., and Nathans, D. (1973). *Virology* **54,** 384.
Brockman, W. W., Lee, T. N. H., and Nathans, D. (1974). *Cold Spring Harbor Symp. Quant. Biol.* **39,** 119.
Brockman, W. W., Gutai, M. W., and Nathans, D. (1975). *Virology* **66,** 36.
Brugge, J. S., and Butel, J. S. (1975). *J. Virol.* **15,** 619.
Bryan, R. N., Gelfand, D. H., and Hayashi, M. (1969). *Nature (London)* **224,** 1019.
Burgoyne, L. A., Hewish, D. R., and Mobbs, J. (1974). *Biochem. J.* **143,** 67.
Butel, J. S., Brugge, J. S., and Noonan, C. A. (1974). *Cold Spring Harbor Symp. Quant. Biol.* **39,** 25.
Campbell, A. (1962). *Adv. Genet.* **11,** 101.
Carbon, J., Shenk, T. E., and Berg, P. (1975a). *Proc. Natl. Acad. Sci. U.S.A.* **72,** 1392.
Carbon, J., Shenk, T. E., and Berg, P. (1975b). *J. Mol. Biol.* **98,** 1.
Carroll, R. B., Hager, L., and Dulbecco, R. (1974). *Proc. Natl. Acad. Sci. U.S.A.* **71,** 3754.
Champoux, J. J., and Dulbecco, R. (1972). *Proc. Natl. Acad. Sci. U.S.A.* **69,** 143.
Chou, J., and Martin, R. G. (1974). *J. Virol.* **13,** 1101.
Chou, J., and Martin, R. G. (1975). *J. Virol.* **15,** 127.
Chou, J., Avila, J., and Martin, R. G. (1974). *J. Virol.* **14,** 116.
Chow, L. T., Boyer, H. W., Tischer, E. G., and Goodman, H. M. (1974). *Cold Spring Harbor Symp. Quant. Biol.* **39,** 109.

Coggin, J. H., Elrod, L. H., Ambrose, K. R., and Anderson, N. G. (1969). *Proc. Soc. Exp. Biol. Med.* **132,** 328.
Cowan, K., Tegtmeyer, P., and Anthony, D. D. (1973). *Proc. Natl. Acad. Sci. U.S.A.* **70,** 1927.
Crawford, L. V. (1969). *Adv. Virus Res.* **14,** 89.
Crawford, L. V., and Black, P. H. (1964). *Virology* **24,** 388.
Crawford, L. V., and Waring, M. J. (1967). *J. Mol. Biol.* **25,** 23.
Crawford, L. V., Follett, E. A. C., and Crawford, E. M. (1966). *J. Microsc. (Paris)* **5,** 597.
Crawford, L. V., Gesteland, R. F., Rubin, G. M., and Hirt, B. (1971). *Lepitit Colloq. Biol. Med.* **2,** 104.
Croce, C. M., and Koprowski, H. (1974a). *J. Exp. Med.* **139,** 1350.
Croce, C. M., and Koprowski, H. (1974b). *Science* **184,** 1288.
Croce, C. M., Girardi, A. J., and Koprowski, H. (1973). *Proc. Natl. Acad. Sci. U.S.A.* **70,** 3617.
Croce, C. M., Huebner, K., Girardi, A. J., and Koprowski, H. (1974a). *Virology* **60,** 276.
Croce, C. M., Huebner, K., Girardi, A. J., and Koprowski, H. (1974b). *Cold Spring Harbor Symp. Quant. Biol.* **39,** 137.
Crumpacker, C. S., Levin, M. J., Wiese, W. H., Lewis, A. M., Jr., and Rowe, W. P. (1970). *J. Virol.* **6,** 788.
Danna, K. J., and Nathans, D. (1971). *Proc. Natl. Acad. Sci. U.S.A.* **68,** 2913.
Danna, K. J., and Nathans, D. (1972). *Proc. Natl. Acad. Sci. U.S.A.* **69,** 3097.
Danna, K. J., Sack, G. H., Jr., and Nathans, D. (1973). *J. Mol. Biol.* **78,** 363.
Das Gupta, R., Shih, D., Saris, C., and Kaesberg, P. (1975). *Nature* **256,** 624.
Davoli, D., and Fareed, G. C. (1974). *Cold Spring Harbor Symp. Quant. Biol.* **39,** 137.
Defendi, V. (1963). *Proc. Soc. Exp. Biol. Med.* **113,** 12.
Del Villano, B. C., and Defendi, V. (1973). *Virology* **51,** 34.
DePamphilis, M. L., and Berg, P. (1975). *J. Biol. Chem.* **250,** 4348.
DePamphilis, M. L., Beard, P., and Berg, P. (1975). *J. Biol. Chem.* **250,** 4340.
Dhar, R., Subramanian, K., Zain, B. S., Pan, J., and Weissman, S. M. (1974a). *Cold Spring Harbor Symp. Quant. Biol.* **39,** 153.
Dhar, R., Zain, B. S., Weissman, S. M., Pan, J., and Subramanian, K. (1974b). *Proc. Natl. Acad. Sci. U.S.A.* **71,** 371.
Dhar, R., Subramanian, K., Pan, J., and Weissman, S. M. (1975). *Fed. Proc., Fed. Am. Soc. Exp. Biol.* **34,** No. 2206. (Abstr.)
Diamandopoulos, G. T., Tevethia, S. S., Rapp, F., and Enders, J. F. (1968). *Virology* **34,** 331.
Dubbs, D. R., Rachmeler, M., and Kit, S. (1974). *Virology* **57,** 161.
Eason, R., and Vinograd, J. (1971). *J. Virol.* **7,** 1.
Easton, J. M., and Hiatt, C. W. (1965). *Proc. Natl. Acad. Sci. U.S.A.* **54,** 1100.
Eron, L., Westphal, H., and Khoury, G. (1975). *J. Virol.* **15,** 1256.
Estes, M. K., Huang, E., and Pagano, J. (1971). *J. Virol.* **7,** 635.
Fareed, G. C., and Salzman, N. P. (1972). *Nature (London), New Biol.* **238,** 274.
Fareed, G. C. Garon, C. F., and Salzman, N. P. (1972). *J. Virol.* **10,** 484.
Fareed, G. C., McKerlie, M. L., and Salzman, N. P. (1973a). *J. Mol. Biol.* **74,** 95.
Fareed, G. C., Khoury, G., and Salzman N. P. (1973b). *J. Mol. Biol.* **77,** 457.
Fareed, G. C., Byrne, J. C., and Martin, M. A. (1974). *J. Mol. Biol.* **87,** 275.
Fiers, W., Danna, K., Rogiers, R., Vandevoorde, A., Van Herreweghe, J., Van Heuverswyn, H., Volckaert, G., and Yang, R. (1974). *Cold Spring Harbor Symp. Quant. Biol.* **39,** 179.

Fiers, W., Rogiers, R., Soeda, E., van de Voorde, A., van Heuverswyn, H., van Herreweghe, J., Volckaert, G., and Yang, R. (1975). Proceedings of the Tenth FEBS Meeting, pp. 17–33. North-Holland, Amsterdam.
Fischer, H., and Sauer, G. (1972). *J. Virol.* **9,** 1.
Fox, R. I., and Baum, S. G. (1974). *Virology* **60,** 45.
Friedman, M. P., Lyons, M. S., and Ginsberg, H. S. (1970). *J. Virol.* **5,** 586.
Fuks, P., Goldblum, T., Cymbalista, S., and Goldblum, N. (1967). *Is. J. Med. Sci.* **3,** 912.
Fuller, W., and Waring, M. J. (1964). *Ber. Bunsenges. Phys. Chem.* **68,** 805.
Ganem, D., Davoli, D., and Fareed, G. C. (1975). *Fed. Proc., Fed. Am. Soc. Exp. Biol.* **34,** No. 1723. (Abstr.)
Garfin, D. E., and Goodman, H. M. (1974). *Biochem. Biophys. Res. Commun.* **59,** 108.
Garigloi, P., and Mousset, S. (1975). *FEBS Lett.* **56,** 146.
Gelb, L. D., Kohne, D. E., and Martin, M. A. (1971). *J. Mol. Biol.* **57,** 129.
Gelfand, D. H., and Hayashi, M. (1969). *Proc. Natl. Acad. Sci. U.S.A.* **63,** 135.
Gerber, P. (1966). *Virology* **28,** 501.
Germond, J. E., Hirt, B., Oudet, P., Gross-Bellard, M., and Chambon, P. (1975). *Proc. Natl. Acad. Sci. U.S.A.* **72,** 1843.
Gerry, H. W., Kelly, J. T., Jr., and Berns, K. I. (1973). *J. Mol. Biol.* **79,** 207.
Gershon, D., Sachs, L., and Winocour, E. (1966). *Proc. Natl. Acad. Sci. U.S.A.* **56,** 918.
Gilden, R. V., Carp, R. I., Taguchi, F., and Defendi, V. (1965). *Proc. Natl. Acad. Sci. U.S.A.* **53,** 684.
Gillespie, D., and Spiegelman, S. (1965). *J. Mol. Biol.* **12,** 829.
Girard, M., Marty, L., and Saurez, F. (1970). *Biochem. Biophys. Res. Commun.* **40,** 97.
Girardi, A. J., and Defendi, V. (1970). *Virology* **42,** 688.
Glaubiger, D., and Hearst, J. E. (1967). *Biopolymers* **8,** 691.
Graham, F. L., and van der Eb, A. J. (1973). *Virology* **52,** 456.
Graham, F. L., Abrahams, P. J., Mulder, C., Heijneker, H. L., Warnaar, S. O., de Vries, F. A. J., Fiers, W., and van der Eb, A. J. (1974). *Cold Spring Harbor Symp. Quant. Biol.* **39,** 637.
Gray, H. B., Jr., Upholt, W. B., and Vinograd, J. (1971). *J. Mol. Biol.* **62,** 1.
Griffith, J. (1975). *Science* **187,** 1202.
Griffith, J., Dieckmann, M., and Berg, P. (1975). *J. Virol.* **15,** 167.
Haas, M., Vogt, M., Dulbecco, R. (1972). *Proc. Natl. Acad. Sci. U.S.A.* **69,** 2160.
Habel, K., and Eddy, B. E. (1963). *Proc. Soc. Exp. Biol. Med.* **113,** 1.
Häyry, P., and Defendi, V. (1968). *Virology* **36,** 317.
Häyry, P., and Defendi, V. (1970). *Virology* **41,** 22.
Hall, M. R., Meinke, W., and Goldstein, D. A. (1973). *J. Virol.* **12,** 901.
Hartley, J. W., Huebner, R. J., and Rowe, W. P. (1956). *Proc. Soc. Exp. Biol. Med.* **92,** 667.
Hashimoto, K., Nakajima, K., Oda, K., and Shimogo, H. (1973). *J. Mol. Biol.* **81,** 207.
Hatanaka, M., and Dulbecco, R. (1966). *Proc. Natl. Acad. Sci. U.S.A.* **56,** 736.
Hedgpeth, J., Goodman, H. M., and Boyer, H. W. (1972). *Proc. Natl. Acad. Sci. U.S.A.* **69,** 3448.
Henry, P., Black, P. H., Oxman, M. N., and Weissman, S. M. (1966). *Proc. Natl. Acad. Sci. U.S.A.* **56,** 1170.
Henry, P., Schnipper, L. E., Samaha, R. V., Crumpacker, C. S., Lewis, A. M., Jr., and Levine, A. S. (1973). *J. Virol.* **11,** 665.
Hirai, K., and Defendi, V. (1971). *Biochem. Biophys. Res. Commun.* **42,** 714.
Hirai, K., and Defendi, V. (1972). *J. Virol.* **9,** 705.

Hirt, B. (1967). *J. Mol. Biol.* **26,** 365.
Hölzel, F., and Sokol, F. (1974). *J. Mol. Biol.* **84,** 423.
Hoggan, M. D., Rowe, W. P., Black, P. H., and Huebner, R. J. (1965). *Proc. Natl. Acad. Sci. U.S.A.* **53,** 12.
Huang, E. Estes, M. K., and Pagano, P. S. (1972). *J. Virol.* **9,** 923.
Huebner, R. J., Chanock, R. M., Rubin, B. A., and Casey, M. J. (1964). *Proc. Natl. Acad. Sci. U.S.A.* **52,** 1333.
Hutchison, C. A., III, and Edgell, M. H. (1971). *J. Virol.* **8,** 181.
Inman, R. B., and Schnös, M. (1970). *J. Mol. Biol.* **49,** 93.
Ishikawa, A., and Aizawa, T. (1973). *J. Gen. Virol.* **21,** 227.
Jackson, D. A., Symons, R. H., and Berg, P. (1972). *Proc. Natl. Acad. Sci. U.S.A.* **69,** 2904.
Jaenisch, R., and Levine, A. (1971). *Virology* **44,** 480.
Jaenisch, R., and Levine, A. (1973). *J. Mol. Biol.* **73,** 199.
Jaenisch, R., Mayer, A., and Levine, A. (1971). *Nature (London), New Biol.* **233,** 72.
Kamen, R., Lindstrom, D. M., Shure, H., and Old, R. W. (1974). *Cold Spring Harbor Symp. Quant. Biol.* **39,** 187.
Keller, W., and Wendel, I. (1974). *Cold Spring Harbor Symp. Quant. Biol.* **39,** 199.
Kelly, T. J., Jr. (1975). *J. Virol.* **15,** 1267.
Kelly, T. J., Jr. and Lewis, A. M., Jr. (1973). *J. Virol.* **12,** 643.
Kelly, T. J., Jr., and Rose, J. (1971). *Proc. Natl. Acad. Sci. U.S.A.* **68,** 1037.
Kelly, T. J., Jr., and Smith, H. O. (1970). *J. Mol. Biol.* **51,** 393.
Kelly, T. J., Jr., Lewis, A. M., Jr., Levine, A. S., and Siegel, S. (1974a). *J. Mol. Biol.* **89,** 113.
Kelly, T. J., Jr., Lewis, A. M., Jr., Levine, A. S., and Siegel, S. (1974b). *Cold Spring Harbor Symp. Quant. Biol.* **39,** 409.
Ketner, G., and Kelly, T. J., Jr. (1976). *Proc. Natl. Acad. Sci. U.S.A.* **73,** 1102.
Khera, K. S., Ashkenazi, A., Rapp, F., and Melnick, J. L. (1963). *J. Immunol.* **91,** 604.
Khoury, G., and Martin, M. A. (1972). *Nature (London), New Biol.* **238,** 4.
Khoury, G., Lewis, A. M., Jr., Oxman, M. N., and Levine, A. S. (1973a). *Nature (London), New Biol.* **246,** 202.
Khoury, G., Martin, M. A., Danna, K. J., Lee, T. N. H., and Nathans, D. (1973b). *J. Mol. Biol.* **78,** 377.
Khoury, G., Bryne, J. C., Takemoto, K. K., and Martin, M. A. (1973c). *J. Virol.* **11,** 54.
Khoury, G., Fareed, G. C., Berry, K., Martin, M. A., Lee, T. N. H., and Nathans, D. (1974). *J. Mol. Biol.* **87,** 289.
Khoury, G., Howley, P., Nathans, D., and Martin, M. A. (1975a). *J. Virol.* **15,** 433.
Khoury, G., Martin, M. A., Lee, T. N. H. and Nathans, D. (1975b). *Virology* **63,** 263.
Khoury, G., Carter, B. J., Ferdinand, F. J., Howley, P. M., Brown, M., and Martin, M. A. (1976). *J. Virol.* **17,** 832.
Kimura, G., and Dulbecco, R. (1972). *Virology* **49,** 394.
Kimura, G., and Dulbecco, R. (1973). *Virology* **52,** 529.
Kimura, G., and Itagaki, A. (1975). *Proc. Natl. Acad. Sci. U.S.A.* **72,** 673.
Kit, S. (1968). *Adv. Cancer Res.* **11,** 73.
Kit, S., Melnick, J. L., Anken, M., Dubbs, D. R., de Torres, R. A., and Kitahara, T. (1967). *J. Virol.* **1,** 684.
Kluchareva, T. E., Shachanina, K. I., Belova, S., Chibisova, V., and Deichman, G. I. 1967). *J. Natl. Cancer Inst.* **39,** 825.

Knowles, B. B., Jensen, F. C., Steplewski, Z. S., and Koprowski, H. (1968). *Proc. Natl. Acad. Sci. U.S.A.* **61,** 42.
Koch, M. A., and Sabin, A. B. (1963). *Proc. Soc. Exp. Biol. Med.* **113,** 4.
Koprowski, H., Jensen, F. C., and Steplewski, Z. S. (1967). *Proc. Natl. Acad. Sci. U.S.A.* **58,** 127.
Kornberg, R. D. (1974). *Science* **184,** 868.
Kornberg, R. D., and Thomas, J. O. (1974). *Science* **184,** 685.
Lai, C.-J., and Nathans, D. (1974a). *Cold Spring Harbor Symp. Quant. Biol.* **39,** 53.
Lai, C.-J., and Nathans, D. (1974b). *Virology* **60,** 466.
Lai, C.-J., and Nathans, D. (1974c). *J. Mol. Biol.* **89,** 179.
Lai, C.-J., and Nathans, D. (1975a). *Virology* **66,** 70.
Lai, C.-J., and Nathans, D. (1975b). *J. Mol. Biol.* **97,** 113.
Laipis, P., and Levine, A. J. (1973). *Virology* **56,** 580.
Lake, R. S., Barban, S., and Salzman, N. P. (1973). *Biochem. Biophys. Res. Commun.* **54,** 640.
Lavi, S., and Winocour, E. (1972). *J. Virol.* **9,** 309.
Lazarus, H. M., Sporn, M. B., Smith, J. M., and Henderson, W. C. (1967). *J. Virol.* **1,** 1093.
Lebowitz, P., and Khoury, G. (1975). *J. Virol.* **15,** 1214.
Lebowitz, P., Kelly, T. J., Jr., Nathans, D., Lee, T. N. H., and Lewis, A. M., Jr. (1974). *Proc. Natl. Acad. Sci. U.S.A.* **71,** 441.
Lee, T. N. H., Brockman, W. W., and Nathans, D. (1975). *Virology* **66,** 53.
Levine, A. J. (1972). *Perspect. Virol.* **8,** 61.
Levine, A. J. (1974). *Prog. Med. Virol.* **17,** 1.
Levine, A. J., and Teresky, A. K. (1970). *J. Virol.* **5,** 451.
Levine, A. J., Kang, H. S., and Billheimer, F. E. (1970). *J. Mol. Biol.* **50,** 549.
Levine, A. S., Oxman, M. N., Henry, P. H., Levin, M. J., Diamandopoulos, G. T., and Enders, J. F. (1970). *J. Virol.* **6,** 199.
Levine, A. S., Levin, M. J., Oxman, M. N., and Lewis, A. M., Jr. (1973). *J. Virol.* **11,** 672.
Lewis, A. M., Jr., and Rowe, W .P. (1970). *J. Virol.* **5,** 413.
Lewis, A. M., Jr., and Rowe, W. P. (1971). *J. Virol.* **7,** 189.
Lewis, A. M., Jr., Baum, S. G., Prigge, K. O., and Rowe, W. P. (1966). *Proc. Soc. Exp. Biol. Med.* **122,** 214.
Lewis, A. M., Jr., Levin, M. J., Wiese, W. H., Crumpacker, C. S., and Henry, P. H. (1969). *Proc. Natl. Acad. Sci. U.S.A.* **63,** 1128.
Lewis, A. M., Jr., Levine, A. S., Crumpacker, C. S., Levin, M. J., Samaha, R. J., and Henry, P. H. (1973). *J. Virol.* **11,** 655.
Lewis, A. M., Jr., Breeden, J. H., Wewerka, Y. L., Schnipper, L. E., and Levine, A. S. (1974). *Cold Spring Harbor Symp. Quant. Biol.* **39,** 651.
Lindberg, V., and Darnell, J. (1970). *Proc. Natl. Acad. Sci. U.S.A.* **65,** 1089.
Lindstrom, D. M., and Dulbecco, R. (1972). *Proc. Natl. Acad. Sci. U.S.A.* **69,** 1517.
Littlefield, J. W. (1964). *Science* **145,** 709.
Livingston, D. M., Henderson, I. C., and Hudson, J. (1974). *Cold Spring Harbor Symp. Quant. Biol.* **39,** 283.
Lobban, P. E., and Kaiser, A. D. (1973). *J. Mol. Biol.* **78,** 453.
Lodish, H. F., Weinberg, R. A., and Ozer, H. L. (1974). *J. Virol.* **13,** 590.
Mantei, N., Boyer, H. W., and Goodman, H. M. (1975). *J. Virol.* **16,** 754.
Martin, M. A., and Axelrod, D. (1969). *Science* **164,** 68.
Martin, M. A., and Chou, J. (1975). *J. Virol.* **15,** 599.

Martin, M. A., Gelb, L. D., Fareed, G. C., and Milstein, J. B. (1973). *J. Virol.* **12,** 748.
Martin, R .G., Chou, J. Y., Avila, J. and Saral, R. (1974). *Cold Spring Harbor Symp. Quant. Biol.* **39,** 17.
May, E., Kpecka, H., and May, P. (1975). *Nucleic Acids Res.* **2,** 1995.
Mayer, A., and Levine, A. J. (1972). *Virology* **50,** 328.
Mertz, J. E. (1975). Ph.D. Thesis, Stanford Univ., Palo Alto, California.
Mertz, J. E., and Berg, P. (1974a). *Virology* **62,** 112.
Mertz, J. E., and Berg, P. (1974b). *Proc. Natl. Acad. Sci. U.S.A.* **71,** 4879.
Mertz, J. E., and Davis, R. W. (1972). *Proc. Natl. Acad. Sci. U.S.A.* **69,** 3370.
Mertz, J. E., Carbon, J., Herzberg, M., Davis, R. W., and Berg, P. (1974). *Cold Spring Harbor Symp. Quant. Biol.* **39,** 69.
Meselson, M., and Yuan, R. (1968). *Nature (London)* **217,** 1110.
Metzgar, R. S., and Oleinick, S. R. (1968). *Cancer Res.* **28,** 1366.
Morrow, J. F., and Berg, P. (1972). *Proc. Natl. Acad. Sci. U.S.A.* **69,** 3365.
Morrow, J. F., and Berg, P. (1973). *J. Virol.* **12,** 1631.
Morrow, J. F., Berg. P., Kelly, T. J., Jr., and Lewis, A. M., Jr. (1973). *J. Virol.* **12,** 653.
Mulder, C., and Delius, H. (1972). *Proc. Natl. Acad. Sci. U.S.A.* **69,** 3215.
Nakajima, K., Ishitsuka, H., and Oda, K. (1974). *Nature (London)* **252,** 649.
Nathans, D. (1976). *Harvey Lect.* **70** (in press).
Nathans, D., and Danna, K. J. (1972). *J. Mol. Biol.* **64,** 515.
Nathans, D., and Smith, H. O. (1975). *Annu. Rev. Biochem.* **44,** 273.
Noll, M. (1974). *Nature (London)* **251,** 249.
Oda, K., and Dulbecco, R. (1968). *Proc. Natl. Acad. Sci. U.S.A.* **60,** 525.
Okazaki, R., Okazaki, T., Sakabe, K., Sugimoto, K., and Sugino, A. (1968). *Proc. Natl. Acad. Sci. U.S.A.* **59,** 598.
Old, R. W., Murray, K., and Roizes, G. (1975). *J. Mol. Biol.* **92,** 331.
Olins, A. L., and Olins, P. E. (1974). *Science* **183,** 330.
Osborn, M., and Weber, K. (1974). *Cold Spring Harbor Symp. Quant. Biol.* **39,** 367.
Osborn, M., and Weber, K. (1975). *J. Virol.* **15,** 636.
Oudet, P., Gross-Bellard, M., and Chambon, P. (1975). *Cell* **4,** 281.
Ozanne, B. (1973). *J. Virol.* **12,** 79.
Ozanne, B., and Sambrook, J. (1971). *Lepetit Colloq. Biol. Med.* **2,** 248.
Ozanne, B., Sharp, P. A., and Sambrook, J. (1973). *J. Virol.* **12,** 90.
Ozer, H. L. (1972). *J. Virol.* **9,** 41.
Patch, C. T., Lewis, A. M., Jr., and Levine, A. S. (1974). *J. Virol.* **13,** 677.
Pollack, R. E., Green, H., and Todaro, G. J. (1968). *Proc. Natl. Acad. Sci. U.S.A.* **60,** 126.
Pope, J. H., and Rowe, W. P. (1964). *J. Exp. Med.* **120,** 121.
Potter, C. W., McLaughlin, B. C., and Oxford, J. S. (1969). *J. Virol.* **4,** 574.
Prives, C. L., Aviv, H., Peterson, B. M., Roberts, B. E., Rozenblatt, S., Ravel, M., and Winocour, E. (1974a). *Proc. Natl. Acad. Sci. U.S.A.* **71,** 302.
Prives, C. L., Aviv, H., Gilboa, E., Ravel, M., and Winocour, E. (1974b). *Cold Spring Harbor Symp. Quant. Biol.* **39,** 309.
Prives, C., Gluzman, Y., Gilboa, E., Revel, M. and Winocour, E. (1976). *Proc. Natl. Acad. Sci. U.S.A.* (in press).
Pulleyblank, D. E., and Morgan, A. R. (1975). *J. Mol. Biol.* **91,** 1.
Qasba, P. K. (1974). *Proc. Natl. Acad. Sci. U.S.A.* **71,** 1045.
Rabson, A. S., O'Conor, G. T., Berezesky, I. K., and Paul F. S. (1964). *Proc. Soc. Exp. Biol. Med.* **116,** 187.

Radloff, R., Bauer, W., and Vinograd, J. (1967). *Proc. Natl. Acad. Sci. U.S.A.* **57,** 1514.
Rapp, F., Kitahara, T., Butel, J. S., and Melnick, J. L. (1964a). *Proc. Natl. Acad. Sci. U.S.A.* **52,** 1138.
Rapp, F., Melnick, J. L., Butel, J. S., and Kitahara, T. (1964b). *Proc. Natl. Acad. Sci. U.S.A.* **52,** 1348.
Rapp, F., Butel, J. S., Feldman, L. A., Kitahara, T., and Melnick, J. L. (1965). *J. Exp. Med.* **121,** 935.
Reed, S. I., Ferguson, J., Davis, R. W., and Stark, G. R. (1975). *Proc. Natl. Acad. Sci. U.S.A.* **72,** 1605.
Reich, P. R., Black, P. H., and Weissman, S. M. (1966). *Proc. Natl. Acad. Sci. U.S.A.* **56,** 78.
Risser, R., and Mulder, C. (1974). *Virology* **58,** 424.
Ritzi, E., and Levine, A. J. (1970). *J. Virol.* **5,** 686.
Robb, J. A. (1973). *J. Virol.* **12,** 1187.
Robb, J. A. (1974). *Cold Spring Harbor Symp. Quant. Biol.* **39,** 277.
Robb, J., and Martin, R. G. (1972). *J. Virol.* **9,** 956.
Robb, J. A., Tegtmeyer, P., Ishikawa, A., and Ozer, H. L. (1974). *J. Virol.* **13,** 662.
Roberts, B. E., Gorecki, M., Mulligan, R. C., Danna, K. J., Rozenblatt, S., and Rich, A. (1975). *Proc. Natl. Acad. Sci. U.S.A.* **72,** 1922.
Roberts, R. J. Myers, P. A., Morrison, A., and Murray, K. (1976a). *J. Mol. Biol.* **102,** 157.
Roberts, R. J., Myers, P. A., Morrison, A., and Murray, K. (1976b). *J. Mol. Biol.* **103,** 199.
Roberts, R. J., Breitmeyer, J. B., Tabachnik, N. F., and Myers, P. A. (1975). *J. Mol. Biol.* **91,** 121.
Rowe, W. P., and Baum, S. G. (1964). *Proc. Natl. Acad. Sci. U.S.A.* **52,** 1340.
Rozenblatt, S., and Winocour, E. (1972). *Virology* **50,** 558.
Rozenblatt, S., Lavi, S., Singer, M. F., and Winocour, E. (1973). *J. Virol.* **12,** 501.
Rundell, K., Lai, C-J., and Tegtmeyer, P. (1976). *J. Virol.* (in press).
Rush, M. G., Eason, R., and Vinograd, J. (1971). *Biochim. Biophys. Acta* **228,** 585.
Sabin, A., and Koch, M. A. (1964). *Proc. Natl. Acad. Sci. U.S.A.* **52,** 1131.
Sack, G. H., Jr., Narayan, O., Danna, K. J., Weiner, L. P., and Nathans, D. (1973). *Virology* **51,** 345.
Salzman, N. P., Sebring, E. D., and Radonovich, M. (1973). *J. Virol.* **12,** 669.
Salzman, N. P., Lebowitz, J., Chen, M., Sebring, E., and Garon, C. F. (1974). *Cold Spring Harbor Symp. Quant. Biol.* **39,** 209.
Sambrook, J., Westphal, H., Srinivasan, P. R., and Dulbecco, R. (1968). *Proc. Natl. Acad. Sci. U.S.A.* **60,** 1288.
Sambrook, J., Sharp, P. A., and Keller, W. (1972). *J. Mol. Biol.* **70,** 57.
Sambrook, J., Sugden, B., Keller, W., and Sharp, P. A. (1973). *Proc. Natl. Acad. Sci. U.S.A.* **70,** 3711.
Sambrook, J., Botchan, M., Gallimore, P., Ozanne, B., Pettersson, U., Williams, J., and Sharp, P. A. (1974). *Cold Spring Harbor Symp. Quant. Biol.* **39,** 615.
Sauer, G., and Kidwai, J. R. (1968). *Proc. Natl. Acad. Sci. U.S.A.* **61,** 1256.
Schell, K., Lane, W. T., Casey, M. J., and Huebner, R. J. (1966). *Proc. Natl. Acad. Sci. U.S.A.* **55,** 81.
Sebring, E. D., Kelly, T. J., Jr., Thoren, M. M., and Salzman, N. F. (1971). *J. Virol.* **8,** 479.
Sebring, E. D., Garon, C. F., and Salzman, N. P. (1974). *J. Mol. Biol.* **90,** 371.

Sen, A., and Levine, A. J. (1974). *Nature (London)* **249,** 343.
Senior, M. B., Olins, A. L., and Olins, D. E. (1975). *Science* **187,** 173.
Sharp, P. A., Sugden, B., and Sambrook, J. (1973). *Biochemistry* **12,** 3055.
Sharp, P. A., Pettersson, U., and Sambrook, J. (1974). *J. Mol. Biol.* **86,** 709.
Shenk, T. E., Rhodes, C., Rigby, P., and Berg, P. (1975). *Proc. Natl. Acad. Sci. U.S.A.* **72,** 989.
Shenk, T. E., Carbon, J., and Berg, P. (1976). *J. Virol.* **18,** 644.
Smith, H. O., and Nathans, D. (1973). *J. Mol. Biol.* **81,** 419.
Southern, E. (1975). *J. Mol. Biol.* **98,** 503.
Subramanian, K. N., Pan, J., Zain, S., and Weissman, S. M. (1974). *Nucleic Acids Res.* **1,** 727.
Subramanian, K. N., Dhar, R., and Weissman, S. M. (1976). *J. Biol. Chem.* (in press).
Sweet, B. H., and Hilleman, M. R. (1960). *Proc. Soc. Exp. Biol. Med.* **105,** 420.
Tai, H. T., Smith, C. A., Sharp, P. A., and Vinograd, J. (1972). *J. Virol.* **9,** 317.
Takemoto, K. K., Kirschstein, R. L., and Habel, K. (1966). *J. Bacteriol.* **92,** 990.
Tegtmeyer, P. (1972). *J. Virol.* **10,** 591.
Tegtmeyer, P. (1974). *Cold Spring Harbor Symp. Quant. Biol.* **39,** 9.
Tegtmeyer, P. (1975). *J. Virol.* **15,** 613.
Tegtmeyer, P., and Ozer, H. L. (1971). *J. Virol.* **8,** 516.
Tegtmeyer, P., Schwartz, M., Collins, J. K., and Rundell, K. (1975). *J. Virol.* **16,** 168.
Tevethia, S. S., Katz, M., and Rapp, F. (1965). *Proc. Soc. Exp. Biol. Med.* **119,** 896.
Tevethia S. S., Diamandopoulos, G. T., Rapp, F., and Enders, J. F. (1968). *J. Immunol.* **101,** 1192.
Tonegawa, S., Walter, G., Bernardini, A., Dulbecco, R. (1970). *Cold Spring Harbor Symp. Quant. Biol.* **35,** 823.
Tournier, P., Cassingena, R., Wickert, R., Coppey, J., and Suarez, H. (1967). *Int. J. Cancer* **2,** 117.
Trilling, D. M., and Axelrod, D. (1970). *Science* **168,** 268.
Uchida, S., and Watanabe, S. (1969). *Virology* **39,** 721.
Uchida, S., Watanabe, S., and Kato, M. (1966). *Virology* **28,** 135.
Uchida, S., Yoshiike, K., Watanabe, S. and Furuno, A. (1968). *Virology* **34,** 1.
Upholt, W. B., Gray, H. B., Jr., and Vinograd, J. (1971). *J. Mol. Biol.* **61,** 21.
van de Voorde, A., Contreras, R., Rogiers, R., and Fiers, W. (1976). *Cell* (in press).
Van Holde, K. E., Sahasrabuddhe, B. R., Shaw, B. R., von Bruggen, E. F. J., and Arnberg, A. (1974). *Biohem. Biophys. Res. Commun.* **60,** 1365.
Vinograd, J., and Lebowitz, J. (1966a). *J. Gen. Physiol.* **49,** 103.
Vinograd, J., and Lebowitz, J. (1966b). "Macromolecular Metabolism," p. 103. Little, Brown, Boston, Massachusetts.
Vinograd, J., Lebowitz, J., Radloff, R., Watson, R., and Laipis, P. (1965). *Proc. Natl. Acad. Sci. U.S.A.* **53,** 1104.
Vinograd, J., Lebowitz, J., and Watson, R. (1968). *J. Mol. Biol.* **33,** 173.
Wall, R., and Darnell, J. E. (1971). *Nature (London), New Biol.* **232,** 73.
Walter, G., Roblin, R., and Dulbecco, R. (1972). *Proc. Natl. Acad. Sci. U.S.A.* **69,** 921.
Wang, J. C. (1971). *J. Mol. Biol.* **55,** 523.
Wang, J. C. (1974). *J. Mol. Biol.* **89,** 783.
Watkins, J. F., and Dulbecco, R. (1967). *Proc. Natl. Acad. Sci. U.S.A.* **58,** 1396.
Weinberg, R. A., Warnaar, S. O, and Winocour, E. (1972). *J. Virol.* **10,** 193.
Weinberg, R. A., Ben-Ishai, Z., and Newbold, J. E. (1974). *J. Virol.* **13,** 1263.
Weiss, M. C., and Green, H. (1967). *Proc. Natl. Acad. Sci. U.S.A.* **58,** 1104.

Werchau, H., Westphal, H., Maass, G., and Haas, R. (1966). *Arch. Gesamte Virusforsch.* **19,** 351.
Westphal, H. (1970). *J. Mol. Biol.* **50,** 407.
Westphal, H., and Dulbecco, R. (1968). *Proc. Natl. Acad. Sci. U.S.A.* **59,** 1158.
Wetmur, J. G., and Davidson, N. (1968). *J. Mol. Biol.* **31,** 649.
White, M., and Eason, R. (1971). *J. Virol.* **8,** 363.
Wilson, G. A., and Young, F. E. (1975). *J. Mol. Biol.* **97,** 123.
Yang, R., Danna, K. J., van de Voorde, A., and Fiers, W. (1975). *Virology* **68,** 260.
Yang, R., Vandevoorde, A., and Fiers, W. (1976). *Eur. J. Biochem.* **61,** 119.
Yoshiike, K., (1968). *Virology* **34,** 391.
Yoshiike, K., Furuno, A., Watanabe, S., Uchida, K., Matsubara, K., and Takagi, Y. (1974a). *Cold Spring Harbor Symp. Quant. Biol.* **39,** 85.
Yoshiike, K., Furuno, A., and Uchida, S. (1974b). *Virology* **60,** 342.

PLANT VIRUS INCLUSION BODIES

Giovanni P. Martelli and Marcello Russo

Istituto di Patologia Vegetale,
University of Bari, Bari, Italy

I. Introduction

Inclusion bodies can be simply defined as intracellular structures produced *de novo* as a result of viral infections. They may contain virus particles, virus-related materials, or ordinary cell constituents in a normal or degenerating condition, either singly or, more often, in various proportions.

By this definition, most, if not all, plant viruses are liable to induce formation of inclusion bodies of some sort. These may be crystalline or amorphous (or both) in the same cell, and may vary in size. Thus, some inclusions are readily seen with the light microscope, as they constitute bulky masses often attaining 10–20 μm along the major axis, whereas others are only visible with the electron microscope.

This is a broad interpretation of the concept of inclusion body. Previously, only those entities observable in the light microscope have been recognized as inclusions (McWhorter, 1965). However, it seems reasonable to agree with Esau (1967) that limiting the definition of inclusion body to structures above a minimal size is impractical. For instance, aggregates of virus particles, which are to be regarded as inclusions, may vary tremendously in dimensions.

Intracellular inclusions arise as a consequence of disturbed cell metabolism following virus infection but are not necessarily a by-product of this process. In some instances, inclusions constitute a more or less inert accumulation of virus particles or excess viral material, but there is mounting evidence that, in many cases, inclusion bodies play a primary role in the synthesis and/or assembly of viral components. Therefore, a knowledge of their fine structure and composition seems fundamental to an understanding of their probable function.

The two basic questions about inclusions are always (1) what are they made of, and (2) what is their significance in the economy of cell–virus interactions? Answers to these questions are being provided by recent studies, but a lot of ground still remains to be covered.

Virus inclusions in plant cells have been reviewed by Smith (1958), Goldin (1963), McWhorter (1965), Thaler (1966), and Rubio-Huertos (1972) in articles containing a great deal of information on the appearance of these inclusions, their staining properties, and their composition, as deduced primarily from light microscope studies. Informative chapters on intracellular inclusions are also present in plant virology textbooks (Bawden, 1964; Matthews, 1970; Bos, 1970) and some of their ultrastructural aspects were dealt with years ago (Matsui and Yamaguchi, 1966; Esau, 1967, 1968). Additional information on these abnormalities can be found in papers reviewing the cytology of virus infections (Bald, 1966; De Zoeten, 1976). Therefore, in this presentation, rather than reviewing plant virus inclusions comprehensively, we will discuss the most significant findings in this field in the context, whenever possible, of an interpretation of the role of inclusions in the infection process.

II. Progress in the Study of Inclusion Bodies: A Short Outline

The notion that virus infections may induce abnormalities of the protoplast resulting in the development of novel structures not seen in healthy cells dates back to the early days of plant virology. The first observations (Iwanowski, 1903) concerned tobacco cells infected with tobacco mosaic virus (TMV), but as studies progressed it became clear that other viruses too were able to induce comparable abnormalities which, in the literature, became known as vacuolate bodies, amoeboid bodies, plasmodium-like masses, spherules or, after Goldstein (1924), X-bodies.

Until the early 1960s investigations on virus-induced intracellular inclusions were dominated by light microscopy. The primary objects of examination were fresh mounts of trichomes and epidermal strips from infected leaves, petioles, and stems, where the inclusions could be readily detected. However, portions of leaves and flowers mounted under vacuum in physiological salt solutions were successfully used (McWhorter, 1951), as well as sectioned tissues (e.g., Esau, 1941, 1944, 1960a,b).

Even at that time, investigators not only directed their efforts toward description of the inclusions and identification of their nature, but also pursued the sequence of events leading to their formation. In this respect, the papers of Sheffield (1931, 1939, 1941) and Zech (1952, 1954, 1960) were of great relevance. These workers followed the *in vivo* development of X-bodies of the aucuba strain of TMV by observing living, infected cells directly under the microscope, or by time-lapse photography (Sheffield, 1931; Singh and Hildebrandt, 1966).

Considerable advances were made in the development of suitable fixation techniques (Bald, 1949a,b; Solberg and Bald, 1964) and of staining procedures that permitted rapid detection of the inclusions and their differentiation from ordinary cell constituents (McWhorter, 1941; Rich, 1948a; Bald, 1949b; Rubio-Huertos, 1950; Littau and Black, 1952; Rawlings, 1957; Robb, 1964).

Despite these contributions, study of virus-induced inclusions evidently had not progressed quite satisfactorily, because McWhorter, in his 1965 review, while emphasizing the diagnostic value of intracellular inclusions and their importance in virus reproduction, sharply criticized the lack of interest in these structures shown by plant virologists.

Since the mid-1960s a considerable body of information on the ultrastructure of virus-infected cells has accumulated. The advent of electron microscopy, however, has not discouraged investigations with the light microscope. Several reports have in fact been published in recent years dealing with light microscopy of inclusions of many different viruses (Mamula and Miličić, 1968; Bos, 1969; Juretić *et al.*, 1970; Miličić and Juretić, 1971; Da Cruz *et al.*, 1972). New rapid-staining procedures were combined with detergent treatment in order to solubilize plastids, thereby facilitating their identification in epidermal strips (Christie, 1967, 1971).

There has also been an increasing tendency to view and compare similar materials with both light and electron microscopes. This has led to a better understanding of both the inclusion body ultrastructure and gross chemical composition by virtue of cytochemical staining reactions and enzyme digestion tests (Rubio-Huertos, 1962, 1968; Kitajima *et al.*, 1968; Warmke, 1969; Harrison *et al.*, 1970; Martelli and Castellano, 1971; Štefanac and Ljubešić, 1971, 1974; Bos and Rubio-Huertos, 1969,

1972; Edwardson *et al.*, 1972; Rubio-Huertos and Bos, 1973; Russo and Martelli, 1975b; Martelli and Russo, 1976a).

At present, selective dyes for the identification of proteins (mercuric bromophenol blue, ninhydrin–Schiff), lipids (Sudan black B, Sudan IV), carbohydrates (periodic acid–Schiff) and nucleic acids (Feulgen, methyl green–pyronine, azure B) are of common use in light microscopy, but the problems of chemical identification are far more severe at the electron microscope level.

The electron microscope has greatly improved our knowledge of viral inclusions by disclosing the occurrence and localization of entities small enough to escape detection by light microscopy, and by providing a more adequate insight into their nature and fine structure. Thus, for example, it was ascertained electron microscopically that the amorphous inclusions of cauliflower mosaic virus, which were formerly thought to consist of dense masses of protoplasm without specific virus content (Rubio-Huertos, 1956), are indeed packed with virions (Fujisawa *et al.*, 1967c); on the other hand, the "pure virus crystals" in cells infected by red clover vein mosaic virus (McWhorter, 1965) are not made up of virus particles at all (Rubio-Huertos and Bos, 1973).

The great potentialities of electron microscopy have been further exploited by techniques like vacuum dehydration (Edwardson *et al.*, 1968), embedding in glycol methacrylate for enzyme digestion (Weintraub *et al.*, 1969), freeze-fracturing (Hatta *et al.*, 1973), freeze-etching (McDonald and Hiebert, 1974a), the use of labeled antibodies (Shepard and Shalla, 1969; Schlegel and DeLisle, 1971), and autoradiography (Hayashi and Matsui, 1967; Kamei *et al.*, 1969c). These techniques have proved valuable in determining what occurs in inclusion bodies as well as where they originate and what they are made of.

A recent most promising approach has used cytological observations at the fine-structure level combined with cell fractionation methods to study how viral inclusions are involved in virus replication (De Zoeten *et al.*, 1974; Matthews, 1975). Further advances in this field can reasonably be expected from the use of plant protoplast–virus systems [for reviews, see Zaitlin and Beachey (1974); Takebe (1975)].

III. Grouping of Inclusion Bodies

In previous reviews, inclusion bodies have been classified according to diverse criteria. Thus, Goldin (1963) followed a matrical approach, grouping inclusions in relation to the botanical families to which the hosts infected by the inducing virus belonged. This system had the advantage of being highly practical for diagnostic purposes but was little concerned with the origin and nature of the pathological structures. Conversely, other authors' classifications (Smith, 1958; McWhorter, 1965; Rubio-Huertos, 1972) were primarily inspired by morphological and topological considerations, based on similarities in the outer appearance of the inclusion bodies and of their location within the cell. McWhorter (1965), in particular, recognized ten different forms of inclusions divided into four groups, but these, in the light of more recent work, appear rather heterogeneous.

At present, therefore, neither of the above classification schemes seems entirely satisfactory. Viral inclusions, in fact, are quite variable and complex. Their final appearance and constitution are the result of an interaction between the host and the pathogen. However, in some instances it is certainly the host plant (i.e., the cell type) that determines both the kind and shape of the inclusions. Thus, it is known that the size of TMV X-bodies varies with the dimensions of the mother cell. Large cells may harbor inclusions up to 30 μm in diameter, whereas in small cells the inclusions may not exceed 5 μm diameter (Bawden, 1964). Esau and Hoefert (1971b) reported the occurrence of massive fibrous aggregates composed almost exclusively of beet yellows virus particles in sieve elements of *Tetragonia expansa* Murr; loose fibrous masses of the same virus intermingled with host components were found in parenchyma cells of the same host (Esau and Hoefert, 1971a). The explanation provided for such a difference was ascribed to the peculiar nature of the protoplast of sieve elements, which, as it contains few organelles and lacks a tonoplast, does not restrict the accumulation of virions to parietal cytoplasm. Similarly, the elongated form of cauliflower mosaic virus inclusions in *Brassica rapa* L. mesophyll was explained as being a consequence of their development in cells with a very narrow parietal strip of cytoplasm that permitted their growth in one direction only (Martelli and Castellano, 1971).

In other instances, it is the virus that plays a primary role in determining the inclusion type. Specificity can be found at the virus group or virus strain level. An example of the former is provided by the proteinaceous cylindrical inclusions of potyviruses, whose specificity for this group is now beyond question (Edwardson, 1974). As to virus strain specificity, one such example was recently illustrated for soil-borne wheat mosaic virus, a virus with divided genome, whose isolates from different geographic regions are known to evoke inclusion bodies differing in size and ultrastructural features (Hibino *et al.*, 1974a,b). Taking soil-borne wheat mosaic virus inclusions as genetic markers, Tsuchizaki *et al.* (1975) showed that RNA within the short virus particles controls inclusion type as well as particle length and serotype (coat protein). An addi-

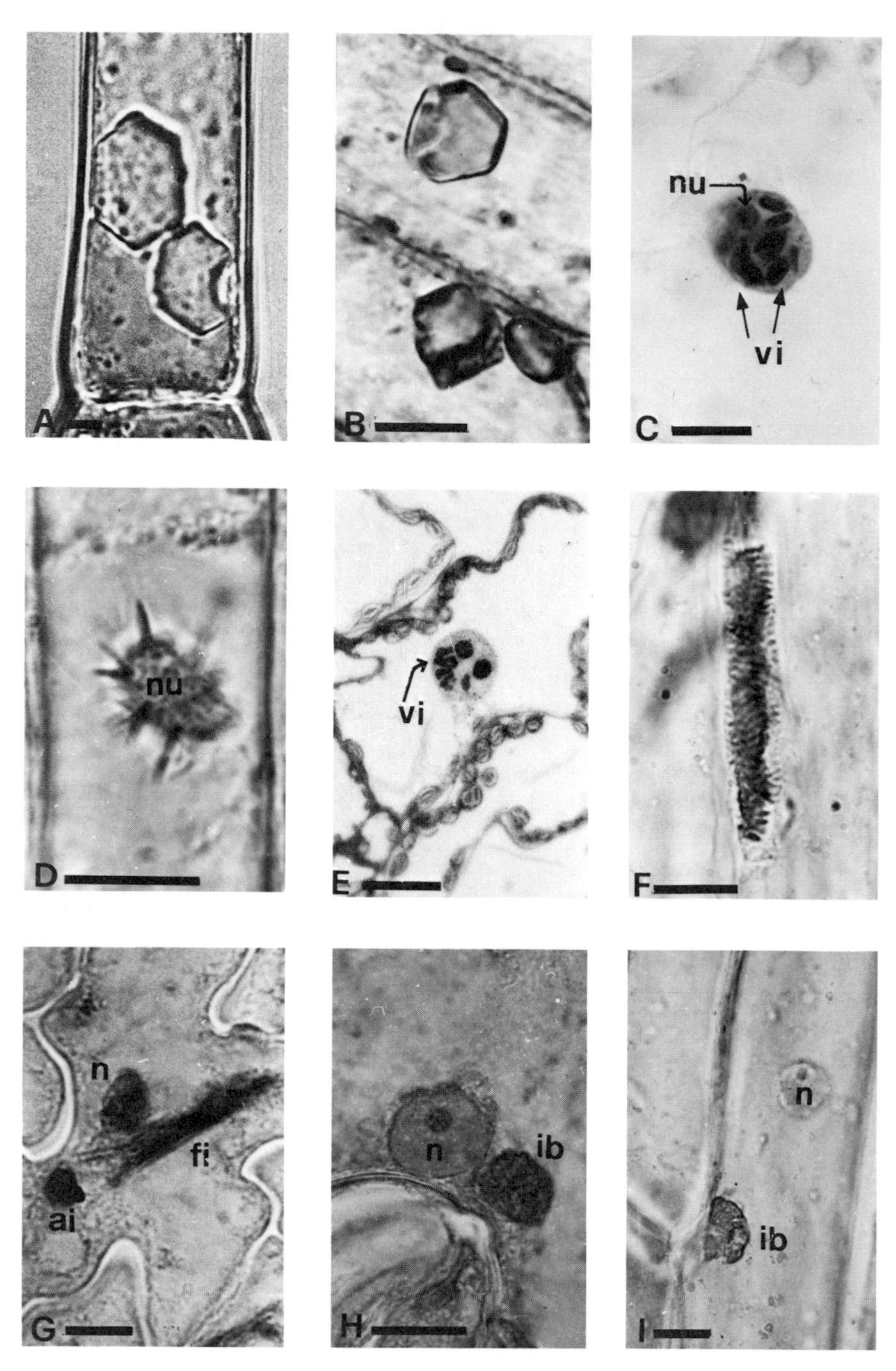

FIG. 1. Light microscopic pictures of various types of inclusions in virus-infected plant cells. (A) Hexagonal crystals of tobacco mosaic virus in a tobacco leaf hair

tional case of virus strain specificity is given by the various forms of cytoplasmic or intranuclear accumulations recorded for alfalfa mosaic virus (AMV). Studying the *in vivo* behavior of 24 AMV strains, Hull *et al.* (1970) recognized four district aggregation forms of cytoplasmic particles. Not all the virus strains (only 13 out of 24) gave rise to aggregation bodies, but those which evoked them did so independently of the host and showed a remarkable consistency in their production, even after repeated transfers from plant to plant. In fact, in mixed infections, the simultaneous occurrence of two AMV strains in the same cell could be clearly established upon detection of the aggregate type that is specific to each strain (Hull and Plaskitt, 1970).

Unfortunately, little is known at present about the significance and function of inclusion bodies. This information, which should come mostly from biophysical and biochemical investigations and cell fractionation studies, would provide a sounder foundation for grouping and classifying the inclusions not only on ultrastructural bases, but also according to the physiological role they play in the infection process—their primary and characterizing feature.

For the time being it is necessary to rely mostly on data from electron microscopy studies, which, although more informative than light microscopic observations, are not sufficient to give a thoroughly satisfactory insight into the significance of inclusion bodies.

Thus, on the basis of morphological similarity, it is tempting to speculate on possible activities of virus inclusions, extrapolating from

---

cell. [From Bos (1970) courtesy of the copyright holder, Pudoc, Wageningen.] (B) Nonviral crystal associated with infection by red clover vein mottle virus. [From Rubio-Huertos and Bos (1973), courtesy of the copyright holder, Netherlands Society of Plant Pathology.] (C) Intranuclear inclusions evoked by tobacco etch virus (vi); nu, nucleolus. (Courtesy of R. G. Christie.) (D) Greatly enlarged nucleolus with many radiating needles in a cell infected by pea necrosis virus. [From Bos and Rubio-Huertos (1969), courtesy of the copyright holder, Academic Press.] (E) Deeply staining perinuclear (apparently intranuclear) inclusions (vi) consisting of accumulations of *Gomphrena* rhabdovirus particles. [From Kitajima and Costa (1966), courtesy of the copyright holder, Academic Press.] (F) Banded inclusion in phloem tissue of beet infected with beet yellows virus (courtesy of L. Bos). (G) Amorphous (ai) and fibrous (fi) inclusions next to the nucleus (n) in a squash cell infected by watermelon mosaic virus. [From Martelli and Russo (1976a), courtesy of the copyright holder, Academic Press.] (H) Amorphous inclusion (ib) and nucleus (n) in a broad bean cell infected with broad bean wilt virus. [From Russo and Martelli (1975b), courtesy of the copyright holder, *Journal of Submicroscopic Cytology*.] (I) Viroplasm (ib) and nucleus (n) in turnip cell infected with cauliflower mosaic virus. [From Martelli and Castellano (1971), courtesy of the copyright holder, Cambridge University Press.] All magnification bars represent 10 μm.

the little biochemical evidence available; however, this requires the assumption that comparable structures afford comparable functions. As a consequence, the grouping that follows is highly tentative and subject to modification as further and more specific information becomes available.

Obvious morphological differences (such as an amorphous or a crystalline appearance, used as differentiating characteristics in the past) are still useful enough to be retained. Likewise, a strict topological criterion may assist for a primary, gross subdivision of inclusion bodies into nuclear and cytoplasmic categories.

IV. Nuclear Inclusions

Nuclear inclusions are of common occurrence throughout the plant kingdom. Therefore, extreme caution must be taken in separating those having a pathological origin, i.e., those which develop following virus infections, from the somewhat similar looking structures which, as will be discussed here, may be ordinary features of healthy cells.

Nevertheless, nuclear inclusions evoked by viruses, which were indeed a rarity when investigations relied only on the light microscope, although more frequently recorded electron microscopically, still appear to be relatively uncommon. Whether or not any of them is directly involved in virus multiplication has not been experimentally ascertained, except in a few instances.

Based on their localization and origin, nuclear inclusions can be classified as nucleolus-related, nucleoplasmic, and perinuclear. However, there are instances in which the whole organelle is transformed into a cytopathological structure.

In those members of the rhabdovirus group whose particles mature at the nuclear envelope, infected nuclei often exhibit marginate chromatin and a large central area with low electron density, where the karyoplasm is uniform, finely granular, or fibrillar. This area has sometimes been referred to as viroplasm-like (Kitajima *et al.*, 1975), but in eggplant mottled dwarf virus it was shown to contain no RNA and to develop as a consequence of the marked decrease of nucleohistones (Russo and Martelli, 1975a). Hence, although such modification implies a direct involvement of the nucleus in virus production, similarity of this alteration with a viroplasmic structure seems difficult to maintain.

The same may not apply to the nuclear inclusion reported for the rhabdovirus infecting a *Dendrobium phalaenopsis* hybrid (Lawson and Ali, 1975), as this inclusion is virtually indistinguishable from the supposed viroplasmic masses found in the perinuclear position. Likewise, the fine structure of orchid nuclei infected by the rhabdovirus-like entities of the KORV type (Lesemann and Doraiswamy, 1975) is intriguingly reminiscent of a viroplasm. Such nuclei display a well-defined central zone with a higher (Kitajima *et al.*, 1974a) or lower (Lesemann and Doraiswamy, 1975) electron opacity than the marginal area, where chromatin is confined. Circumstantial evidence suggests that these inclusion-like areas may be factories for the inner components (nucleocapsids) of virus particles, which after being formed migrate to the periphery of the nucleus. There they become attached to the nuclear membrane, eliciting the development of bleb-like evaginations which may act as carriers of these viral structures to the cytoplasm.

Profound modifications of the nuclear structure have also been recorded in connection with pea enation mosaic virus (PEMV) infections, a virus which is suspected to multiply in the nucleus (Shikata and Maramorosch, 1966).

A. *Nucleolus-Related Inclusions*

Nucleolus-related inclusions may be accumulations of materials, conceivably synthesized at the nucleolar level, which remain *in situ* appearing as structures morphologically distinct from the organelle itself and which may assume an amorphous or a crystalline organization; or such inclusions may be complex entities originating from modification of the whole organelle.

Some of the most peculiar nucleolar inclusions are the satellite bodies elicited by beet mosaic virus (BMV). These abnormalities have been encountered at different times in four different hosts infected by BMV, so that it is highly probable that they represent reactions specifically induced by this virus. The first record of these peculiar structures is apparently that of Reitberger (1956), who with the light microscope observed anatomical deviations of nucleoli of BMV-infected sugarbeet plants; these anomalies bear a striking resemblance to the modifications later investigated with the electron microscope in *Gomphrena globosa* L. and *Chenopodium quinoa* Willd. and called satellite bodies (Martelli and Russo, 1969a,b; Russo and Martelli, 1969). The "sprouting" condition of nucleoli of BMV-infected pea cells (Bos, 1969) can be considered similar to these anomalies.

Satellite bodies consist of accumulations of intensely electron-opaque proteinaceous material, seen as dense masses having an amorphous or finely granular texture, localized at the periphery of the nucleolus. The bodies are not uniform in size or appearance, for they enlarge as the infection progresses, and contain one or more electron-clear cavities lined with ribosome-like particles. A change in their structure takes place as the infection ages. The granular matrix becomes organized into rod-like, rigid-looking units 18–20 nm in diameter, possessing a tubular nature. As a consequence, the satellite bodies become progressively disrupted. Both the amorphous matrix and the tubules are composed of protein that is believed to be synthesized *de novo* (Martelli and Russo, 1969a), but it is not known whether this protein is a quite separate material or whether it bears any relationship to virus coat protein or to the protein constituting cylindrical inclusions. Satellite bodies contain a small amount of RNA (probably ribosomal) and no DNA.

A second type of nucleolar inclusion is the crystalline formations induced in several hosts by bean yellow mosaic virus (BYMV) and related viruses. These were first observed by McWhorter (1941), who described them as isometric, isotropic crystals 0.2–3 μm in diameter, with a cubic or octahedral shape. Their appearance in the nucleoli before external symptoms were visible was pointed out by Rich (1948b), who also demonstrated their proteinaceous nature by cytochemical staining (Rich, 1948c). Virus strain–dependent differences in the size, shape, and occurrence of these crystals were reported by Mueller and Koenig (1965), who successfully used this specificity as an indicator of cross-protection between different BYMV strains; their usefulness for distinguishing filamentous legume viruses was pointed out by Edwardson *et al.* (1972).

The closeness of the relationship between crystals and nucleolus was unraveled by thin sectioning. Weintraub and Ragetli (1966) reported that in BYMV-infected broad bean cells in the early stages of infection the nucleolus was nearly always enlarged and contained, embedded in its matrix, one or more crystalline inclusions which were sometimes large enough to distort the shape of the organelle. The inclusions appeared as intensely electron-opaque bodies with a rounded or, more often, angular outline, forming squares or hexagons with an area ranging from 0.3 to 4 μm^2, and were often lined by spherical ribosome-like particles.

The fine structure of the nucleolar crystals was resolved as striations with a repeating pattern of about 7 nm. Enzyme digestion tests confirmed previous findings (Rich, 1948c) in that BYMV crystals proved to be composed entirely of pepsin and tripsin-sensitive material, with no evidence of the occurrence of nucleic acids (Weintraub and Ragetli, 1968; Weintraub *et al.*, 1969). Nucleolar inclusions virtually identical to those induced by BYMV in outer appearance, size, fine structure, and possibly chemical nature have been recorded in pea mosaic (Cousin *et al.*, 1969), sweet vetch crinkly mosaic (Russo *et al.*, 1972), and clover yellow vein (Pratt, 1969) viruses, and in an unnamed virus from virginia crabb apple (Weintraub and Ragetli, 1971).

A peculiar deviation from the pattern just described is represented by the nucleolar inclusions elicited by a virus causing pea necrosis, related but not identical with BYMV (Bos, 1969). In this case the nucleoli were much enlarged and, under the light microscope, appeared as if an array of protruding needles radiated from them (Bos, 1969; Bos and Rubio-Huertos, 1969). In thin sections of needle-containing nuclei, no organized nucleoli or nucleolar material could be observed. The needles proved to be large, electron-opaque radiating bars with a crystalline shape but no visible ultrastructure. They were large enough so as to induce deformation of the nucleus by pushing outward the nuclear envelope, without actually puncturing and breaking through it (Bos and Rubio-Huertos, 1969). Evidence from light microscopy indicates that, indeed, these inclusions begin to develop at the nucleolar level (Bos and Rubio-Huertos, 1969), but it is not known whether they represent the end product of a transformation of the nucleolus or are structures synthesized *de novo*. Their nature is also obscure, although their very appearance is suggestive of a protein composition.

Additional nucleolus-related crystalline entities have been reported as induced by an unidentified potyvirus of broad bean (Edwardson, 1974) and by two Brazilian strains of potato virus Y [PVY (Kitajima *et al.*, 1968)]. The PVY crystals are reminiscent of those evoked by BYMV, having a polygonal shape and a pattern of repeating striations; presumably they are made up of proteins, but it is not known if they bear any relationship to the cytoplasmic proteinaceous inclusions occurring in the same cells. The function of all these crystalline bodies is also unknown. In this they do not differ from crystals of elongated tubular elements with hexagonal packing found in nuclei, nearly always closely associated with the nucleolus, of tobacco cells infected with the Caldy strain of AMV (Hull *et al.*, 1970). The nature of these crystalline formations, or that of the nucleolus-related loose aggregates of long rod-shaped particles evoked by the 15/64 strain of the same virus, has not been elucidated, although the possibility exists that they may be composed of viral protein (Hull *et al.*, 1970). The same authors have also recorded an interesting though unexplained connection of type 4 aggregates of AMV particles with the nucleolus.

There are instances in which the nucleolus as a whole, because of profound modifications in its appearance and contents, becomes so abnormal as to constitute an inclusion body itself. One such example is proved by globe amaranth cells infected with *Gomphrena* rhabdovirus. As reported by Kitajima and Costa (1966), the nucleoli become greatly enlarged (10–15 μm in diameter) and vacuolated. They contain, in intimate association, fibrillar material and tubular particles, either scattered or in a paracrystalline array, identified as the internal component (nucleocapsid) of the virus. The remarkable increase in size was believed to be due to nucleolar activity during virus synthesis. A similar hypertrophic condition of nucleoli was also found associated with infections by the agent of white ear of wheat (Kitajima *et al.*, 1971), but the significance of this observation remains to be established.

B. Nucleoplasmic Inclusions

Nucleoplasmic inclusions represent an array of structures differing in size, appearance, and constitution, but all occurring in the karyoplasm of infected cells. It is by no means certain that some of them do not have a nucleolar origin. For instance, BMV satellite bodies can also be found in the nucleoplasm, far away from the site of their production (Martelli and Russo, 1969a), and the same applies to BYMV crystals (Weintraub and Ragetli, 1966). The nature of some of the nucleoplasmic inclusions is known, but their function is in most cases still obscure.

Spherical inclusions composed of discrete units and having an overall granular appearance are characteristically present in the karyoplasm of beet curly top virus–infected nuclei. Although their nature and role are unknown, the inclusions seem to disappear as the virus, which is present only in the nucleus where it arises, increases in quantity (Esau and Hoefert, 1973). Hence, they may be connected with virus multiplication.

Rounded amorphous inclusions, granular in texture and much less electron-opaque than the nucleolus, have also been recorded in the nucleoplasm of celery cells infected by celery mosaic virus, but here again no information is available on their composition and function (Edwardson, 1974). Similarly, very little is known about the nucleoplasmic inclusions of the amorphous type, recorded in cotton infected with leaf crumple virus (Tsao, 1963).

Fibrous inclusions are so called because their constituents are threadlike or are elongated tubular units aggregated in bundles of variable size. Such inclusions are resolved by electron microscopy only. They predominate in cells infected by filamentous viruses, and very often they are actually composed of aggregates of virus particles. This is the case with beet yellows (Cronshaw *et al.*, 1966; Esau, 1968; Esau and Hoefert, 1971a), beet yellow stunt (Hoefert *et al.*, 1970), and rice "hoja blanca" viruses (Kitajima and Galves, 1973), as well as with several carlaviruses (Thaler *et al.*, 1970; Turner, 1971) and potyviruses (Edwardson *et al.*, 1972; Cadilhac *et al.*, 1972; Koenig and Lesemann, 1974).

The massive intranuclear aggregates of particles of an unnamed virus from *Lantana horrida* H.B.K. and of its supposed replication intermediates (Arnott and Smith, 1968) are very striking. So are the fibrous bundles observed in the karyoplasm of nuclei infected with the rhabdovirus causing striate mosaic of wheat and interpreted as uncoiled nucleoprotein components of the virions (Vela and Lee, 1974).

Viruses with rigid rod-shaped particles like TMV and barley stripe mosaic sometimes accumulate in the nucleus, giving rise to structures resembling fibrous inclusions (Esau and Cronshaw, 1967; Esau, 1968; Gardner, 1967; Granett and Shalla, 1970). In other instances, conspicuous nucleoplasmic fibrous inclusions can be composed of whorls of threadlike proteinaceous structures which resemble but are not identical with virus particles (Kitajima and Costa, 1976). They can also consist of aggregates of tubules 30–40 nm in diameter, reminiscent of ordinary plant microtubules, but whose nature and significance is not known (Weintraub *et al.*, 1973).

Up to now intranuclear membranous inclusions have been reported in four instances. In cells infected with cowpea mosaic virus (CoMV) the nuclei often contain discrete accumulations of vesicular membranous bodies forming a rather compact structure (Van der Scheer and Groenewegen, 1971; Langenberg and Schroeder, 1975). The role of such an inclusion, which is unique among the comoviruses investigated so far, is not known, but it does not seem to be the result of simple nuclear degeneration. On the other hand, the intranuclear membranous inclusions of pelargonium leaf curl (Martelli and Russo, 1972) and of artichoke mottled crinkle viruses [AMCV (Martelli and Russo, 1973)] are visible inside both viable and necrotic nuclei and may be either a cause or a consequence of nuclear degeneration. These inclusions consist of groups of tightly packed membranous cisternae arranged in bundles of parallel elements, forming complex configuration in which stacked tubules and rows of rounded vesicles (i.e., profiles of cisternae in longitudinal and transverse section) are discernible.

Although the role of such membranous inclusions remains obscure, at least those induced by CoMV seem to be derived from the nuclear envelope (Langenberg and Schroeder, 1975). A similar origin could be inferred for membranous inclusions of AMCV, though no connection was observed between membranous vesicles and the nuclear envelope. Conversely, the intranuclear membranes elicited by carnation etched ring virus infection in *Saponaria vaccaria* do not seem to be connected with the nuclear envelope and may originate from *de novo* synthesis. Their development is thought to be related to the infection process by Lawson and Hearon (1974), who speculate that such membranes may be involved both in the production and storage of coat protein for particle assembly.

Nucleoplasmic crystalline inclusions are basically of two kinds: (1) protein crystals like those typically induced by tobacco etch virus (TEV); (2) crystalline arrays of virus particles.

TEV inclusions have a three-dimensional morphology resembling that of a truncate four-sided pyramid (Matsui and Yamaguchi, 1964) and may be equally present in the nucleus and cytoplasm. The consensus is that their place of origin is the nucleus, but it is not known whether the nucleolus has a bearing in their formation. In this connection, it is worth mentioning that Sheffield (1941) suggested the possibility that the crystals could develop at the expense of the nucleolus but could not substantiate this by chemical tests. Subsequent electron microscopic investigations have confirmed the occasional topological relationship between TEV crystals and the nucleolus without providing evidence that this organelle is the actual site of their formation. For these reasons, TEV crystals are presently regarded as nucleoplasmic inclusions.

The intranuclear occurrence of these entities was first reported by Kassanis (1939), who, together with Sheffield (1941), studied accurately their development with the light microscope. Their chemical nature was ascertained to be basically proteinaceous, and it was therefore concluded that nuclear crystals were probably not constituted by virus (McWhorter, 1965). Except for one report claiming the viral nature of these inclusions (Rubio-Huertos and Garcia-Hidalgo, 1964), electron microscopic investigations and light microscopic cytochemistry fully substantiated these early findings (Matsui and Yamaguchi, 1964; Takahashi, 1962; Hooker and Summanwar, 1964). In particular, use of tritiated uridine and leucine (Hayashi and Matsui, 1967) and enzyme digestion tests with RNase and proteases (Shepard, 1968) proved unequivocally that TEV nuclear crystals are made up of protein and do not contain viral nucleic acid. The constituent is a distinct protein possessing an antigenic specificity totally different from that of TEV-induced cylindrical inclusions or of assembled or dissociated TEV coat protein. The lack of immunochemical cross-reactivity among the above types of TEV-induced proteins was ascertained both *in situ* with ferritin-labeled antibodies (Shepard *et al.*, 1974) and *in vitro* using purified preparations of intranuclear inclusions (Knuhtsen *et al.*, 1974).

The substructure of TEV crystals has been recently investigated by freeze-etching. These studies have revealed that the inclusions are formed by vertical stacking of rectangular laminae, each composed of a sheet of doughnut-shaped subunits arranged in a rectangular array and spaced about 14.5 nm apart in the two orthogonal directions (McDonald and Hiebert, 1974a).

Karyoplasmic crystalline formations of unknown nature but resembling proteinaceous bodies, have been detected in nuclei of cells infected by sharka (Van Bakel and Van Oosten, 1972), sweet-potato mosaic (Kitajima and Costa, 1974a), and russet crack (Lawson *et al.*, 1971) viruses.

Intranuclear virus crystals have been recorded for several polyhedral viruses. Thus, distinct crystalline arrays of virus particles were visualized in the karyoplasm of cells infected by the cowpea strain of southern bean mosaic [SBMV (Weintraub and Ragetli, 1970a)], tomato bushy stunt (Russo and Martelli, 1972), beet western yellows (Esau and Hoefert, 1972a,b), maize streak (Bock *et al.*, 1974), and eggplant mosaic (Hatta, 1976) viruses. Except for maize streak and eggplant mosaic viruses, whose relationship with the host cells has not yet been thoroughly investigated, in all other instances the suggestion was made that the nucleus and specifically the nucleolus are involved in virus multiplication. If so, the virions that crystallize in the nucleus are formed *in situ* and do not come from the cytoplasm via nuclear pores. This would negate the idea of a redistribution of particles to equilibrate virus concentration throughout the cell, as suggested by De Zoeten and Gaard (1969b) for intranuclear SBMV.

Admittedly, the contention that these viruses are synthesized or assembled in the nucleus is based on visual evidence, but time-course studies have revealed striking similarities in the three aforementioned instances, such as: (1) plentiful occurrence of nuclear virions in young infections concomitant to a relative scarcity of particles in the cytoplasm; (2) an obvious and strict topological relationship of the virus particles with the nucleolus; (3) apparent integrity of the nuclear envelope. These features can indeed be taken as evidence that the nucleus may be more than a mere site of particle accumulation.

Crystalline arrays of empty capsids have been visualized within nuclei of cells infected by turnip yellow mosaic virus (TYMV) and three additional tymoviruses: eggplant mosaic, okra mosaic, and desmodium mosaic (Hatta, 1976). As no nucleoprotein particles accumulate in such nuclei, the presence of empty capsids has been considered as evidence that the nucleus may be the site of their assembly (Matthews, 1975).

Intranuclear crystals induced by TMV have also been reported, though in rare instances (Woods and Eck, 1948; Goldin, 1963). Although the shape and outer appearance of the crystalline bodies suggest that, by analogy with the similar looking structures found in the cytoplasm, they are aggregates of virus particles, yet no electron microscopic evidence of this is available. True intranuclear TMV crystals have not been detected so far in tissues examined with the electron microscope (Esau, 1968).

Other virus inclusions in the nucleus are those occasionally reported for some como- and nepoviruses. Particle-containing tubules of CoMV (Van der Scheer and Groenewegen, 1971) and grapevine fanleaf virus (Peña-Iglesias and Rubio-Huertos, 1971) were in fact seen in the karyoplasm of infected cells. Conceivably, their intranuclear occurrence is accidental and may be due to transport from the cytoplasm or to incorporation during cell division (Esau and Gill, 1969).

C. Perinuclear Inclusions

Perinuclear inclusions arise from accumulation of materials between the two lamellae of the nuclear envelope, which as a consequence may become widely separated. Such a peculiar localization is explained by the fact that these materials are in the process of being transferred to the cytoplasm, or vice-versa. Thus, perinuclear inclusions are often transitory and may appear only in certain stages of the infection.

The inclusions can be composed of virus particles, virus-related products, or cell constituents. The first is the case of most plant rhabdoviruses, i.e., those which mature at the nuclear envelope and are discharged in the perinuclear space [for review, see Howatson (1970); Francki (1973)]. Resulting accumulations of virions may be large enough to become visible with the light microscope (Kitajima and Costa, 1966; Christie *et al.*, 1974). In one case, perinuclear rhabdovirions were intermingled with masses of finely granular matrix that was identified as viroplasm because it appeared to give rise to the inner nucleoprotein component of the particles (Lawson and Ali, 1975). If this interpretation is correct, the perinuclear area can function as a site for virus multiplication as well as virus accumulation.

Perinuclear accumulations of virus particles were also recorded in pelargonium leaf curl virus and, to a lesser extent, in SBMV (Martelli and Castellano, 1969; Weintraub and Ragetli, 1970a).

Inclusions composed of vesicular elements have been reported for pea enation mosaic virus (De Zoeten *et al.*, 1972; Vovlas *et al.*, 1973). The vesicles are produced from the inner nuclear membrane and are electron-lucent except for a fine network of DNase-digestible fibrils (De Zoeten *et al.*, 1972). Such vesicles may represent a major step in the synthetic pathway leading to virus production, for PEMV–RNA polymerase appears to be associated with them (G. De Zoeten, personal communication).

Perinuclear localization of vesicles containing a fibrillar network identified as possible viral nucleic acid has been reported by Esau and Hoefert (1972a,b) for beet western yellow virus. The authors envisage a developmental sequence by which the vesicles arise in the cytoplasm, are surrounded by endoplasmic reticulum, and are then carried to the nucleus, where the enveloping membrane fuses with the outer nuclear membrane so that the vesicles finally become perinuclear. It is suggested that this is the way by which viral nucleic acid and perhaps also protein are transferred to the nucleus, where virus assembly takes place. The origin of the vesicles is unknown; certainly it does not appear to be dictyosomal (Esau and Hoefert, 1972b).

A somewhat similar situation has been observed in tobacco protoplasts infected with cowpea chlorotic mottle virus. However, in this instance the fibril-containing perinuclear vesicles were interpreted as being a stage of a budding process leading to formation of double membrane-bounded small vacuoles and to their subsequent release into the cytoplasm (Burgess *et al.*, 1974).

Perinuclear vesicles are also found in the very early stages of infection of lettuce necrotic yellow virus. These vesicles originate in the inner lamella of the nuclear envelope and may act as carriers of viral RNA from the nucleus, where it is synthesized and/or accumulates, to the perinuclear area and cytoplasm, where virus assembly takes place (Wolanski and Chambers, 1971).

None of the above interpretations of the direction of movement of membrane-bound fibrillar material (viral nucleic acid?) is substantiated by evidence other than circumstantial. However, they are consistent with the final distribution of virus particles. The movement of material is postulated to be from the cytoplasm to the nucleus for beet western yellows virus, which may be produced in the nucleus, and from the nucleus to the cytoplasm for cowpea chlorotic mottle and lettuce necrotic yellows viruses, which are entirely cytoplasmic.

The perinuclear localization of crystalline bodies related to virus infection has been recorded at least in one case, i.e., in PEMV-infected cells, where crystals of nuclear origin but of unknown nature and significance were detected between the parted lamellae of the nuclear envelope (De Zoeten *et al.*, 1972).

D. Nonviral Nuclear Inclusions

Inclusions in the nuclei of healthy plant cells have been observed and studied with the light microscope since the last century (Zimmermann, 1893) and, more recently, have been extensively reinvestigated electron microscopically (Schnepf, 1964; Barton, 1967; Villiers, 1968; Weintraub *et al.*, 1968; Wergin *et al.*, 1970; Unzelman and Healey, 1972). In a most informative review on this subject, Thaler (1966) reports that nuclear inclusions have been observed throughout the plant kingdom in more than 200 species, including, of course, representatives of monocotyledons and dicotyledons. Such inclusions may have localization and outward appearance intriguingly similar to those produced by viruses. Thus, the only safe discriminating criteria are the failure to transmit an infectious agent from inclusion-bearing plants (Weintraub *et al.*, 1968) and the lack of recognizable virus particles or virus-related structures in such plants.

As illustrated by Wergin *et al.* (1970), nuclear inclusions of healthy plants display a variety of shapes and sizes. Although amorphous and fibrous bodies are often encountered, those with a crystalline lattice predominate (Amelunxen and Giele, 1968; Weintraub *et al.*, 1968; Villiers, 1968; Wergin *et al.*, 1970; Unzelman and Healey, 1972). The localization of the inclusions is also variable, for they are found in the karyoplasm, perinuclear space, and occasionally in the cytoplasm (where they may be transferred from the nucleus) but more frequently are contiguous to the nucleolus. It is generally agreed that such a strict association between nucleolus and inclusions may have a metabolic basis, and also that the nucleolus may be the site of origin of the crystals, which represent protein storage structures (Frey-Wyssling and Mühlethaler, 1965). Indeed, the crystalline bodies are composed of different kinds of protein, as confirmed by recent cytochemical staining reactions (Wergin *et al.*, 1970; Unzelman and Healey, 1972) and enzyme digestion tests (Weintraub *et al.*, 1971), and do not seem to contain either insoluble carbohydrates or nucleic acids (Weintraub *et al.*, 1971; Unzelman and Healey, 1972).

V. Cytoplasmic Inclusions

Cytoplasmic inclusions, which have been observed in most known groups of plant viruses, differ tremendously in size, location, composition, and ultrastructural organization and have diverse origins and functions. According to their gross morphology, the inclusions can be divided into crystalline and amorphous types. Both types may be induced by the same virus and may occur side by side in the same cell. Similarly, different strains of the same virus may produce either crystalline or amorphous inclusions, or both.

As mentioned earlier, in the past cytoplasmic inclusions have been interchangeably called amoeboid or vacuolated bodies, or (more commonly) X-bodies, owing to the apparent similarity with the round, granular, and vacuolated inclusions originally observed in TMV-infected cells by Goldstein (1924). However, this terminology seems inappropriate. McWhorter (1965) sought to restrict use of the term *X-body* to structures of unknown function, and he recommended use of terms like *amorphous* and *crystalline* for inclusions associated with virus formation. More recently, Warmke (1969) has proposed using the term *X-body* strictly in the original sense, i.e., for the amorphous inclusions characteristically associated with TMV infections. Esau and Hoefert (1971c), in agreement with Warmke's suggestion, point out that "The continued use of X-body with reference to TMV infections is justifiable only on historical grounds." Therefore, the expressions *amorphous inclusions* or *amorphous inclusion bodies* will be used in the present review to designate noncrystalline aggregates of materials—whatever their origin—arising in the cytoplasm of the host cells as a consequence of virus infection.

A. Aggregates of Virus Particles

Viruses multiplying at a high rate produce a tremendous number of particles; these may accumulate in the cytoplasm, giving rise to the simplest possible inclusions, i.e., pure or nearly pure aggregates of virions. In such accumulations, virus particles may be disposed at random with no orderly orientation, or they may be aligned side by side and/or end to end in a two-dimensional array, or they may be highly organized in a three-dimensional lattice. Accordingly, the resulting inclusions will be fibrous, paracrystalline, or crystalline. Since the boundary between fibrous and paracrystalline viral aggregates is virtually impossible to define, these two kinds of inclusions will be discussed together.

1. Fibrous and Paracrystalline Inclusions

The most striking of these inclusions occur in cells infected by anisometric viruses. Thus, "fibrous masses" composed of bundles of filamentous particles have been seen in many hosts infected with potexviruses (Kikumoto and Matsui, 1961; Purcifull *et al.*, 1966; Amelunxen and Thaler, 1967; Kozar and Sheludko, 1969; Zettler *et al.*, 1968; Thaler *et al.*, 1970; Turner, 1971; Appiano and Pennazio, 1972; Shalla and Shepard, 1972; Štefanac and Ljubešić, 1974; Doraiswamy and Lesemann, 1974), carlaviruses (Lyons and Allen, 1969; Tu and Hiruki, 1970; De Bokx and Waterreus, 1971; Bos and Rubio-Huertos, 1971, 1972; Rubio-Huertos and Bos, 1973; Hiruki and Shukla, 1973; Shukla and Hiruki, 1975), potyviruses [for review, see Edwardson (1974)], and closteroviruses (Price, 1966; Esau, 1968; Hoefert *et al.*, 1970; Esau and Hoefert, 1971a,b,c; Chen *et al.*, 1971, 1972; Kitajima *et al.*, 1971; Kitajima and Galves, 1973).

Sometimes, these virus aggregates represent relatively small, sinuous fascicles of loose particles, whereas in other cases viral bundles are much larger and compact, with particles tightly packed side by side in a paracrystalline array. In these instances bulky accumulations of virions may occupy a great deal of the ground cytoplasm.

Conspicuous virus aggregates, having a periodicity arising from the orderly alignement of particles in stacked layers, form "banded" inclusions like those characteristic of beet yellows virus infections (e.g., Esau, 1968) or of members of the potexvirus group (Weber and Kenda, 1952; Thaler, 1956; Amelunxen, 1956, 1958). Under the electron microscope the banded bodies appear to be composed almost entirely of virus particles aligned in horizontal tiers whose width coincides with the particle length of the virus. Successive layers are not continuous with one another, being usually separated by thin cytoplasmic strands containing ribosomes, which are likely to be trapped within the inclusion in the course of its formation (Purcifull *et al.*, 1966; Esau, 1968; Thaler *et al.*, 1970; Esau and Hoefert, 1971c; Pennazio and Appiano, 1975). Although virus layers may sometimes be accurately aligned, in no instance has evidence been obtained of an end-to-end abutment of virus particles (which, presumably, would give rise to a true three-dimensional crystalline structure as in TMV; see Section V, A, 2). In potexviruses, fibrous masses correspond to the spindle-shaped inclusions seen under the light microscope (Miličić, 1954; Goldin, 1963; Amelunxen and Thaler, 1967; Štefanac and Ljubešić, 1974; Giri and Chessin, 1975). In a detailed study of cactus virus X-induced spindles, Amelunxen and Thaler (1967) described the substructure of these inclusions as being largely composed of virus particles in a parallel orientation, but also containing cytoplasmic strands with ribosomes, lipid droplets, and electron-lucent gaps. The fibrillar appearance of the spindles in the light microscope was explained by these authors as being due to the presence of furrows of variable depth, grooving the surface lengthwise.

In TMV-infected cells, a variety of inclusion bodies deriving from linear aggregation of the virions can be found. Under the light microscope these inclusions appear as long fibrous structures sometimes extendinig the whole length of a cell in a straight or looped condition, or as very refractive spindle-shaped or spike-like bodies (e.g., Kassanis and Sheffield, 1941; Goldin, 1963). All these various kinds of inclusions are paracrystalline virus aggregates in which the particles are regularly packed side by side in a hexagonal array. As pointed out by Warmke and Edwardson (1966b), such aggregates are capable of unlimited growth in length but not in width, which obviously explains their elongate needle-like appearance.

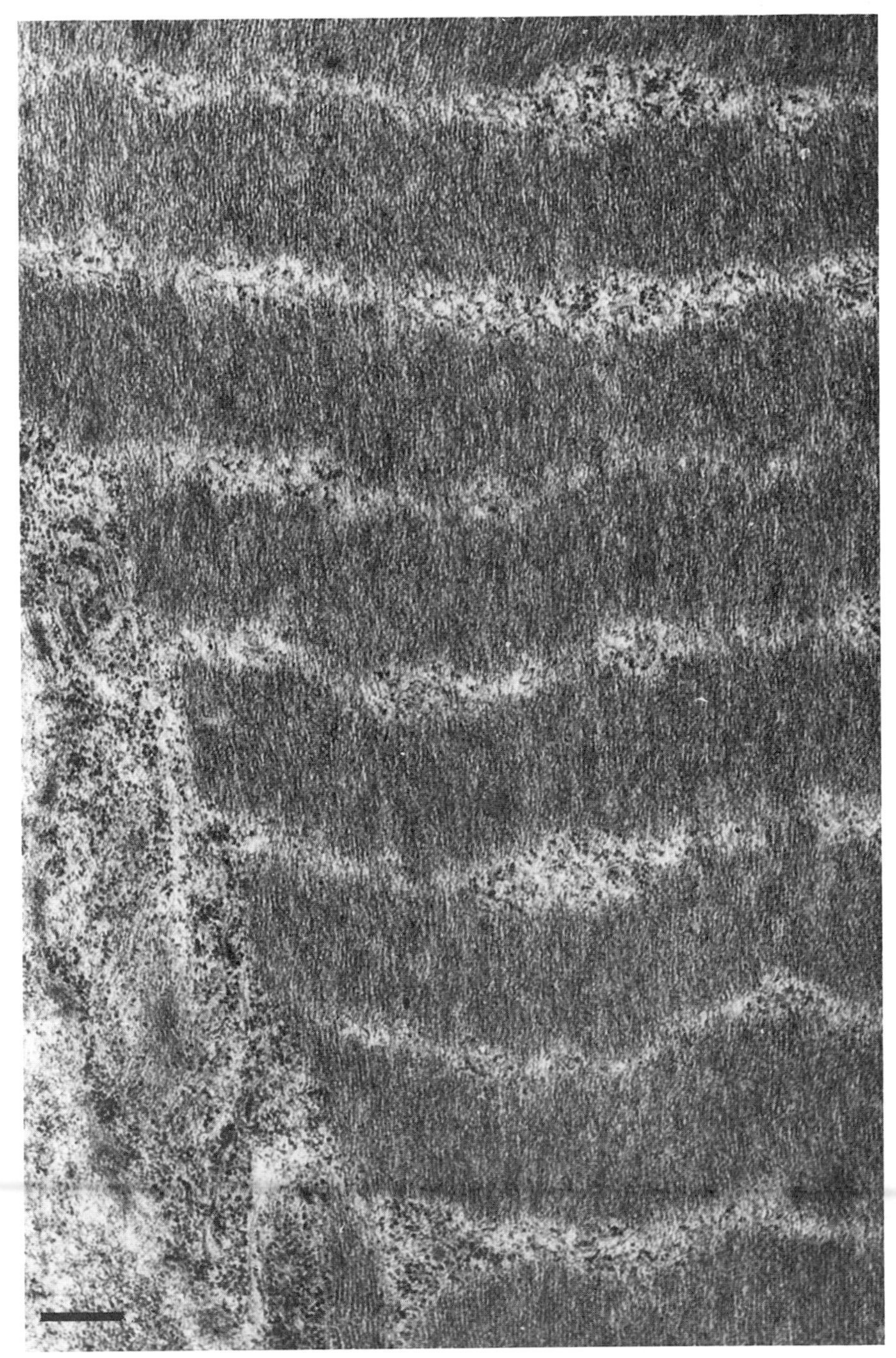

FIG. 2. Ultrastructural aspects of a banded inclusion composed of potato virus X particles in stacked layers. (Courtesy of A. Appiano.) Magnification bar represents 250 nm.

Based on the fact that formation of paracrystalline TMV inclusions can be induced by various treatments in cells containing three-dimensional crystals only and that they seem to arise concomitantly with the degradation of these crystals, Warmke and Edwardson (1966b) suggested that paracrystalline bodies are a secondary aggregation product of the breakdown of preexisting crystalline structures. These authors envisage a transformation sequence by which virus particles of successive layers of true crystals fuse end-to-end across the boundary of the tiers, forming elements several particles long. At the same time the side-to-side attractive forces lessen, thus allowing the crystal to split into needle-like bundles. The paracrystalline needles may remain together, or—if the original crystal collapses—they may separate, dispersing into the cytoplasm and vacuole. According to Warmke and Edwardson (1966b), a similar mechanism is likely to operate under natural conditions, producing breakdown of hexagonal crystals in cells of older leaves with long-standing infections in which elongated paracrystalline forms of TMV prevail.

It should be noted, however, that needle-like inclusions do not necessarily represent degradation products of larger crystalline bodies, for they normally occur, as revealed by light microscopy, in cells containing "angled layer" aggregates of TMV particles (Granett and Shalla, 1970). These aggregates are another type of paracrystalline inclusions that were reported independently and at about the same time by different workers as being characteristically induced by some strains of TMV such as aucuba (Warmke, 1967, 1968), U5 (Shalla, 1968; Granett and Shalla, 1970) and a pepper isolate (Herold and Munz, 1967). More recently they were also found with a tomato strain of TMV (Chen, 1974) and with beet necrotic yellow vein virus (Tamada, 1975), which has particles morphologically but not serologically related to TMV. The most detailed information on these unusual viral aggregates derives from the studies of Warmke (1967, 1968, 1969, 1974), who was the first to term them angled layer aggregates by analogy with similar-looking aggregates of detached bacterial pili previously so labeled by Brinton (1965).

Angled layer TMV inclusions are indeed very peculiar and differ from the three-dimensional crystals and all other known paracrystalline bodies induced by this virus. The particles are arranged in stacks of parallel elements oriented across the long axis of the aggregate. Each layer is placed over the next at an angle of 60° in the aggregates formed by aucuba, U5, and pepper strains (Warmke, 1968) and of 30° in the tomato strain aggregate (Chen, 1974). We agree with Warmke (1968) that the 90° displacement between virus layers reported for the pepper strain (Herold and Munz, 1967) may not have been correctly interpreted.

Consequent to this diversity of orientation of the virus layers, angled layer aggregates appear as three-way crosshatched figures in cross section and as short parallel lines interspersed with rows of dots when longitudinally cut. Oblique sections of 30° aggregates reveal more complex feather-like patterns (Chen, 1974). Angled layer aggregates are slightly twisted along the major axis; the presence of a shift in the angle of the particles in successive cross sections (about 6.5° from one section to the next) led Warmke (1974) to conclude that these bodies form a constant left-handed helix.

Angled layer aggregates seem to be more stable than regular three-dimensional TMV crystals, as the former withstand processing for electron microscopy much better. The reasons leading to their formation are, nevertheless, obscure. Warmke (1968) speculated that a difference in the spatial arrangement or in the chemical properties of the protein subunits between aucuba and common TMV strains could account for the production of angled layer aggregates instead of ordinary crystals. This hypothesis, however, although intriguing, seems difficult to reconcile with the occurrence of angled layer configurations within ordinary crystalline structures (Warmke, 1968) and with the contemporary presence of both kind of inclusions in the same cell, unless this results from mixed infection by two different TMV strains, which in the reported cases is unlikely (Warmke, 1968; Chen, 1974).

Paracrystalline monolayers of parallel virus particles are encountered with most plant rhabdoviruses, both around the nucleus and in the cytoplasm. Although conspicuous because of the large size of the virions, these aggregates are usually built up of a relatively small number of particles in a two-dimensional hexagonal packing [for review, see Francki (1973); Martelli and Russo (1977)].

The situation with alfalfa mosaic virus, which also possesses bacilliform particles although of a smaller size and diverse constitution, is considerably different. Different AMV strains can form a variety of strain-specific paracrystalline bodies (De Zoeten and Gaard, 1969; Hull *et al.*, 1969, 1970). Comparing 24 strains of AMV, Hull *et al.* (1970) recognized four distinct aggregation forms: (1) short rafts of particles arranged in a hexagonal array; (2) long bands of particles packed side by side, sometimes in a stacked layer configuration; (3) complex aggregates consisting of two or more parallel rows of particles in a side-by-side arrangement, appearing packed in a rhomboid or hexagonal lattice when transversely cut; (4) complex bodies consisting of centers of aggregation connected to one another by groups of particles and revealing a variety of diverse profiles according to the plane of sectioning. These aggregates develop in the early stages of infection, sometimes at the same time as the first appearance of particles, but the cause of aggregation remains unexplained. Noticing that in type 2, 3, and 4 aggregates there appears to be some restriction in the two-dimensional formation of the hexagonal lattice, Hull *et al.* (1970) argued that in these cases the aggregation mechanism is unlikely to be chemical, as it is in type 1 aggregates (i.e., reduction of the net surface charge on the particles). They postulated a physical mechanism, such as aggregation on a template preexisting in the cell, but found no evidence of it.

Paracrystalline aggregation forms of isometric viruses can be recognized in particle-containing tubules, like those present in reovirus-like plant viruses, comoviruses, nepoviruses, ilarviruses (Gerola *et al.*, 1969a), and in parsnip yellow fleck virus (Murant *et al.*, 1975). These tubules (see Sections V, F and V, G) may occur singly or, as in the case of nepoviruses, in bundles. A close examination of intracellular arrays of nepoviruses reveals that they seldom show the cubic or pseudocubic packing in crystalline structures that is customary for other small icosahedral plant viruses (see the following subsection). Instead, virus aggregates appear as stacked rows of particles in straight or, more often, curved stratification (De Zoeten and Gaard, 1969b; Gerola *et al.*, 1969b; Peña-Iglesias and Rubio-Huertos, 1971; Roberts *et al.*, 1970; Vovlas *et al.*, 1971; Halk and McGuire, 1973; Šarić and Wrischer, 1975). It is therefore possible that these aggregates originate from a close packing of particle-containing tubules. This interpretation seems compatible even with the odd spherical aggregates of arabis mosaic virus (Gerola *et al.*, 1965b, 1966a).

If this is so, virus particles are prevented from interacting to form true crystals unless the tubule walls dissolve, as may occur in the instances where regular crystalline bodies have been observed (Gerola *et al.*, 1965b, 1966a; Roberts *et al.*, 1970; Yang and Hamilton, 1974). Interestingly, the elongated aggregates of the type and "datura quercina" strains of tobacco streak virus (Edwardson and Purcifull, 1974) may also perhaps be built of closely packed particle-containing tubules, as their overall aspect and organization seem to suggest.

2. *Crystalline Inclusions*

Intracellular crystalline bodies formed by orderly arrays of virus particles can be induced both by rod-shaped and isometric viruses.

Thus, for instance, in infected plant tissues or in tissue culture cells (Goldin, 1963; Chandra and Hildebrandt, 1965) members of the tobamovirus group produce crystalline inclusions appearing as hyaline plates with a hexagonal or rounded shape. Hexagonal plates typically occur with infections by the common strain of TMV and many of its variants [for reviews, see Goldin (1963); Bawden (1964); McWhorter (1965); see also Miličić and Juretić (1971)], whereas rounded plates are formed by some TMV strains (Resconich, 1961) but mostly by other viruses such as ribgrass mosaic (Goldin, 1953; Miličić, 1968; Miličić *et al.*, 1968), sunn hemp mosaic (Miličić and Juretić, 1971), cucumber green mottle [CGMV (Miličić, 1969)], and *Odontoglossum* ringspot [ORSV (Miličić and Štefanac, 1971)]. CGMV and ORSV may also form thin plates which, although having the tendency to assume an hexagonal shape, differ from those of typical TMV (Miličić and Juretić, 1971).

TMV crystals have been the object of intensive investigations since their discovery (Iwanowski, 1903). A number of studies were carried out in the past with the light microscope to determine their nature and composition (Goldstein, 1924; Beale, 1937; Goldin 1938a,b; Sheffield, 1939; Wilkins *et al.*, 1950); but it was the classical work by Steere and Williams (1953) with the electron microscope that provided final evidence that TMV crystals are composed largely of infective virus rods. Many reports on intracellular crystalline TMV have now accumulated (Rubio-Huertos, 1950; Brandes, 1956; Steere, 1957; Wehrmeyer, 1960a,b; Shalla, 1964; Kolehmainen *et al.*, 1965; Milne, 1966b; Warmke and Edwardson, 1966a,b; Warmke and Christie, 1967; Esau, 1968; Granett and Shalla, 1970; Langenberg and Schroeder, 1972, 1973a; Willison, 1976), so that a considerable body of information concerning the structural organization of the crystals is available. This progress was made possible by the development of embedding and fixing procedures which retained the structural integrity of the crystalline inclusions *in situ*. Thus, excellent micrographs of whole intracellular TMV crystals were produced by Warmke and co-workers either by short fixation of tissue samples in weak water or acetone solutions of potassium permanganate (Warmke and Edwardson, 1966a,b) or by a modification of the ordinary osmium tetroxide technique (Warmke and Christie, 1967). Subsequently, buffered chromic acid–formaldehyde fixatives proved very useful in preserving the structure of all TMV crystals present in an entire tissue block (Langenberg and Schroeder, 1972, 1973a). Fixation with glutaraldehyde (alone or in mixture with formaldehyde) followed by osmium tetroxide, although capable of revealing the crystalline structure of TMV (Esau, 1968), seems much less effective than the aforementioned methods (Langenberg and Schroeder, 1973a).

In thin sections, TMV crystals appear to be made up of stacked layers of virions in which individual rods are closely aligned in a parallel array. Each layer has the depth of a single virus particle. Particles may be

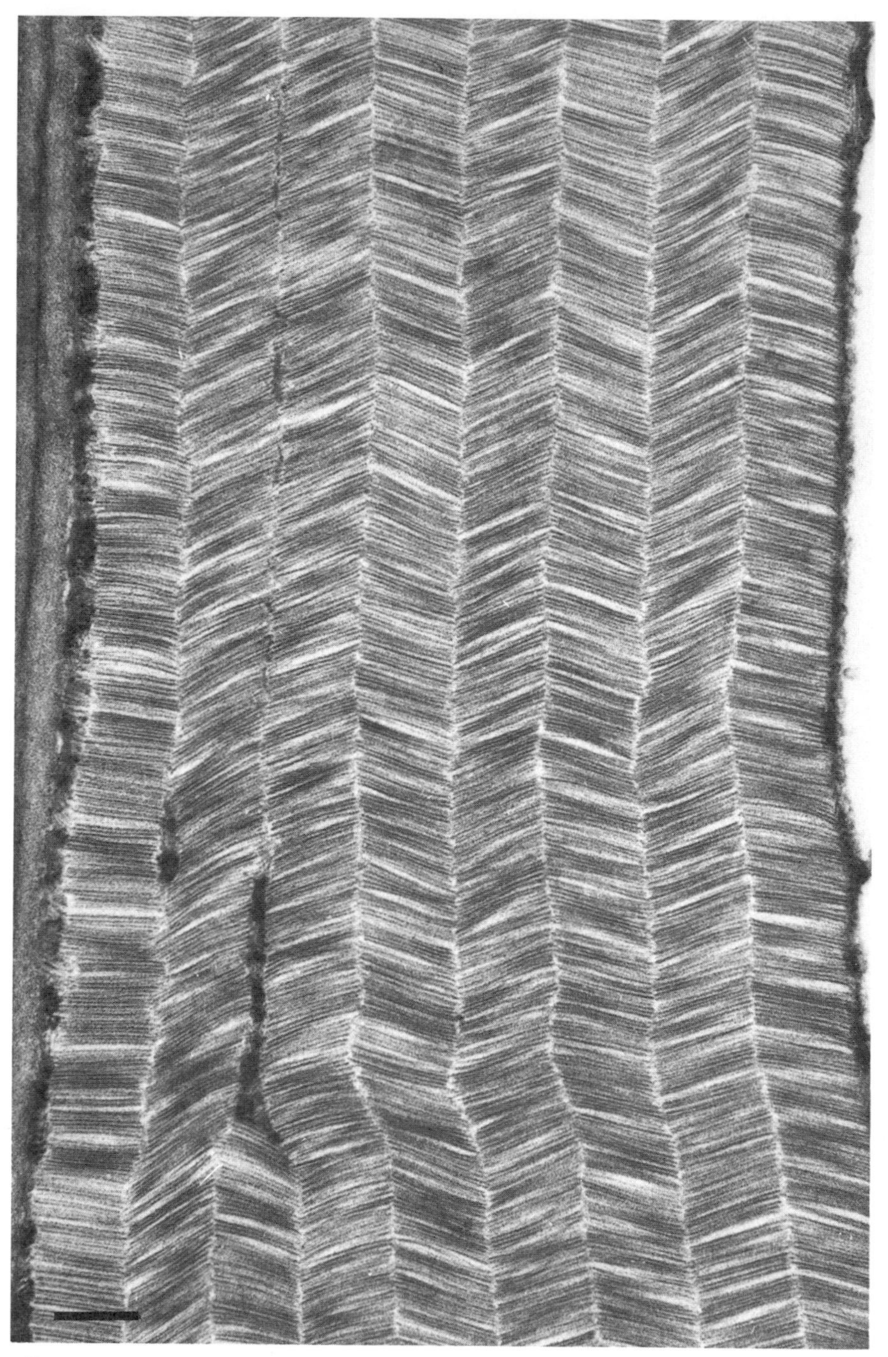

FIG. 3. Crystalline inclusion of tobacco mosaic virus showing virus particles arranged in a herringbone pattern. (Courtesy of H. E. Warmke and the copyright holder, Academic Press.) Magnification bar represents 250 nm.

accurately aligned in successive rows or may be slightly tilted, thus forming an angle and giving rise to the characteristically familiar herringbone pattern (Steere, 1957; Warmke, 1968; Warmke and Christie, 1967; Willison, 1976). In the crystal, virus particles are oriented perpendicularly, or nearly so, to the hexagonal faces, and parallel to the lateral faces. In transverse section, virions appear to be arranged in a hexagonal pattern. The interparticle spacing is about 24 nm, owing to the hydrated condition of the crystals, which according to recent estimates contain 45% crystallizable water (Willison, 1976).

The development of TMV crystals has been followed by conventional and phase-contrast microscopy in living French bean or tobacco cells (Resconich, 1961; Bald and Solberg, 1961; Solberg and Bald, 1962) and by electron microscopy (Warmke and Edwardson, 1966b) in tobacco leaf and epidermal hair cells. Warmke and Edwardson (1966b) envisaged a developmental sequence by which the growth of the crystal begins with virus particles accumulating in the cytoplasm to form small aggregates of parallel rods with ends aligned. These aggregates increase in size and become variously extended monolayers which have the capacity of unlimited repetition in two dimensions. Such viral tiers are likely to correspond to the "gray plates" indicated by Bald and co-workers as being an early stage in the formation of crystals. Subsequent aggregation and stacking of several monolayers gives way to the three-dimensional crystalline structure.

In TMV crystals the virions are very densely packed, so that in general [exceptions are reported by Esau (1968)] little room is left for foreign material to become entrapped within the inclusions. This is not always the case with the crystalline bodies known as "rounded plates" that are induced by other tobamoviruses; being less compact than those of TMV, these bodies are more likely to trap components of the host cytoplasm. Rounded plates, like TMV crystals, are composed of layers of virus rods in parallel arrays (Hatta and Ushiyama, 1973). However, the layers in the stack are not all the same size, as they are in TMV. The middle tiers are larger than the outer ones, so that the inclusion ends up with a rounded, oval, or lens-shaped appearance (Goldin, 1953; Miličić, 1968; Miličić and Juretić, 1971).

Despite their outward diversity the crystalline inclusions of TMV and other members of the tobamovirus group are structurally almost identical. This may imply that they develop in a similar manner, but why a comparable, if not identical, developmental process should result in the formation of differently shaped bodies remains to be explained.

Light microscope records of true cytoplasmic crystals composed of isometric virus particles are few. Perhaps the best examples available are given by the beautifully shaped polyhedra and elongate platelike inclusions associated with infection by several strains of broad bean wilt virus (Rubio-Huertos, 1962; Juretić *et al.*, 1970; Miličić *et al.*, 1974). These crystalline bodies, as demonstrated by electron microscopy, are indeed entirely made up of regularly arranged virions (Rubio-Huertos, 1968; Miličić *et al.*, 1974).

Since the early report of apparent *in situ* crystallization of tomato bushy stunt virus (Smith, 1956), ultrastructural observations have revealed the widespread tendency of isometric viruses to crystallize. The number of these records has grown to such an extent that it would be virtually impossible to provide a complete list of them. For instance, among the single-stranded RNA viruses with particles 25–30 nm in diameter, intracellular crystalline aggregates have been reported for representatives of bromoviruses (De Zoeten and Schlegel, 1967a; Hills and Plaskitt, 1968; Paliwal, 1970, 1972), comoviruses (Milne, 1967a; Štefanac and Ljubešić, 1971; Hooper *et al.*, 1972; Kim and Fulton, 1972; Kitajima *et al.*, 1974b; Langenberg and Schroeder, 1975), cucumoviruses (Gerola *et al.*, 1965a; Honda and Matsui, 1968; Russo and Martelli, 1973; Otsuki and Takebe, 1973; Honda *et al.*, 1974), ilarviruses (Edwardson and Purcifull, 1974), luteoviruses (Esau and Hoefert, 1972a; Amici *et al.*, 1975), nepoviruses (Gerola *et al.*, 1965b, 1966a; Roberts *et al.*, 1970; Yang and Hamilton, 1974), tombusviruses (Russo *et al.*, 1968; Rubio-Huertos and Garcia-Hidalgo, 1971; Martelli and Russo, 1972, 1973; Russo and Martelli, 1972), tymoviruses (Gerola *et al.*, 1966b; Rubio-Huertos *et al.*, 1967; Ushiyama and Matthews, 1970; Allen, 1972; Granett, 1973; Moline and Fries, 1974; Hatta and Matthews, 1974; Hatta, 1976), and, in addition, for southern bean mosaic and related viruses [cocksfoot mottle, cocksfoot mild mosaic, and phleum mottle (Weintraub and Ragetli, 1970a,b; Chamberlain and Catherall, 1976)], tobacco necrosis and satellite (Edwardson *et al.*, 1966; Kassanis *et al.*, 1970), potato leaf roll (Kojima *et al.*, 1969), sowbane mosaic (Milne, 1967a), pea enation mosaic (Shikata and Maramorosch, 1966), Brazilian eggplant mosaic (Kitajima and Costa, 1974), and rice stripe (Kiso *et al.*, 1974) viruses.

Interestingly, crystalline formations occur also in cells infected with reovirus-like plant viruses, which have particles 79 nm in diameter containing double-stranded RNA (e.g., Shikata and Maramorosch, 1965, 1967a,b; Gerola *et al.*, 1966c,d; Giannotti *et al.*, 1968; see also review by Milne and Lovisolo, 1977). Beside the nucleus, as mentioned before, isometric virus crystals are found in the cytoplasm and vacuole of infected cells. Vacuolar crystals may arise *in situ* following disruption of the tonoplast or, as in tombusviruses, may be driven from the cytoplasm

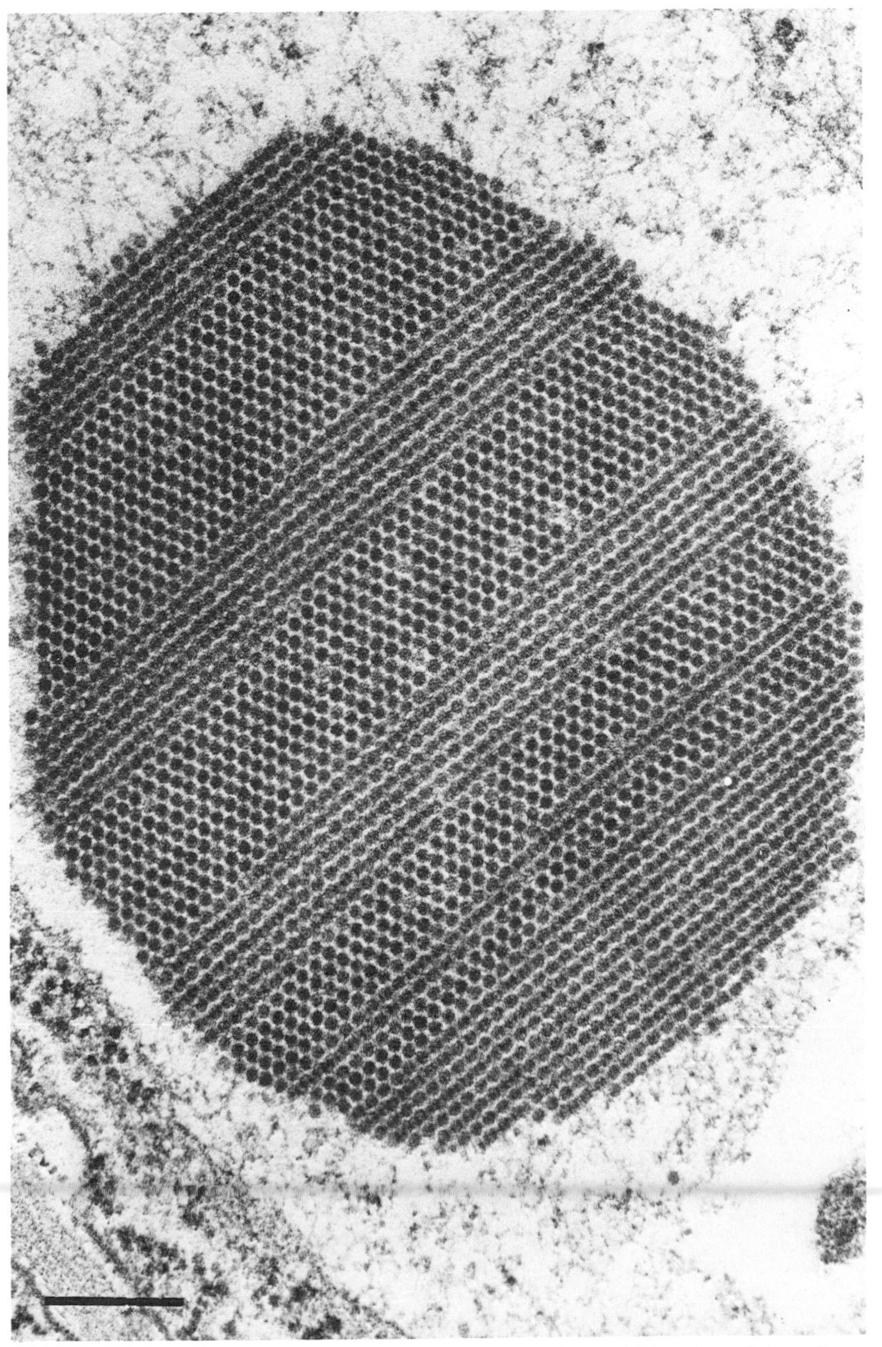

FIG. 4. A large vacuolar crystal of artichoke mottled crinkle virus. [From Russo *et al.* (1968), courtesy of the copyright holder, Academic Press.] Magnification bar represents 250 nm.

through an active transport mechanism performed by membranous, bubble-like evaginations of the tonoplast (Russo *et al.*, 1968).

The crystals are tremendously variable in size, ranging from minute particle aggregates, detectable only with the electron microscope, to conspicuous structures attaining 15–20 μm in length. Judging from published micrographs, they are nearly always made up entirely of virus particles. If foreign material is incorporated within the structure, it must occur in a nonparticulate form, as it is seldom detectable and represents a rather small proportion of the inclusion.

The composition of the crystals is also variable, for it reflects that of the particle population of the parent virus. Thus, crystals of a single-component virus, like SBMV, are built up of elementary units which are sensitive both to protease and RNase digestion, i.e., the constituents are nucleoprotein particles (Weintraub and Ragetli, 1970b). Conceivably, the crystals of other single-component viruses (tombusviruses), or of other viruses with a multipartite genome but having no accessory particles deprived of nucleic acid (e.g., bromo- and cucumoviruses), have a comparable composition. Instead, crystals of multicomponent viruses can be built of empty protein shells alone or in a mixture with nucleoprotein components as, for instance, in tymoviruses (Markham and Smith, 1949; Hatta, 1976). Also the "viroplasm with a gridlike appearance" seen in CoMV-infected cells (Langenberg and Schroeder, 1975) could actually be a large crystalline aggregate in which empty capsids predominate over full particles. Whereas empty shells and viral nucleoprotein particles may be readily recognized in crystalline structures where nearly all elements are of one type or the other (Hatta, 1976; Hatta and Matthews, 1976), it is not always easy to assess their relative proportion in a mixed crystal. Hatta (1976) showed with densitometer scans that it is virtually impossible to identify the type of individual constituents (i.e., empty shells or nucleoproteins) of TMV crystals when the sections are thick enough to comprise more than one row of particles in depth. However, the use of thinner sections makes the distinction between full and empty particles easier, as, for example, with broad bean wilt virus (R. G. Milne, personal communication).

As to the structural organization of crystalline inclusions, X-ray diffraction studies of wet crystals of tomato bushy stunt virus (Carlisle and Donberger, 1948; Caspar, 1956), TYMV (Klug *et al.*, 1957a,b; Klug and Finch, 1960) and tobacco ringspot virus (Klug and Caspar, 1960) have indicated that all have cubic symmetry. Specifically, the crystal lattice is body-centered cubic, that is, the unit cell is a cube containing virus particles at each corner and one in the center. Conversely, recent X-ray analysis of SBMV crystals showed that, with this virus, particle ar-

FIG. 5. Two vacuolar crystals of cucumber mosaic virus. Virus particles are arranged in a square and hexagonal pattern in the upper and lower crystal, respectively. [From Russo and Martelli (1973), courtesy of the copyright holder, Mediterranean Phytopathological Society.] Magnification bar represents 250 nm.

rangement differs from the above, for the unit cell is rhombohedral or orthorhombic (Johnson *et al.*, 1974; Akimoto *et al.*, 1975).

Such a variety of crystalline structures revealed by X-ray studies was not confirmed by early electron micrographs of carbon replicas of dried crystals, which, owing to the drastic manipulations necessary for electron microscopy, always showed a cubic close-packed organization (Labaw and Wyckoff, 1957; Klug and Caspar, 1960). However, whether the same applies to thin-sectioned intracellular virus crystals remains to be established. In fact, there are indications that, in thin sections of embedded material, crystalline structures are better preserved than they are after drying and replication. For instance, Johnson *et al.* (1974) have produced impressive micrographs of a SBMV aggregate sectioned perpendicularly to the crystallographic 3-fold axis, where the open hexagonal arrangement of virus particles does not show appreciable signs of collapse. Rather, it exhibits a remarkable similarity with the X-ray diffraction model. Likewise, Hatta (1976) in detailed studies of crystalline inclusions of intracellular TYMV, has convincingly identified linear, hexagonal, and tetragonal arrays of particles as being views down the 2-, 3-, and 4-fold crystallographic axes of the cubic unit cell, respectively. The similarity of patterns seen in intracellular crystals of a number of other viruses (Russo *et al.*, 1968; Paliwal, 1970; Rubio-Huertos and Garcia-Hidalgo, 1971; Kassanis *et al.*, 1970; Štefanac and Ljubešić, 1971; Allen, 1972; Kim and Fulton, 1972; Granett, 1973; Russo and Martelli, 1973; Yang and Hamilton, 1974; Kitajima and Costa, 1974b; Kitajima *et al.*, 1974b; Langenberg and Schroeder, 1975) may perhaps indicate that a comparable situation also occurs in these instances. However, when adjacent virus layers are shifted with respect to each other, individual particles can be out of phase in relation to the plane of sectioning. This produces the variety of patterns (herringbone, block, scalar) observed, for example, in SBMV crystals; Weintraub and Ragetli (1970a) were able to interpret and reproduce such patterns artificially by superimposing images of an orderly hexagonal particle array and then tilting and/or shifting them in lateral directions along the x and y axis.

A most peculiar form of particle aggregation is represented by the tubular crystals induced by some fungal (Yamashita *et al.*, 1973) and plant viruses. Such structures were first reported by Rubio-Huertos (1968) in cells infected with the petunia strain of broad bean wilt virus (BBWV). Subsequently the crystals were found associated with infections by additional viruses related to BBMV (Sahambi *et al.*, 1973; Hull and Plaskitt, 1974; Miličić *et al.*, 1974; Russo and Martelli, 1975b; Weidemann *et al.*, 1975). The tubules are present in the cytoplasm, singly or in aggregates. In broad bean cells they are almost invariably seen in parallel arrays lining the cytoplasmic side of the tonoplast or the outer side of the bounding membrane of secondary vacuoles, so that they are often constrained within cytoplasmic bridles separating such vacuoles (Russo and Martelli, 1975b). When occurring in bundles, the tubules give rise to conspicuous structures, which under the light microscope appear as hyaline, rather compact spindles several micrometers long (Miličić *et al.*, 1974). The ultrastructure of the tubular crystals reveals that they are constructed of a single sheet of virus particles in an hexagonal arrangement (Sahambi *et al.*, 1973), tilted to give a primary helix with a pitch angle of about 20° (Hull and Plaskitt, 1974). Their diameter ranges from 60 to 80 nm, and 9 to 10 particles are contained in their circumference. The building units of the tubes can be empty shells and whole virions (Rubio-Huertos, 1968; Sahambi *et al.*, 1973; Russo and Martelli, 1975b). Tubular aggregates have been reported in which 95% of the constituents were empty viral capsids (Sahambi *et al.*, 1973).

Crystalline aggregates comparable but not quite identical to those just described are produced by the parsley strain of BBWV. In this case virus particles are aggregated into sheets which are either rolled into scrolls or into large, irregular tubes 200–250 nm in diameter, having 25–30 particles in their circumference (Frowd and Tomlinson, 1972; Hull and Plaskitt, 1974).

In healthy plant cells, particles do not occur in crystalline configurations except (as will be discussed in Section V, B, 4) for phytoferritins, whose arrays are infrequent and limited to the inside of chloroplasts. Ribosome aggregates are rare, and whenever present they appear as polysome-like complexes with a helical configuration (e.g., Russo and Martelli, 1969; Barnett *et al.*, 1971).

Therefore, when virus particles are arranged in a regular lattice rather than being scattered at random, the level of confidence in their identification rises considerably. In this respect, virus crystals help us to distinguish between ribosomes and small polyhedral viruses (Milne, 1967a; Weintraub and Ragetli, 1970a); however, their utility for particle size determination has recently been questioned by Hatta (1976), who in TYMV crystals found considerable discrepancies between the measured and expected interparticle distance.

There are cases in which some strains of a certain virus (or some viruses of the same group) induce formation of intracellular crystals whereas others, even though closely related or serologically identical to them, do not (Weintraub and Ragetli, 1970a; Martelli and Russo, 1972; Kitajima *et al.*, 1974b). This differential behavior is likely to remain unexplained until the mechanism(s) leading to crystal production are elucidated. At present, virtually nothing is known about this. In fact, there

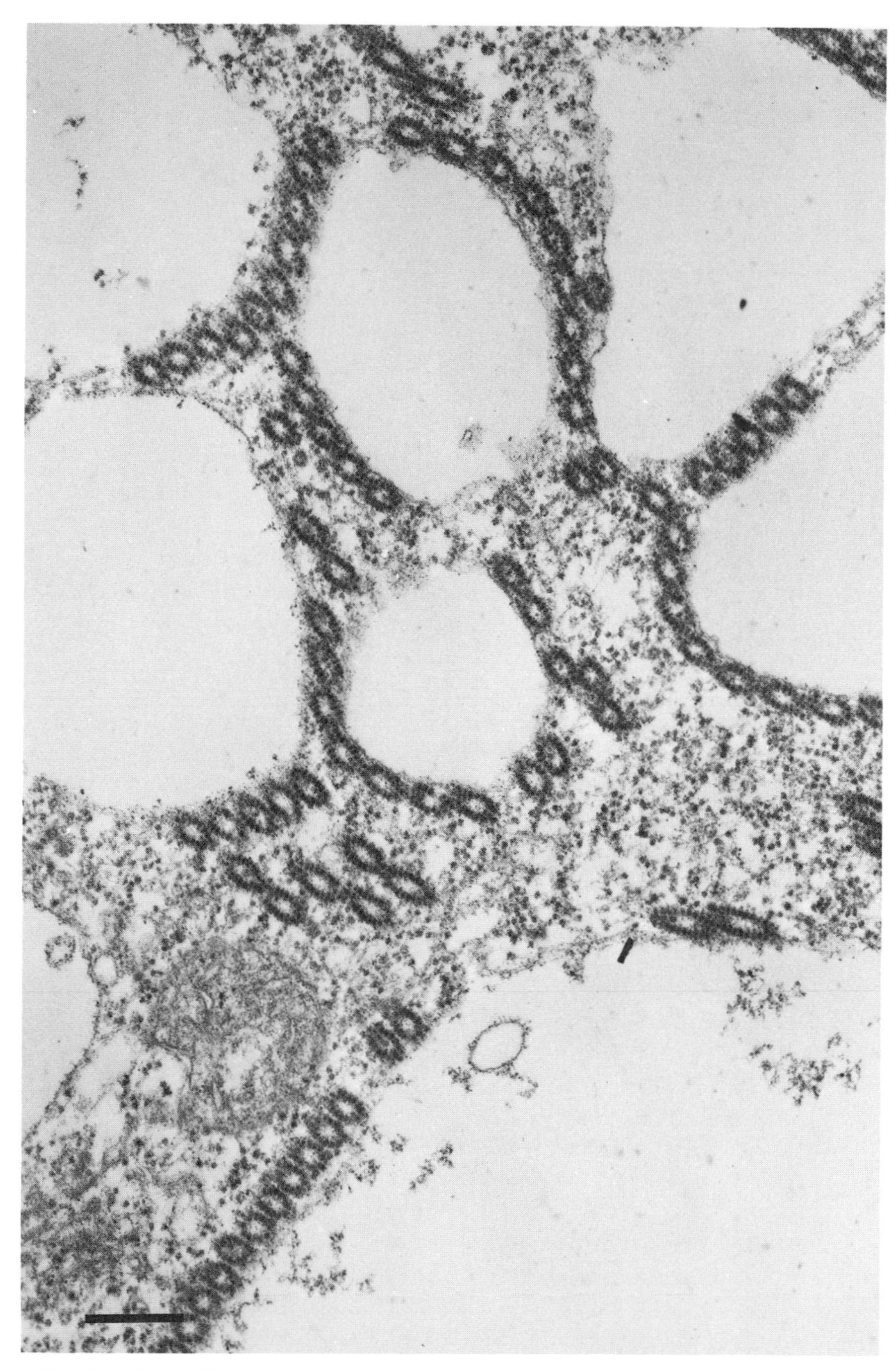

FIG. 6. Rows of tranversely sectioned crystalline tubules made up of broad bean wilt virus particles. The tubules line the bounding membrane of secondary vacuoles. [From Russo and Martelli (1975b), courtesy of the copyright holder, *Journal of Submicroscopic Cytology*.] Magnification bar represents 250 nm.

is not even consistency between the *in vivo* and *in vitro* behavior of many viruses with respect to crystallization.

Thus, for instance, test tube production of TMV crystals similar to intracellular ones has been achieved only in the presence of basic proteins (Milne, 1967b,c). Under the influence of lowered pH, or high salt concentrations, or of a number of diverse substances (Bernal and Fankuchen, 1937; Bawden and Sheffield, 1939; Dudman, 1966), purified TMV preparations form paracrystalline aggregates that bear only a pale resemblance to the orderly crystalline structures visible *in vivo*. Furthermore, cytoplasmic crystals have not been observed in cells infected by chicory yellow mottle virus (G. P. Martelli and M. Russo, unpublished data), a virus which crystallizes spontaneously during purification (Quacquarelli *et al.*, 1972; Vovlas *et al.*, 1974), whereas viruses that occur naturally in crystalline form within the cell (e.g., tomato bushy stunt, southern bean mosaic, and tobacco ringspot) need treatment with ammonium sulfate in order to crystallize *in vitro*.

Klug and Caspar (1960) recognized that the main factors determining the rate of crystallization of purified virus *in vitro* are (1) virus concentration; (2) ionic strength and salt composition; and (3) pH. Under the conditions of the negative staining–carbon technique (Horne and Pasquali Ronchetti, 1974), it is the ambient temperature at which drying of icosahedral plant virus suspensions is carried out and the pH of the suspension in presence of the negative stain that governs particle packing into square or hexagonal arrays (Horne *et al.*, 1975). Whether or not all of this also applies to crystallization in a cellular environment is unknown. However, based on these notions and on the fact that cytochrome *c* and histones act as powerful crystallizing agents for purified TMV (Milne, 1967b,c) and alfalfa mosaic virus (De Zoeten and Gaard, 1969a, 1970), the latter authors suggested that decompartmentalization of divalent cations or of basic proteins due to membrane breakdown induces an environment favorable to AMV crystallization. Aside from these causes, removal of water appears to be the principal single factor accounting for intracellular crystal formation. In fact, crystalline arrays of virions can be artificially induced by wilting leaf tissues prior to fixation (Milne, 1967a; Ushiyama and Matthews, 1970), by ultracentrifugation of whole leaves (Favali and Conti, 1971), or by plasmolysis with concentrate sucrose solutions (Hatta and Matthews, 1974). It would therefore be interesting to investigate the water balance of those infected cells where "spontaneous" crystallization of virus particles has occurred. For instance, crystalline packets of TMV particles, which are not seen in relatively undamaged cells of *Nicotiana glutinosa* local lesions, become evident in necrotic cells, possibly due to shrinkage in cell volume during necrosis (Milne, 1966c; Weintraub *et al.*, 1967).

B. Proteinaceous Inclusions

Under the general name of *proteinaceous inclusions*, several quite different types of intracellular bodies are here included. Their common feature is a similarity in chemical constitution, for all contain protein as the basic component, although this may be of different origin.

Some of these inclusions do not have a specific organization and lack an obvious substructure. Others are highly organized into well-defined forms.

1. Amorphous Granular Bodies

The "antigenic inclusions" of clover yellow mosaic (CYMV), a member of the carlaviruses, belong to the first group. These inclusions have been found in early stages of infection by CYMV in the cytoplasm and central vacuole of broad bean cells, appearing as electron-opaque rounded bodies filled with amorphous material which becomes heavily tagged when exposed to virus-specific ^{125}I-labeled antibodies. Because their size decreases with time concurrently with an increase in the number of virus particles, it has been suggested that these inclusions represent a pool of excess virus protein being slowly incorporated into complete virions (Schlegel and DeLisle, 1971). Similar-looking inclusions have been detected in the cytoplasm of sugarbeet cells infected by curly top virus; they appear as bulky amorphous aggregates, sometime close to the nucleus and partially surrounded by endoplasmic reticulum strands, which are evidently related to virus infection, but whose nature and function have not been ascertained (Esau and Hoefert, 1973).

In contrast, it is known that protein is the major component of the electron-opaque rounded or elongated, amorphous bodies occurring in the cytoplasm of watermelon mosaic virus–infected cells (Martelli and Russo, 1976a). These inclusions are sensitive to protease and stain positive for proteins. They also show, in the immediate vicinity or immersed in the matrix, filamentous profiles believed to be virus particles. The plentiful presence of these particles perhaps explains why the inclusions also stain positive for RNA. It is not known whether they represent accumulations of excess virus-coat protein or protein not directly related to the virus. As cylindrical inclusions occur normally in cells with or without amorphous inclusions, it seems unlikely that the latter arc made up of cylindrical inclusion protein in a nonorganized form.

It is noteworthy that these protein bodies have been encountered in watermelon mosaic virus isolates coming from different countries (Martelli and Russo, 1976a). Their general aspect recalls that of the satellite bodies of beet mosaic virus (Martelli and Russo, 1969a), which were also detected in the cytoplasm of *Gomphrena* cells, although only in rare in-

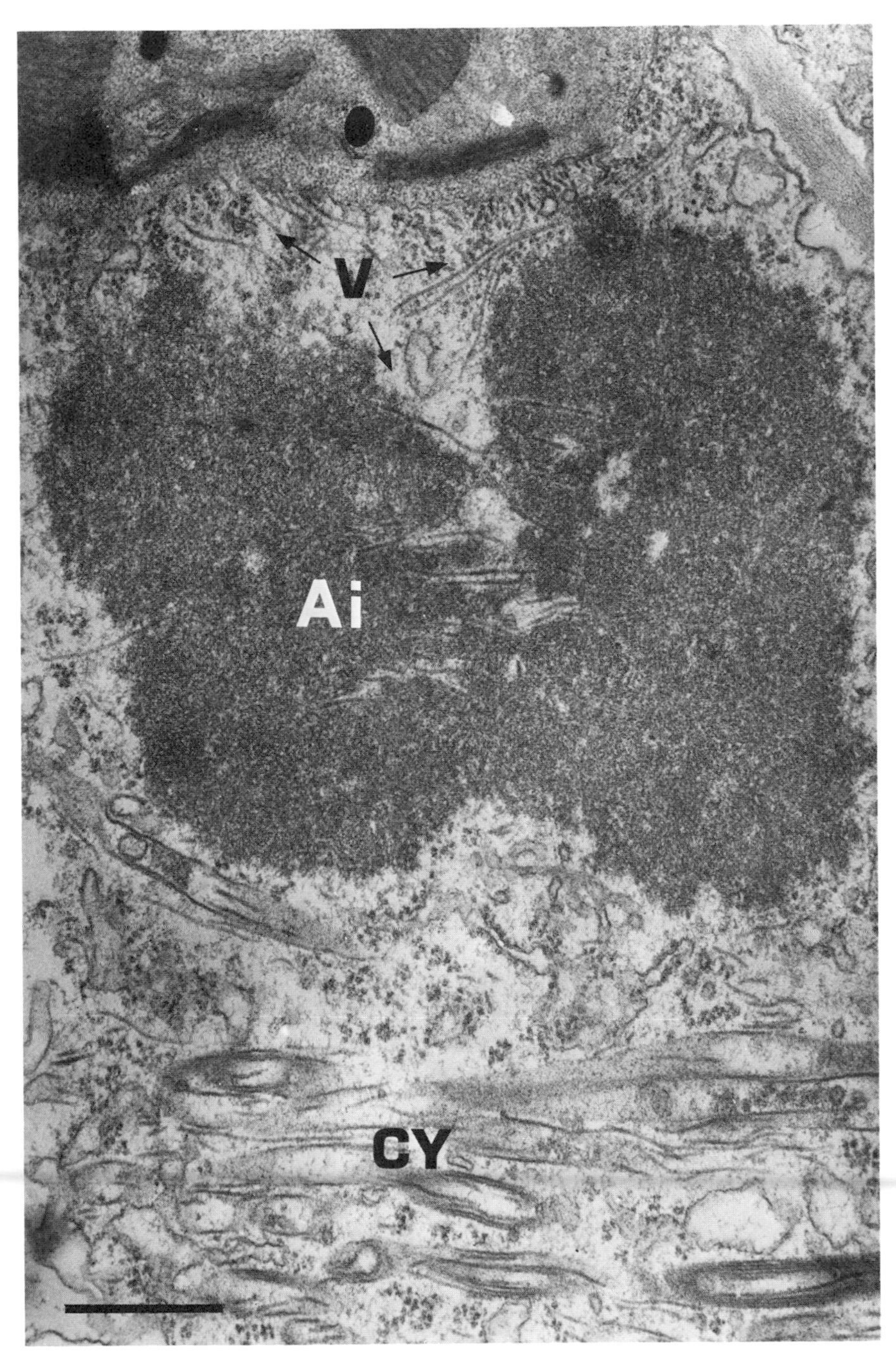

FIG. 7. Amorphous cytoplasmic inclusion (Ai) and cylindrical inclusions (CY) in a squash leaf infected by watermelon mosaic virus. Profiles of virus particles (V) are seen inside and around the amorphous inclusion. Magnification bar represents 250 nm.

stances. Similar inclusions had been noticed before in cells infected by a Florida strain of watermelon mosaic virus and the serologically related papaya ringspot virus (Edwardson, 1974), but in neither case were they investigated further.

2. *Cylindrical Inclusions*

Cylindrical inclusions are complex, three-dimensional structures typically induced by all the members of the potyvirus group (particles 700–900 nm long) that have so far been investigated at the ultrastructural level [for review, see Edwardson (1974)]. These inclusions are also found in cells infected with some viruses that have shorter filamentous particles (Rubio-Huertos and Vela-Cornejo, 1966; Weintraub and Ragetli, 1971; Inouye, 1970, 1971; Peña-Iglesias and Ayuso-Gonzales, 1972) or longer ones (Hooper and Wiese, 1972; Langenberg and Schroeder, 1973c). They form intracellularly, irrespective of whether the inducing virus occurs singly or in mixed infections (e.g., McWhorter and Price, 1949; Fujisawa *et al.*, 1967a; Rubio-Huertos *et al.*, 1968a; Kamei *et al.*, 1969a; Castro *et al.*, 1971; Peña-Iglesias *et al.*, 1972; Lee and Ross, 1972; Russo and Martelli, 1973; Russo *et al.*, 1975).

The expression *cylindrical inclusions* was first coined by Edwardson (1966a; see also Rubio-Huertos and Lopez-Abella, 1966). It was defined in later papers by Edwardson and co-workers (Purcifull and Edwardson, 1967; Edwardson *et al.*, 1968), who, from serial sections, reconstructed the inclusion bodies associated with infections by tobacco etch virus (TEV) and watermelon mosaic virus and obtained a basically cylindrical configuration. However, Andrews and Shalla (1974) reported that in TEV those bodies which appear as pinwheels when cross sectioned are cone shaped, at least during the early stages of formation, with arms tapering towards the apex of the structure. The results of more recent investigations involving mathematical determination of the shape of inclusions produced by wheat streak mosaic virus indicate that they approximate an elliptic hyperboloid. The actual shape of these pinwheels is a flat-ended cylinder gradually sloping toward one end, where the slope rapidly rounds to a cone (R. L. Mernaugh and W. S. Gardner, personal communication). Hence, the term *cylindrical inclusions* does not seem to be accurate; nevertheless, in this review it will be used to designate broadly all kinds of inclusions induced by potyviruses, in synonymy with the term *pinwheel inclusions,* now widely accepted and firmly established in the literature.

If properly stained, cylindrical inclusions can be detected with the light microscope in epidermal strips of infected plants; they are seen more easily after treating tissues with Triton X-100 (Christie, 1971). According to their intracellular location, in epidermal strips the inclusions appear as intensely stained, cone-shaped structures abutting on the cell wall (e.g., Edwardson, 1974), or as a series of straight, thin elements radiating from a central point (e.g., Edwardson *et al.*, 1968, 1972), or as elongate to spindle-shaped fibrous bodies with a loose texture (e.g., Zettler *et al.*, 1967; Martelli and Russo, 1976a).

The first electron microscopic record of pinwheels seems to be due to David-Ferreira and Borges (1958) who observed them in PVY-infected cells. Subsequently a variety of intracellular structures were reported in association with infection by different potyviruses; these structures were variously interpreted (Cremer *et al.* 1960; Matsui and Yamaguchi, 1964; Rubio-Huertos and Garcia-Hidalgo, 1964; Lee, 1965) until Edwardson (1966a,b) demonstrated that they represent different aspects of the same type of inclusion.

Presently, the consensus is that pinwheels are real entities evoked by viral infections. Data from freeze-etching, negative staining, and from use of different fixatives firmly support the conclusion that these bodies do not constitute artifacts arising during tissue processing for electron microscopy (Edwardson, 1974; McDonald and Hiebert, 1974a,b; Andrews and Shalla, 1974). Contrary to these views, Wilson and co-workers (Wilson and Israel, 1972; Wilson *et al.*, 1974a,b,c) reported the occurrence of pinwheel inclusions in cultured "healthy" carrot cells; their conclusion that pinwheels in plant cells may be produced by means other than a virus is based primarily on failure to recover infectivity from cultured tissues using current virus-detection methods. However, it is not always easy to demonstrate viruses in infections by members of the potyvirus group. Moreover, only sketchy details are available on how Wilson and colleagues analyzed their tissue for virus. Finally, these authors have not convincingly excluded the possibility that the carrots used in their studies were infected with viruses of the PVY type, like those found in Brazil (Camargo *et al.*, 1971) or the United States (E. Hiebert, personal communication). Therefore, in view of the overwhelming evidence provided by the current literature on potyvirus infections (see Edwardson, 1974), the findings of Wilson and co-workers warrant further investigation.

In another instance, intracellular inclusions comparable to, although not identical with, the "classical" ones were recorded in supposedly healthy *Kalanchoe* plants. These, when more accurately observed, proved to contain a filamentous virus of the PVY type (Kapil *et al.*, 1975).

In cross section, cylindrical inclusions appear as pinwheels, scrolls, or laminated aggregates, whereas the respective profiles in longitudinal section are bundles for pinwheels, tubes for scrolls, and laminated aggregates (these remain unchanged). Pinwheels are composed of curved plates radiating from a central axis constituted by a cylinder with a solid core; scrolls are cylinders constructed of a rolled-up plate; and laminated aggregates are formed by a series of stacked laminae. Virus particles may be associated with all types of inclusions, either loosely (e.g., Edwardson *et al.*, 1968) or rather intimately [as, for instance, in turnip mosaic virus (TuMV) infections (Kamei *et al.*, 1969a; Russo *et al.*, 1972; Edwardson, 1974)].

Pinwheel inclusions are highly organized and show a distinct substructure that has been investigated in detail in various ways. For instance, the vacuum-dehydration technique (Edwardson *et al.*, 1968) has shown that the plates constituting the inclusions, when cross sectioned, appear as linear arrays of subunits with a spacing of about 5 nm. Similar results were obtained by observing negatively stained inclusions from tissue extracts. These exhibited clear-cut unidirectional striations along the major axis, again with an average periodicity of 5 nm (Purcifull and Edwardson, 1967; Edwardson *et al.*, 1968).

In negatively stained preparations, the rectangular plates showing striations in various directions (indicating a multilayered condition) were assumed to be either scrolls or parts of pinwheels (curved plates at the central portion), whereas the rectangular or, more commonly, triangular plates exhibiting unidirectional striations were interpreted as profiles of laminated aggregates (Edwardson *et al.*, 1968; Purcifull *et al.*, 1970). Triangular plates can also be identified as pinwheel arms (Andrews and Shalla, 1974).

Using freeze-etching techniques, McDonald and Hiebert (1974b) demonstrated that the plates of cylindrical inclusions of TEV, PVY, and TuMV are composed of globular subunits arranged in a rectangular array with a spacing of about 6 nm in the two orthogonal directions. These authors also suggested that the subunits revealed by freeze-etching correspond to those isolated in polyacrylamide gel electrophoresis after disruption of the inclusions (Hiebert and McDonald, 1973). The substructure revealed by freeze-etching contrasts with the unidirectional striation seen in negatively stained plates. This discrepancy was explained by assuming that in the latter type of preparations the stain is able to penetrate the gap between the rows of subunits on one axis but not on the other, either because of real physical limitation or because of shrinking due to exposure to vacuum in the electron microscope (McDonald and Hiebert, 1974b).

The proteinaceous nature of cylindrical inclusions was first established by means of staining reactions by Rubio-Huertos (1950) and then

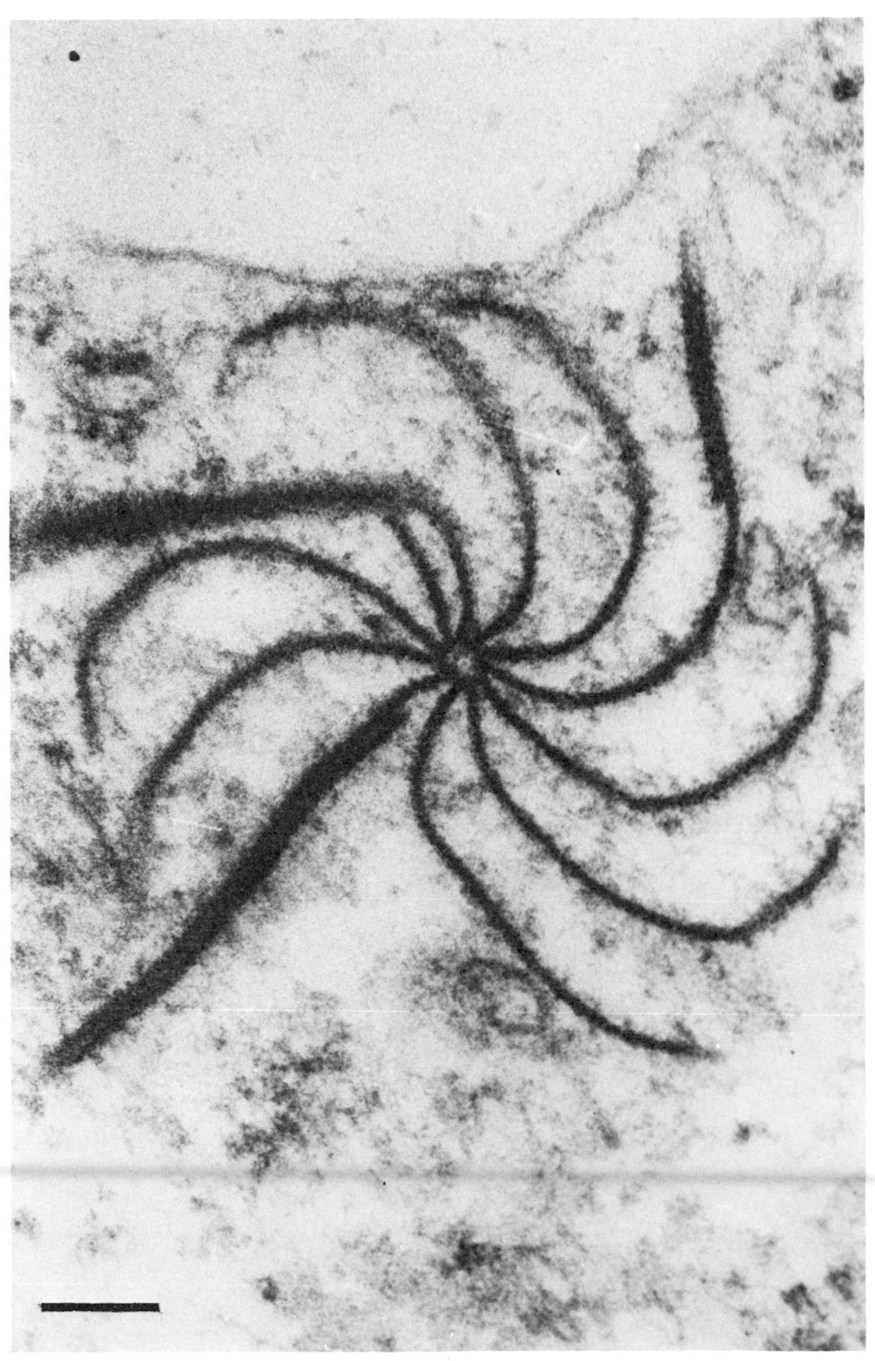

FIG. 8. A pinwheel induced by beet mosaic virus. Magnification bar represents 100 nm.

confirmed by enzyme digestion tests (Shepard, 1968; Weintraub *et al.*, 1969; Hoefert, 1969) and chemical analysis (Hiebert and McDonald, 1973). Sodium dodecyl sulfate–dissociated inclusions yielded subunits that migrated as a single electrophoretic species with a molecular weight in the range of 67,000 to 70,000, i.e., almost double the molecular weight of viral capsid subunits estimated to be about 33,000 daltons (Hiebert and McDonald, 1973).

The serological behavior of cylindrical inclusion proteins has been investigated in detail. Shepard and co-workers (Shepard and Shalla, 1969; Shepard *et al.*, 1974) demonstrated by ferritin tagging that TEV-induced inclusions were serologically unrelated to the parent virus. This was soon confirmed for other potyviruses by using antisera to purified inclusion preparations (Hiebert *et al.*, 1971). In addition it was found that pinwheels evoked by different viruses are not related serologically to one another (Purcifull *et al.*, 1973) nor to host proteins. In conclusion, cylindrical inclusions, although coded by their causal virus, are not serologically related to its coat protein. Furthermore, viruses that are serologically unrelated induce unrelated inclusions, whereas those which show a close serological relatedness elicit inclusions that are also closely related (McDonald and Hiebert, 1975).

As mentioned earlier, pinwheels, scrolls, and laminated plates are inclusions of the same kind, although of different shape, which are produced by the same group of viruses. However, not all members of this group elicit the formation of all three types of inclusions at the same time. Actually, different viruses yield inclusion bodies which differ in appearance. Since this is considered a virus-specific characteristic, a separation of potyviruses into subdivisions has been attempted on the basis of the configuration of their inclusions. Thus, Edwardson (1974) has proposed a classification of potyviruses into three groups according to the presence of the following structures attached to the central portion of the pinwheel inclusion: (a) scrolls or tubes—subdivision I, with 21 members; (b) laminated plates—subdivision II, with 18 members; (c) scrolls and laminated plates together—subdivision III, with 25 members. As pointed out by Edwardson (1974), mixed infections with viruses belonging in subdivisions I and II could result in erroneous assignment to subdivision III.

This grouping is based on the constancy of the inclusions induced by a single virus irrespective of the influence of the host. The electron microscopic evidence accumulated so far is in favor of such constancy, although there are indications that inclusions may vary in the course of infection. This variation, however, refers to the number, size, and relative preponderance of the type of inclusion, rather than to a change from one kind to another. For instance, in young infections of beet mosaic virus in *G. globosa* mesophyll cells, small pinwheels with a few laminated plates prevail, whereas in older infections, pinwheels are less numerous but much larger and with conspicuous laminated inclusions. In very old infections, pinwheels disappear but laminated plates are still found (Russo and Martelli, 1969). Thus, irrespective of the age of infection and of the changes in the general appearance, beet mosaic virus would always be attributed to Edwardson's subdivision II. A comparable situation has been reported for sweet potato russet crack virus (Lawson *et al.*, 1971; Nome *et al.*, 1974), TEV (Andrews and Shalla, 1974), and watermelon mosaic virus (Martelli and Russo, 1976a). In sweet potato russet crack virus, pinwheels prevail in young infections whereas tubes, scrolls, and laminated plates are plentiful in older infections. These observations, and the finding of apparent transitional forms between pinwheels and tubes and between pinwheels and laminated inclusions, led Lawson *et al.* (1971) and Nome *et al.* (1974) to conclude that cylindrical inclusions undergo conformational changes during the course of infection, a phenomenon already noticed for beet mosaic virus (Russo and Martelli, 1969). Additional evidence that this is a common feature in potyvirus infection is given by Andrews and Shalla (1974), who report that pinwheels in TEV-infected root tips change gradually by addition and growth of arms, fusion of arms, and opening of the central cores. The suggestion of Nome *et al.* (1974) that tissues of comparable age of infection be used in studies aimed at comparing the form of inclusions evoked by different potyviruses therefore seems justified.

It should also be noted that certain types of inclusions, e.g., laminated aggregates, seem to be poorly preserved by conventional glutaraldehyde–osmic acid fixation, tending to disappear because of the lytic activity of cellular enzymes during tissue processing. As a consequence, final tissue preparations used for electron microscopy may not entirely reflect the original intracellular conditions with respect to relative occurrence of different inclusion forms (Langenberg and Schroeder, 1973b, 1974a).

Other components may be lost because of insufficient fixation or staining of sections. Thus, Begtrup (1976) reports an intriguing association of microtubular structures with the pinwheel inclusions of carnation vein mottle virus–infected cells. These tubules, however, were only seen when samples had been subjected to prolonged osmic acid treatment and stained with alcoholic uranyl acetate solutions. They were not visible in conventionally treated tissues infected with the same virus (Weintraub and Ragetli, 1970c; Begtrup, 1976).

The development of pinwheels has been investigated to some extent. Russo and Martelli (1969), Lawson *et al.* (1971), and Andrews and

Shalla (1974) studied the developmental sequence leading to pinwheel formation in beet mosaic, sweet potato russet crack, and tobacco etch viruses, respectively. In all instances it was noted that the early profiles of the inclusions were a series of bundle aggregates abutting on the plasma membrane. These, in sweet potato russet crack and tobacco etch viruses, were interpreted as views of longitudinally cut, complete pinwheels, centered at and radiating from plasmodesmata, with which they were consistently associated (Lawson and Hearon, 1971; Lawson *et al.*, 1971; Andrews and Shalla, 1974). A similar interpretation was offered for some of the profiles seen in BMV-infected cells. In particular, Russo and Martelli (1969) proposed that the lamellae constituting pinwheel arms develop as a series of parallel elements perpendicular to the plasma membrane, then converge at their free extremity around a central axis to form pinwheels, which detach themselves from the cell wall and move into the cytoplasm. Lawson *et al.* (1971), instead, reported that pinwheel cores may derive from plasmodesmata around which the inclusion arms converge. This view was shared in part by Andrews and Shalla (1974), who did not find evidence, however, that the central axis of the inclusions originates from extended desmotubules. These authors too (Lawson *et al.*, 1971; Andrews and Shalla, 1974) admit a separation of cylindrical inclusions from the cell wall. Therefore, whatever the developmental sequence is, it appears that inclusions of some potyviruses originate at plasmodesmata, in strict connection with the cell walls, and then move into the cytoplasm and change their size and shape in the course of infection.

A developmental model differing from the above was constructed for wheat spindle streak mosaic virus, which has filamentous particles of extremely variable length, extending to more than 1000 nm, but was assigned to subdivision III of potyviruses by Edwardson (1974). In this case, pinwheels originated from a membranous inclusion arising from the endoplasmic reticulum (Hooper and Wiese, 1972; Langenberg and Schroeder, 1973c). In particular, membranous sheets connected either with the nucleus or the plasma membrane were produced in the endoplasmic reticulum, giving rise to tubes and convoluted plates. The tubes developed into pinwheel structures which continued to grow as the membranous body enlarged and was invaded by ribosomes. It is not clear whether a relationship exists between pinwheel formation and viral synthesis, but virus particles were often seen in close connection with developing inclusions and became trapped between the pinwheel arms (Langenberg and Schroeder, 1973c). Similar profiles of convoluted endoplasmic reticulum were reported in cells infected by pea seed-borne mosaic virus, but these were not identified as sites of pinwheel formation (Hampton *et al.*, 1973).

The constant association of pinwheel inclusions with viruses of the PVY group and the fact that both their morphological and serological properties are virus specific and host independent strongly support the concept that pinwheels are a product of the viral genome. With but a single exception [see Hibino and Schneider (1971), who however, refer to pear vein yellowing, a poorly characterized virus, to date], none of the infections by potyviruses investigated at the ultrastructural level has been found without inclusion development in the host. Hence, pinwheel inclusions may have an essential function in potyvirus pathogenesis. But this function is still unknown.

Current knowledge of their structure and composition has ruled out all previous speculations that referred to pinwheels as being aggregates of virus particles, or of virus-coat protein, or of normal host protein (see Edwardson, 1974). A possible correlation between the presence of pinwheels in epidermal tissues and virus transmission by aphids was envisaged (Zettler *et al.*, 1967) but was not further investigated. In this connection, it is worth pointing out that ryegrass mosaic and agropyron mosaic viruses are not aphid transmitted although both induce pinwheels.

An interesting point was raised by Chamberlain (1974, 1975), who suggested that pinwheel inclusions may be linked with tolerance to virus infection. In particular, she found that in tolerant ryegrass genotypes, ryegrass mosaic virus pinwheels were present in larger groups and were of greater size than in sensitive genotypes. However, the percentage of cells containing inclusions was not related to tolerance. Conversely, in tolerant wheat varieties, agropyron mosaic virus produced larger groups of pinwheels of a similar size, compared with those of the sensitive cultivars, but the number of cells producing them was higher in the tolerant genotypes (Chamberlain, 1975). Based on these observations, Chamberlain (1974, 1975) speculated that pinwheels are produced by plants in response to virus infections, as part of their tolerance mechanism; pinwheels reduce the detrimental effect of virus on the host cell by absorbing free viral components or by restricting the movement of virus particles.

The association of pinwheels with plasmodesmata led Andrews and Shalla (1974) to suggest that they could play some regulatory role in the intracellular transport of virions and various metabolites, for instance, delaying systemic distribution of virions, thereby allowing time for the activation of other host defences. It is undoubtedly conceivable that by plugging up plasmodesmata, pinwheels could impair cell-to-cell transport of virus particles. In fact, whenever virions of potyviruses have been seen within plasmodesmata, these either did not show any obvious association with pinwheel inclusions (e.g., Russo and Martelli, 1969; Weintraub *et al.*, 1974) or, if such an association were seen (Nome *et al.*, 1974), the spatial relationship between pinwheel and virus-containing plasmodesma could not be established with certainty. It is then peculiar that such a powerful mechanism for secluding the virus and preventing its systemic spread should only operate for a short time in the early stages of infection. Besides, as pointed out by Hiebert (1975, and personal communication), the concept that pinwheels are a defense mechanism against potyvirus infections is not easily reconciled with evidence indicating that these inclusions are virus specific and independent of the propagating host. It would be more sensible to assume that it is the virus they serve, not the host plant. Thus, Hiebert (1975) suggested that a clue to the possible role of these inclusions could be provided by their structural organization. In fact, the pinwheel structure maximizes the surface exposure of a body composed of repeating protein subunits to its environment. Thus, the function of the inclusions may be related to some surface phenomenon (Hiebert, 1975, and personal communication).

3. Laminate Inclusions

Laminate inclusions are multilayered bundles of proteinaceous sheets which have been found consistently associated with infections of potato virus X [PVX (Milne, 1968; Kozar and Sheludko, 1969; Stols *et al.*, 1970; Shalla and Shepard, 1972; Doraiswamy and Lesemann, 1974; Pennazio and Appiano, 1975)]. Together with aggregates of virus particles, these structures constitute large intracellular inclusion bodies with a more or less granular appearance in the light microscope and no specific localization within the cell, but always in the cytoplasm (Goldin, 1963; Bawden, 1964; Shalla and Shepard, 1972).

These inclusions have been studied in detail at the ultrastructural level by several workers. Kozar and Sheludko (1969) interpreted the tenuous lines seen in thin sections, which average 3 nm in thickness, as RNA filaments to which ribosomes were attached to form huge polysomes involved in virus synthesis. Shortly afterwards, Stols *et al.* (1970) demonstrated through serial sectioning that the "filaments" were in fact transectional views of membranous structures.

Shalla and Shepard (1972) were the first to interpret correctly the conformational architecture of the inclusions and also to elucidate their chemical constitution. These authors recognized three prominent types of

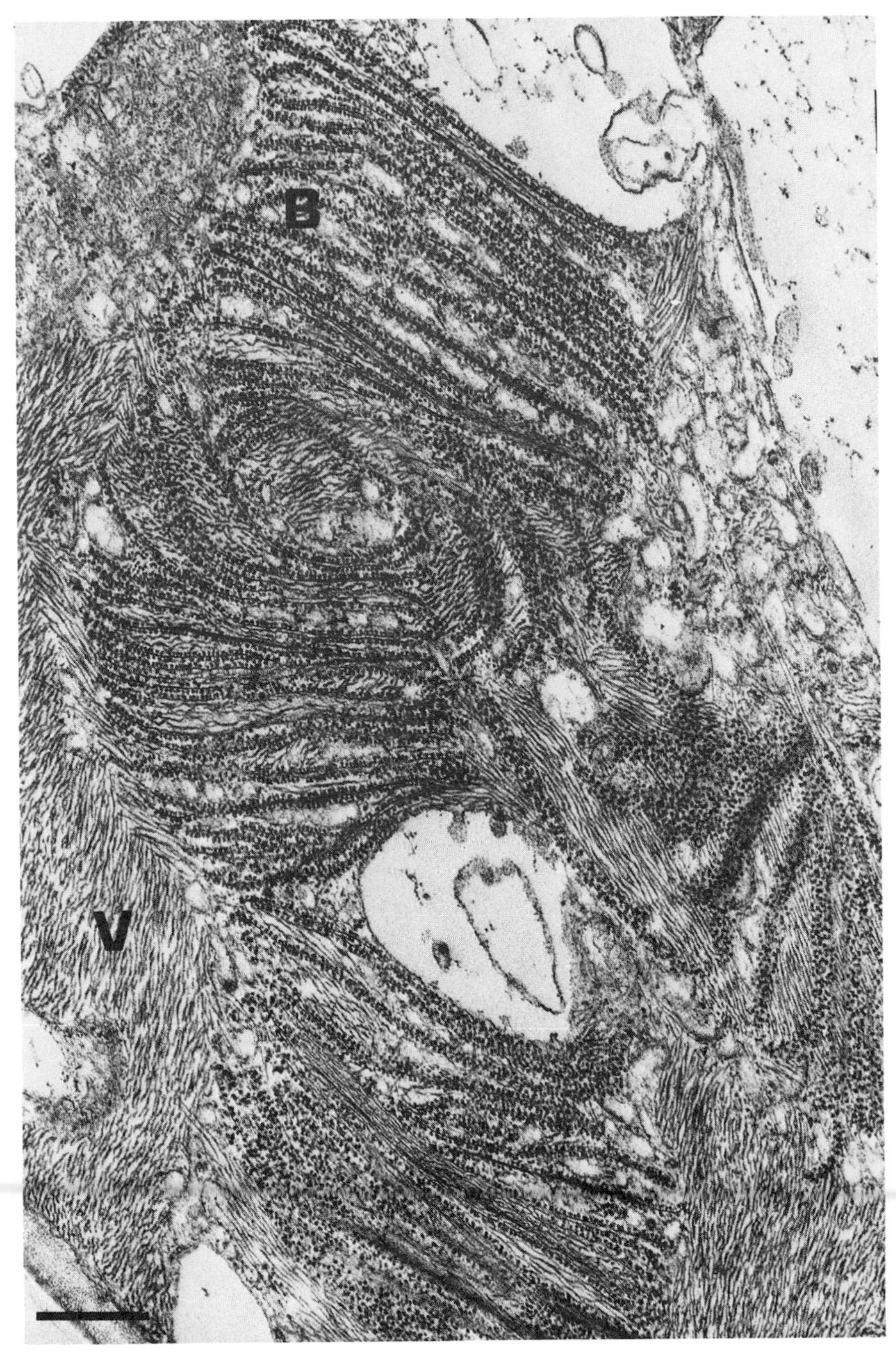

FIG. 9. Beaded sheets (B) and virus particles (V) constituting a complex inclusion induced by potato virus X. [From Shalla and Shepard (1972), courtesy of the copyright holder, Academic Press.] Magnification bar represents 500 nm.

laminate inclusions: (1) sheets mostly free of virus but heavily studded on both sides with beads about 14 nm in diameter (beaded sheets); (2) beaded sheets with virus particles interspersed between the collateral layers; (3) smooth-surfaced sheets with numerous virus particles between them. They also provided experimental evidence by means of differential enzymic digestion that sheets and beads are composed of proteins (although possibly of a different type) which are not related antigenically to PVX. Ferritin-labeled antibodies specific for either whole PVX or depolymerized PVX proteins (D-protein) failed to tag any of the inclusion components (beads and sheets) but heavily tagged virus particles associated with the inclusions (Shalla and Shepard, 1972). Therefore, beaded sheets are not sites of virus or virus-coat protein synthesis as suggested by Kozar and Sheludko (1969).

Identical inclusions are also present in isolated plant protoplasts infected with PVX (Shalla and Petersen, 1973; Otsuki *et al.*, 1974; Honda *et al.*, 1975); they begin to develop as small aggregates of fine granules having the size of the beads of mature inclusions and never precede the appearance of progeny virus particles. The agreement on the last finding is general and applies both to protoplast infections and to expanding local lesions in *G. globosa* (Allison and Shalla, 1974). In infected protoplasts the inclusions appear to develop late in relation to detection of PVX antigens (Shalla and Petersen, 1973) or of whole virus particles (Honda *et al.*, 1975). In expanding local lesions, virus particles are invariably found further away from the center than are inclusion body components (Allison and Shalla, 1974). These observations add strength to the conclusion that laminate inclusions are formed later than virus and therefore are very unlikely to be sites of virus infection.

The function of laminate inclusions is unknown. Shalla and Shepard (1972) postulated that they are analogous to the pinwheel inclusions of potyviruses, whose significance is not clear either. Although this may well be so, laminate inclusions certainly differ from cylindrical inclusions in that they do not constitute a diagnostic feature for the potexvirus group as a whole. So far, they have been detected in connection with PVX infections, but not in cells infected with any of the other viruses of the same group which have been investigated to date: clover yellow mosaic (Purcifull *et al.*, 1966), papaya mosaic (Zettler *et al.*, 1967), narcissus mosaic (Turner, 1971; Štefanac and Ljubešić, 1974), cactus X, and cymbidium mosaic (Doraiswamy and Lesemann, 1974).

4. Crystalline Nonviral Inclusions

Cytoplasmic crystalline inclusions that occur in virus-infected cells but are not made up of virus particles can be roughly grouped into three major categories: (1) crystals consistently associated with infections by certain viruses and never found in healthy cells; (2) crystalline material normally present in plant cells which may be influenced in various ways by virus infections; (3) ordinary crystalline structures whose occurrence is independent of virus infections.

A classic example of the first group is given by the crystals induced by some strains of red clover vein mosaic virus (RCVMV). These were first observed with the light microscope by Porter and McWhorter (1952) and later described as having basically the shape of a thick hexagonal prism (McWhorter, 1965). The inclusions were regarded as highly diagnostic for the parent virus, and in fact they were thought to be composed entirely of virus particles. Their positive reaction to stains for proteins and RNA supported this idea (McWhorter, 1965). However, recent investigations with the light and electron microscopes have failed to confirm the viral nature of such crystals (Rubio-Huertos and Bos, 1973). The filamentous particles of RCVMV were present in the cytoplasm in amorphous bulky aggregates but did not enter the constitution of the crystalline inclusions. These appeared as rather compact structures with unidirectional striations having a periodicity of about 11 nm in diameter. Thus, convincing evidence has been provided that RCVMV crystals are not composed of virus particles. However, their significance and chemical nature remains to be established, although, by analogy with rather similar inclusions induced by some potyviruses, Rubio-Huertos and Bos (1973) suggested that the crystalline bodies are composed of proteins.

Spherical units about 10 nm in diameter, packed in a regular array to form discrete crystalline structures, have been observed by Lawson *et al.* (1971) in cells infected by sweet potato russet crack virus. The authors were unable to ascertain whether the crystals are composed of spherical or elongate particles, but, as the inclusions occurred in infected cells only, it is reasonable to assume that they consisted of virus-related products.

Many additional members of the potyvirus group evoke formation of nonviral crystalline bodies. Thus, for instance, protein crystals seemingly identical with the nucleolar ones that have been previously described were found in the cytoplasm of cells infected with bean yellow mosaic and related viruses (see Edwardson, 1974), as well as in association with infections by clover yellow vein virus (Singh and Lopez-Abella, 1971) and pea seed-borne mosaic virus (Hampton *et al.*, 1973). Furthermore, crystalline material of unknown nature occurs in infections by the viruses of sharka (Bovey, 1971), wheat spindle streak mosaic (Hooper and Wiese, 1972), and henbane mosaic (Harrison and Roberts, 1971; Kitajima and Lovisolo, 1972). In the last case, the inclusions were described as hexagonally arranged tubules in tightly packed bundles with a periodicity of about 12 nm. The crystals were approximately spindle shaped and up to about 10 μm long (Harrison and Roberts, 1971; Kitajima and Lovisolo, 1972).

Records of nonviral cytoplasmic crystals are indeed a rarity with isometric viruses. Milne (1967a) reported the occurrence of regular crysstals of densely stained units hexagonally packed about 6 nm apart in cells of *Chenopodium amaranticolor* Coste et Reyn. infected by cowpea mosaic virus. These inclusions were not seen in healthy cells of the same host and apparently were never recorded again in any of the hosts harboring the same virus which were investigated in later studies (Van der Scheer and Groenewegen, 1971; De Zoeten *et al.*, 1974; Langenberg and Schroeder, 1975). As suggested by Milne (1967a), it is possible that these structures, like all the other nonviral crystalline inclusions, are no more than pathological side-products of virus infections.

There are cases in which ordinary crystalline cell constituents are much more abundant in infected than in healthy cells. This was reported, for example, for crystal-containing microbodies in the zone peripheral to TMV lesions (Israel and Ross, 1967; Ross and Israel, 1970), and in cells infected by AMV (De Vecchi and Conti, 1975). Evidently, these are consequences of deranged cell metabolism or, as suggested for TMV, may be expressions of a resistance mechanism (Israel and Ross, 1967). Thus, in both these instances, the increased number of crystalline structures represents an indirect effect of virus activity upon the cell. A similar explanation was provided by Esau (1975), who related the higher number of intraplastidial protein crystals found in beet curly top–infected spinach cells, as compared with that of healthy samples, to an effect of the causal agent on the metabolism and sugar translocation in diseased leaves. Instead, the pseudocrystalline bodies observed in the chloroplasts of detached healthy leaves and isolated protoplasts (e.g., Ragetli *et al.*, 1970; Otsuki *et al.*, 1972; Milne, 1972; Nagata and Yamaki, 1973) seem to arise following plasmolysis and dehydration of the cells (e.g., Gigot *et al.*, 1975).

In the cytoplasm of perfectly healthy cells a tremendous array of crystalline inclusions can be found that bear no relation to pathogenic agents. These are known to occur throughout the plant kingdom in a great number of species and have been extensively reviewed by Thaler (1966). The inclusions display a variety of sizes and shapes, such as the spindle-shaped crystalline P-protein (e.g., Wergin and Newcomb, 1970), the more or less globular "extruded nucleoli" (e.g., Deshpande and Evert, 1970), the angular vacuolar crystals of *Lilium tigrinum* (Thaler and Amelunxen, 1975), the irregular crystalline bodies contained within spherosomes (e.g., Frederick *et al.*, 1968), pastids (e.g., Newcomb, 1967; Esau, 1975; Freeman and Duysen, 1975) and mitochondria (e.g., Leak, 1968).

In most cases these crystals represent protein storage structures, but they can also be accumulations of fraction I protein (Pyliotis and Gooldchild, 1975) or enzymes, as in microbodies. Cases in which they have been mistaken for virus-induced structures are rare. However, on several occasions, even recently (Kuppers *et al.*, 1975), intraplastidial crystalline arrays of phytoferritin have been wrongly interpreted as crystalline virus (see Craig and Williamson, 1969; Esau and Hoefert, 1973). Phytoferritins are iron–protein complexes which tend to accumulate in plastids of plants growing in suboptimal conditions; hence, it is not surprising that they occur in crystalline form rather frequently in cells whose metabolism is altered by viral infections (Craig and Williamson, 1969). Nevertheless, phytoferritins should be relatively easy to distinguish from icosahedral virus particles because of the distinct appearance of the complexes, their smaller size, and the different system in which they crystallize (Amelunxen *et al.*, 1970). Besides, no intraplastidial crystals of any known plant virus have been recorded so far.

C. Aggregates of Modified Cell Organelles or Host-Derived Material

Following virus infection, cytological disturbances of various types ensue. Different organelles may become modified and, in the end, can be so profoundly altered as to lose their identity, thus giving rise to novel structures with the appearance of inclusion bodies. Several such examples have been illustrated in the literature, and most involve modifications of mitochondria as, for instance, with some strains of tobacco rattle virus (Harrison *et al.*, 1970), cucumber green mottle virus (Hatta *et al.*, 1971; Hatta and Ushiyama, 1973; Sugimura and Ushiyama, 1975), a potyvirus from apple (Weintraub and Ragetli, 1971), and an unidentified virus from begonia (Petzold, 1972).

The inclusion bodies of tobacco rattle virus are well defined, rather compact and, in addition to ribosomes and other constitutents of ground cytoplasm, may contain small packets of virus particles. As illustrated by Harrison *et al.* (1970) the inclusions begin to develop as small aggregates of relatively unaltered mitochondria and grow, as infection progresses, by addition of new mitochondria, to attain the average size of about 17 μm. Remarkable features of these when fully developed are the apparent lack of virus-coat protein, a high RNA content, and the presence of an unusual and heavy production, within mitochondria, of small vesicles containing finely fibrillar material resembling nucleic acid.

A similar mitochondrial vesiculation has been observed in association with infections by cucumber green mottle virus, both in intact leaf tissues of several cucurbit hosts (Hatta *et al.*, 1971; Hatta and Ushiyama, 1973) and in isolated tobacco protoplasts (Sugimura and Ushiyama, 1975).

With this virus, the first detectable sign of infection was the presence of mitochondria with peripheral vesicles containing fine-stranded material which increased with time after inoculation. Strikingly similar profiles of modified mitochondria have been recorded by Petzold (1972) in begonia plants infected by an unidentified virus.

The aforementioned studies provide plenty of circumstantial evidence that mitochondrial vesiculation is intimately related to virus infection. Hatta and Ushiyama (1973) think that these vesicles may represent a step of the replication process of cucumber green mottle virus, rather than the result of a secondary degeneration after virus infection. On the other hand, Harrison *et al.* (1970) suggest that mitochondria are specifically involved in the synthesis of tobacco rattle virus RNA. Thus, it would be tempting to speculate that the fibrillar material contained within mitochondrial vesicles, by analogy with what is known about plastidial vesicles of tymoviruses (discussed next), represents the replicative form of viral RNA or a related product. However, no experimental proof of such an intriguing possibility is presently available.

Inclusions formed from altered chloroplasts are typical of turnip yellow mosaic virus infections and may be a characterizing feature of other tymoviruses too. The inclusions consist of more or less compact clumps of chloroplasts, and were first observed with the light microscope in Chinese cabbage leaves by Rubio-Huertos (1950). Subsequently, they were thoroughly investigated with both the light and electron microscopes (e.g., Rubio-Huertos, 1956; Gerola *et al.*, 1966b; Chalcroft and Matthews, 1966; Rubio-Huertos *et al.*, 1967; Miličić and Štefanac, 1967; Laflèche and Bové, 1968; Ushiyama and Matthews, 1970; Hatta and Matthews, 1974), so that an impressive body of knowledge is now available on their development, gross morphology, and ultrastructural organization. Moreover, *in situ* labeling, cell fractionation, and biochemical studies have answered most of the questions referring to the significance of the inclusions and their role in TYMV multiplication (e.g., Ralph and Clark, 1966; Laflèche and Bové, 1968; Bové, 1975; Hatta and Matthews, 1976). Although it seems possible that all TYMV strains induce chloroplast clumping (Miličić and Štefanac, 1967) there is certainly a strain-dependent difference in the way in which these organelles aggregate. Thus, according to the virus strain and the severity of infection, chloroplasts may be arranged in rows of tightly appressed units (which do not lose their identity) or in progressively more compact aggregation forms, which, with the very virulent strains, are represented by amorphous masses where single organelles are no longer discernible (Chalcroft and Matthews, 1967a,b; Matthews, 1973).

The sequential development of chloroplast abnormalities and clumping has recently been investigated at the fine-structure level by Hatta and Matthews (1974). These authors visualize a stepwise process involving seven different stages: from a relatively unaltered condition (stage A), the organelles become progressively swollen and vesiculated until they clump together (stages E–G). Some remarkable pathological features are connected with these intracellular modifications, as follows: (1) After stage A, periplastidial flask-shaped vesicles are observed that contain finely stranded material. The vesicles are bounded by a membranous wall and have necks pointing to and opening at the chloroplast surface [for more structural detail, see Hatta *et al.* (1973)]. As first suggested by Gerola *et al.* (1966b), the vesicles originate from localized invaginations of the chloroplast limiting membrane. (2) Also visible are endoplasmic reticulum strands which, during stage C of the infection, provide an intimate connection between nucleus and chloroplasts. (3) An electron-lucent material overlies clusters of plastidial vesicles.

As well as in TYMV infection, vesicular structures are characteristically present in all tymovirus infections investigated so far with the electron microscope (Harrison and Roberts, 1970; Allen, 1972; Granett, 1973; Gibbs and Harrison, 1973; Moline and Fries, 1974; Hatta and Matthews, 1976). Several lines of evidence, provided by the studies of R. E. F. Matthews and co-workers and J. M. Bové and co-workers (summarized by Matthews, 1973) strongly indicate that chloroplast vesicles contain the enzyme–template complex for replication of TYMV–RNA. It has been suggested that RNA plus strands, once synthesized, move to the neck of the vesicles and that here they specfically combine with pentamers and hexamers of coat protein (i.e, the aforementioned electron-lucent material), whereby complete virus particles are formed (Hatta and Matthews, 1976). Just where virus-coat protein is produced remains unknown, but both the cytoplasm (Bové, 1975; Hatta and Matthews, 1976) and nucleus (Hatta and Matthews, 1976) have been suggested as possible sites of synthesis. Whatever its place of origin, viral protein appears to accumulate in the cytoplasm around chloroplasts, acting as a sort of glueing matter that causes the organelles to clump (Hatta and Matthews, 1976).

If this is so, as the available evidence seems to indicate, inclusions formed by chloroplast aggregates are a secondary product of TYMV multiplication rather than pathological structures primarily and specifically involved in virus synthesis. In fact, the individual "infected" plastid is the supporting structure for carrying through part of the events leading to virus production. However, why the severity of the modifications suffered by chloroplasts and their clumps varies with the virus strain remains to be established.

Somewhat similar though smaller aggregates of chloroplasts have occasionally been observed in barley stripe mosaic virus-infected cells (Carroll, 1970). Also in this instance the organelles were swollen and exhibited peripheral vesicles containing stranded material reminiscent of but not structurally identical to those of tymoviruses (C. R. McMullen, W. S. Gardner, and G. A. Meyers, personal communication). Furthermore, the rigid rodlike virus particles were often attached "head on" to the external surface of the chloroplast limiting membrane (Shalla, 1966; Gardner, 1967; Carroll, 1970) and filled the interplastidial area within the aggregates (Carroll, 1970). Whether these structures are comparable in function to those of tymoviruses and whether their formation is mediated by the virions surrounding the individual plastids has not been ascertained.

Accumulations of host-derived material representing specific cytopathological effects caused by viruses are exemplified by the clusters of vesicular structures associated with pea enation mosaic virus infections. These peculiar aggregates were first recorded in PEMV-invaded pea cells (De Zoeten *et al.*, 1972) and soon afterwards in the same host and in broad bean plants infected by different isolates of the same virus (Vovlas *et al.*, 1973; G. P. Martelli and M. A. Castellano, unpublished data).

The membranous vesicles originate from a distension and pinching off of the inner lamella of the nuclear envelope and are released first in the perinuclear space and then in the cytoplasm, where they may give rise to massive accumulations occupying a great deal of the cell lumen. The vesicles contain a network of DNase-sensitive fibrillar material, tentatively identified as DNA by De Zoeten *et al.* (1972). In later studies infected nuclei were shown to contain a PEMV–RNA hybridizable product of virus multiplication and PEMV–RNA polymerase (De Zoeten, 1975; and personal communication). PEMV–RNA polymerase was also found associated with cell fractions containing membranous vesicles only (G. A. De Zoeten, personal communication). It would then be tempting to speculate that the nuclear vesicles act as carriers of polymerase and viral RNA complements from the nucleus to the cytoplasm where virus assembly could take place. However, this would not be in line with previous observations (Shikata and Maramorosch, 1966) pointing to the nucleus as a site of PEMV multiplication.

A very consistent ultrastructural feature of infections by members of the tombusvirus group is the occurrence of odd-looking cytoplasmic inclusions having the size of small plastids and made up of three major components: (1) a peripheral enveloping membrane; (2) an electron-opaque finely granular or fibrillar matrix; (3) many globose to ovoid vesicles containing a network of stranded material resembling nucleic acid. The matrical matter may be very abundant and is sometimes arranged in a form resembling protein crystalloids, thus conferring a considerable electron density upon the inclusions. The vesicles line the internal boundary of the envelope and, when present, a central vacuolar area (Russo and Martelli, 1972; Martelli and Russo, 1972, 1973).

The origin of these abnormalities is still obscure, although it seems unlikely that they arise from degenerative changes of ordinary cell organelles. Participation of mitochondria and microbodies in the formation of the inclusions has been suggested (Russo and Martelli, 1972), but further findings indicate that a most probable origin of the bodies lies in the reorganization of preexisting cellular components (e.g., endoplasmic reticulum strands, or dictyosomal vesicles) into a pseudoorganellar form. The function of these inclusions is unknown, although their formation as an early event of the infection process may indicate a direct relationship to virus multiplication.

Multivesicular bodies bearing some resemblance to those just described have been reported from carrot and aster explants affected by aster yellows disease but perhaps also containing an unidentified small isometric virus (Jacoli and Ronald, 1974). These inclusions have an unknown origin and might even represent degenerated mitochondria. They are constructed of a membrane-bound aggregate of vesicles containing the usual threadlike material resembling nucleic acid. It seems plausible that such bodies have a pathological significance, but it is not known whether their origin should be attributed to the virus or the mycoplasma.

D. Clumps of Normal Cell Organelles

Peculiar associations between virions and apparently normal mitochondria causing the organelles to clump have been observed in studies of infections by viruses with rod-shaped and, at least in one instance, isometric particles.

Mitochondrial aggregates have been reported for Brazilian isolates of tobacco rattle virus (TRV), both in intact leaves (Harrison and Roberts, 1968; Kitajima and Costa, 1969) and in isolated tobacco protoplasts (Kubo *et al.*, 1975), for potato virus Y (Borges and David-Ferreira, 1968; Edwardson, 1974) and henbane mosaic virus (Kitajima and Lovisolo, 1972). With TRV the aggregation occurs because the end of the virions becomes attached to the mitochondria, thus linking several together, whereas with PYV and henbane mosaic virus the particles are apposed sideways to the mitochondrial membrane. Hence, in the aggregates induced by the last two viruses, the mitochondria are tightly packed. Interestingly, with TRV only long particles associate with the mitochondrial surface (Harrison and Roberts, 1968; Kitajima and Costa, 1969; Silberschmidt *et al.*, 1970; Kubo *et al.*, 1975). This preferential association was explained on the basis of a difference in the mean surface charge of long and short particles (Cooper and Mayo, 1972), but it is not known whether this has any biological significance. Also unclear is the significance of the small aggregates of mitochondria (probably clumped by particles of the petunia strain of broad bean wilt virus) which occur in monolayers between adjacent organelles (Hull and Plaskitt, 1973/1974).

E. Paramural Bodies and Cell Wall Modifications

Records are becoming increasingly frequent of structural modifications at the cell wall–plasmalemma interface connected with virus infections. The abnormalities originate either from membrane proliferation or from deposition of newly synthesized cell wall material that, respectively, lead to formation of multivesicular structures, which we will refer to with the general term *paramural bodies* (Marchant and Robards, 1968), and to cell wall outgrowths. Neither of these alterations seems to be exclusively associated with any one taxonomic group of plant viruses for one or the other (or both) have been encountered in infections by caulimoviruses (Kitajima and Lauritis, 1969; Conti *et al.*, 1972; Brunt and Kitajima, 1973; Lawson and Hearon, 1974), nepoviruses (Halk and McGuire, 1973; Jones *et al.*, 1973; Yang and Hamilton, 1974; Tsuchizaki, 1975), comoviruses (Kim and Fulton, 1971, 1973a, 1975; Van der Scheer and Groenewegen, 1971; Kim *et al.*, 1974), potexviruses (Allison and Shalla, 1974), tobamoviruses (Ross and Israel, 1970; Spencer and Kimmins, 1971), carlaviruses (Tu and Hiruki, 1971; Hiruki and Tu, 1972; Shlukla and Hiruki, 1975), and reovirus-like plant viruses (Gerola and Bassi, 1966) and, in addition, in oat necrotic mottle (Gill, 1974), parsnip yellow fleck (Murant *et al.*, 1973), oak ringspot (Kim and Fulton, 1973b) Brazilian eggplant mosaic (Kitajima and Costa, 1974b), barley stripe mosaic (C. R. McMullen, W. S. Gardner, and G. A. Myers, personal communication) viruses and the citrus exocortis viroid (Semancik and Vanderwoude, 1976). It is not even known for certain whether, despite the outward resemblance, their origin and functions are always the same in all virus-host combinations or vary with different pathogens.

Paramural bodies consist of variously extended accumulations of vesicles and convoluted membranes adjacent to the cell wall and separated from the ground cytoplasm by the plasmalemma which surrounds them. They may or may not be associated with cell wall outgrowths (thickenings and/or protrusions). When they are, it is thought that paramural bodies take part in the production of new cell wall material to be incorporated in the growing protrusions. If so, the vesicles could be expected to have a dictyosomal origin as Golgi apparatus is known to be specifically involved in cell wall production (Mollenhauer and Morré, 1966). However, in bean pod mottle virus infections, no apparent connection between dictyosomes and paramural bodies has been found by conventional electron microscopy [see discussion in Kim and Fulton (1973)], thus favoring the idea that the vesicles arise from plastic activity of the plasma membrane. Also with cauliflower mosaic virus, autoradiographic experiments with tritiated glucose failed to show a clear-cut relationship between Golgi bodies and vesicles but, since the label was heavily present on cell wall protrusions and associated vesicles, demonstrated that the latter indeed contain cell wall building material (Bassi *et al.*, 1974a).

For bean pod mottle virus, pramural bodies represent one of the earliest responses to virus infection. They even precede the development of cell wall protrusions and disappear, with the latter, as the infection ages (Kim and Fulton, 1973a).

An early appearance of paramural bodies, or plasmalemmasomes, is also characteristic of citrus exocortis viroid infections. These inclusions are produced by membrane proliferations and form rather compact masses situated near the cell surface. They are not connected with production of cell wall modifications but rather with viroid RNA metabolism. Semancik and Vanderwoude (1976) regard these membranous structures as likely sites of pathogenic RNA synthesis or accumulation, based on the notion that viroid RNA was found to be associated with a component of the cell membrane system (Semancik *et al.*, 1976).

Cell wall modifications seem to fall into two diverse structural forms: (1) extensive callose depositions along the preexisting wall; (2) localized finger-like projections. Callose deposits occur in local lesions induced by anisometric viruses (Ross and Israel, 1970; Spencer and Kimmins, 1971; Tu and Hiruki, 1971; Allison and Shalla, 1974), and it has been suggested by these authors that the abnormal wall thickenings of the cells surrounding the lesions act as a barrier restricting virus movement.

Finger-like wall projections that do not become lignified (Bassi *et al.*, 1974a,b) occur both in localized and systemic infections. They are associated with plasmodesmata to such an extent that, in the literature, they have often been referred to as "extended plasmodesmata." Their number and size vary tremendously according to the inducing virus. Instances have been reported in which these protrusions extend far into the cytoplasm (Murant *et al.*, 1973; Kim and Fulton, 1973a). Customarily these elongated protrusions are not "empty" but contain, more or less in a central position, plasmodesma-like channels (Bassi *et al.*, 1974a) which often harbor rows of virus particles (e.g., Gerola and Bassi, 1966; Kim and Fulton, 1971; Yang and Hamilton, 1974; Murant *et al.*, 1975) or, as in carrot mottle virus, densely staining material interpreted as possible nucleocapsids (Murant *et al.*, 1973). The significance of these abnormalities is obscure. According to some authors they may be a secondary pathological response of the cell to the tubules (Murant *et al.*, 1973), whereas for other workers they may favor cell-to-cell spread of viruses providing them with many channels of communication between adjacent cells (Bassi *et al.*, 1974a). A similar concept was put forward by Kim and Fulton (1973a), who recognized a similarity between bean pod mottle virus–infected cells with wall outgrowths and ordinary plant "transfer cells" (Gunning *et al.*, 1968; Gunning and Pate, 1969).

According to this hypothesis, in the early stages of infection, cell wall projections and related structures (paramural bodies) could be actively engaged in absorption or secretion as well as in facilitating cell-to-cell movement of materials originating from viral infection, including virus particles (Kim and Fulton, 1973a). Whatever their significance, based on present knowledge it appears unlikely that cell wall protrusions represent defense mechanisms which impair intracellular virus transport.

Another kind of inclusion, sometimes found associated with cell wall–plasmalemma disturbances, is represented by extracytoplasmic aggregates of entangled virus particles. Such virus bundles were first observed in wheat streak mosaic virus infections (Lee, 1965) and subsequently were reported with increasing frequency, in cells harboring a number of different viruses with filamentous particles (Krass and Ford, 1969; Russo and Martelli, 1969; Macovei, 1971; Barnett *et al.*, 1971; Kim and Fulton, 1973b; Allison and Shalla, 1974; Weintraub *et al.*, 1974; Horvat and Verhoyen, 1975a,b). The virus aggregates are in paramural position, surrounded by the plasma membrane which cuts them off from the ground cytoplasm. They are sometimes partially embedded in the cell wall and often are connected with plasmodesmata. It is not clear whether the peculiar localization of these viral aggregates is to be related to cell-to-cell transport of viruses or, as suggested by Allison and Shalla (1974), to a mechanism of entrapment by which intracellular virus movement is prevented.

F. Complexes of Viral Products and Host Components

The cytopathological structures that are certainly the most complex among virus-induced cytoplasmic inclusions are viral product–host component aggregates. Rather than being mere accumulations of by-products of altered cell metabolism, they seem to represent active sites of virus production.

The inclusions are usually massive and thus easily identified by light microscopy, and often lie next to the nucleus. They are found in several groups of plant viruses and have a diversified ultrastructural organization. All of them have in common the occurrence of small membranous vesicles containing fine-stranded material reminiscent of nucleic acid, which in some instances has indeed been identified as the replicative form of viral RNA.

1. *Tobamoviruses*

Some members of the tobamovirus group evoke formation of cytoplasmic amorphous inclusions for which, as discussed earlier, the name of X-bodies seems appropriate. Those inclusions induced by common strains of TMV are by far the most studied owing to the steady interest they have aroused in plant virologists ever since their discovery (Iwanowski, 1903).

X-bodies were so termed by Goldstein (1924, 1926), who studied them in detail. Afterwards, Sheffield (1931, 1934) showed that they were more stable than crystals and stained strongly for protein. She was able to isolate whole inclusions by microdissection and proved that were highly infectious, containing virus particles plus other materials of unknows origin (Sheffield, 1939, 1946). The validity of Sheffield's observations in relation to classical TMV X-bodies was recently questioned by Warmke (1969), however; by comparative light and electron microscopy, the latter author demonstrated that the large vacuolate inclusions seen under the light microscope in TMV aucuba-infected plants are, for the greatest part, angled layer virus aggregates rather than X-bodies.

Details on the fine structure of X-bodies became available with the improvement of electron microscopic techniques. Thus, Shalla (1964) found TMV-infected cells to contain heavily stained thick filaments and suggested that they were major components of the X-bodies. Areas of cytoplasm rich in filaments also harbored virus particles and accumulations of normal cell constituents like ribosomes, endoplasmic reticulum, and mitochondria. Shalla's (1964) findings relative to the presence of filamentous structures were soon confirmed by Kohlemainen *et al.* (1965) and Milne (1966a,b), who ascertained that these filaments were made out of tubules occurring singly or, more often, in pairs or triplets. Kohlemainen *et al.* (1965) suggested that these filament-rich regions could be involved in virus synthesis or assembly but did not relate them to X-bodies. On the other hand, Milne (1966a) suggested that the filaments could not be developmental forms of TMV.

In comparative studies with the light and electron microscopes, Esau and Cronshaw (1967) found that the filamentous structures, which they called X-tubules or X-components, occurred in the X-bodies. In particular, they suggested that X-tubules were a developmental modification of granular material (X-material) occurring in aggregates in the cytoplasm of infected cells, which became organized into tubular form (Esau, 1968). Esau (1968) proposed that the tubules were polymerized X-protein, an anomalous proteinaceous material biochemically identical with virus-coat protein, found in TMV-infected tissue extracts by Takahaski and Ishii (1953).

Thus, according to Esau's (1968) studies, TMV X-bodies are aggregates of host components such as ribosomes, endoplasmic reticulum and small vacuoles or vesicles, viral protein (in granular or tubular form), and small pockets of virus particles. Next to the inclusions and partly intermingled with them, there may be voluminous accumulations of virions bulging into the central vacuole (e.g., Esau, 1968).

X-bodies of TMV aucuba differ from the classical ones in that they are made up primarily of angled layer aggregates and thus have a high virus content, but they also contain areas where filaments, vesicles, and ribosomes accumulate, as in X-bodies of common TMV strains (Warmke, 1969). Additional differences in ultrastructure can be found in X-bodies induced by TMV mutants—as, for example, in strains Ni 118 (Kassanis and Milne, 1971) and PM_2 (Kassanis and Turner, 1972), where large masses of amorphous, very electron-dense material have been interpreted as accumulations of insoluble virus-coat protein. However, in neither instance was direct proof of the nature of this material obtained, and evidence for its identification as viral protein is purely circumstantial. Similar-looking osmiophilic material found abundantly in cells infected by yellow TMV strains has been interpreted, again without cytochemical evidence, as being depositions of lipids (Kohlemainen *et al.*, 1965; Liu and Boyle, 1972).

Some TMV strains, like TMV U5, do not form X-bodies. Consistent with the idea that X-protein accumulates in such bodies, X-protein could not be isolated from U5-infected tissues (Granett and Shalla, 1970). However, in the same comparative studies, the amount of X-protein extracted from tissues infected with two U1 strains, which showed considerable differences in the production of X-bodies, was not related to X-body frequency. This was taken as strong evidence against X-bodies containing X-protein as their main component (Granett and Shalla, 1970). Interestingly, Shalla (see Granett and Shalla, 1970) demonstrated by ferritin tagging that X-tubules are serologically unrelated to X-protein.

Even though the interpretation of the nature of some of the structural

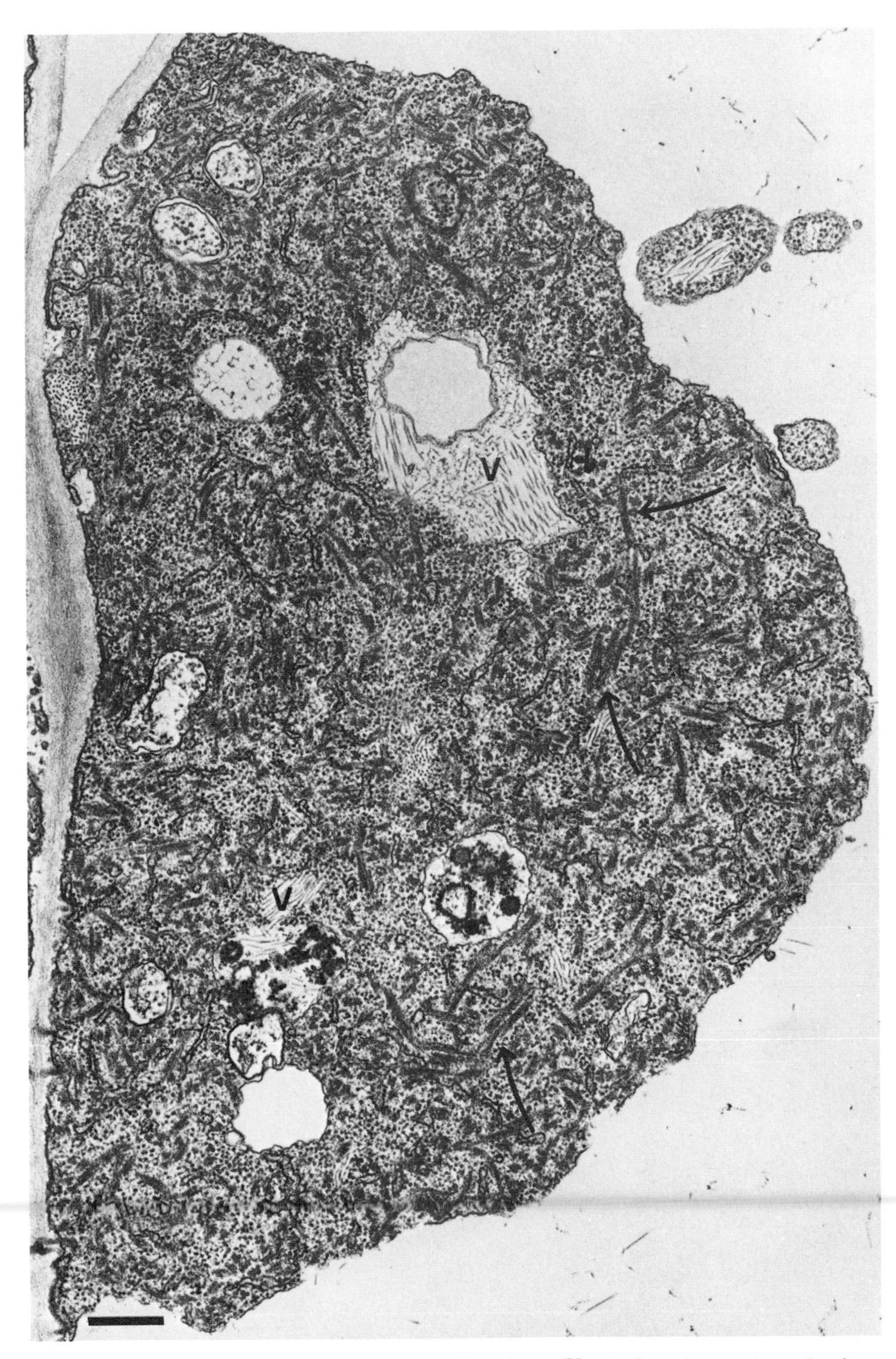

FIG. 10. An X-body of tobacco mosaic virus. X-tubules (arrows) and virus pockets (V) intermingled with host material are visible. [From Esau (1968), courtesy of the copyright holder, University of Wisconsin Press.] Magnification bar represents 500 nm.

components of TMV X-bodies is controversial or obscure, there are strong indications that these inclusions may play a role in virus synthesis. Incorporation of tritiated uridine in X-bodies has been reported by Smith and Schlegel (1965), whereas several workers (Ralph *et al.*, 1971; Zaitlin *et al.*, 1973; Nilsson-Tillgren and Bekke, 1975) have found an intriguing association of virus-induced RNA replicase, double-stranded RNA, and complementary viral RNA with membranous components. These were found by Ralph *et al.* (1971) to be similar to the membranous vesicles of X-bodies, and therefore these inclusions were suggested as being a possible site of TMV–RNA replication.

It has also been proposed that X-bodies are regions of virus-directed protein synthesis (Esau, 1968), but the results of other studies seem to point at the nucleus as a site for virus protein production (Shalla and Amici, 1967; Langenberg and Schlegel, 1967, 1969; Reddi, 1972).

2. *Potyviruses*

In addition to proteinaceous pinwheel inclusions, infections by most if not all members of the potyvirus group are accompanied by the production of complex amorphous inclusions in the cytoplasm. These can be seen with the light microscope, although with difficulty, and appear as granular bodies of variable size often next to the nucleus (e.g., Kunkel, 1922; Sheffield, 1934; McWhorter, 1940; Bawden and Sheffield, 1944; Berkeley and Weintraub, 1952; Hollings, 1957; Hollings and Nariani, 1965; Štefanac and Miličić, 1965; Schmelzer and Miličić, 1966; Purcifull and Shepard, 1967; Bos, 1969; Bos and Rubio-Huertos, 1969; Kamei *et al.*, 1969b; Zettler, 1969). These inclusions are likely not to withstand exposure to Triton X-100, and ought not to be confused with the fibrous, needle-like, or star-shaped bodies which are prominent in detergent-treated epidermal strips and represent light microscopic views of pinwheels [for review, see Edwardson (1974)].

Excellent electron microscopic illustrations of these altered cytoplasmic areas are reported in Edwardson's (1974) review of potyviruses. Ultrastructurally, they appear as aggregates of normal cell constituents, such as ribosomes and endoplasmic reticulum strands, together with membranous vesicles, pinwheels, and often virus particles in bundles or in paracrystalline arrangement. The inclusions are sometimes surrounded by mitochondria but are never bounded by limiting membranes. The vesicular structures may be either electron-lucent or may contain materials with various degrees of electron density or display a network of tiny fibrils recalling nucleic acid strands. As to their origin, the vesicles seem to develop either from dictyosomes (Kim and Fulton, 1969; Russo and Martelli, 1969; Camargo *et al.*, 1971; Martelli and Russo, 1976a) or from the endoplasmic reticulum (Martelli *et al.*, 1969; Krass and Ford, 1969; Langenberg and Schroeder, 1974b). Whatever their source, they appear in the early stages of infection and may represent, as in the case of watermelon mosaic virus (Martelli and Russo, 1976a), one of the very first cytological signs of infection.

The significance of these inclusion bodies is not known, but should the vesicles prove to contain viral RNA, as suggested by Krass and Ford (1969), their implication in virus multiplication would become obvious.

A noteworthy exception to the foregoing pattern are the inclusion bodies reported by Fujisawa *et al.* (1967b) associated with sugar beet mosaic infections. These structures are unlike any of those known for members of the potyvirus group studied so far at the ultrastructural level, as they are very compact, denser than the nucleus, and entirely made up of vesicular structures. Certainly, nothing of the kind was observed elsewhere in cells infected by beet mosaic virus (Hoefert, 1969; Russo and Martelli, 1969; Martelli and Russo, 1969a,b). These discrepancies, and the fact that Fujisawa *et al.* (1967b) did not report observing pinwheel inclusions in their material, make questionable the identification of beet mosaic virus as the inducer of the abnormalities they described.

Secondary vacuolation is another characteristic feature of potyvirus infections that is often associated with cytoplasmic inclusions. This plastic activity of intracellular membranes induces a peculiar type of inclusion, consisting of monolayers of virus particles in parallel array, restrained within narrow cytoplasmic strands delimited by two membranes. These membranous virus-containing structures may bridge secondary vacuoles or extend in a convoluted form into the central vacuole (Hoefert, 1969; Kim and Fulton, 1969; Russo and Martelli, 1969; Martelli *et al.*, 1969; Weintraub and Ragetli, 1970c; Lawson *et al.*, 1971; Hampton *et al.*, 1973; Kitajima and Costa, 1974a; Horvat and Verhoyen, 1975b; Martelli and Russo, 1976a). Their biological significance is not clear, but it has been suggested (Russo and Martelli, 1969; Kitajima and Lovisolo, 1972; Horvat and Verhoyen, 1975b) that their formation may be part of a defense reaction which serves to eject particles by moving them through the cytoplasm toward the vacuole.

3. Closteroviruses

Amorphous cytoplasmic inclusions have been reported for beet yellows virus (Esau, 1968; Esau and Hoefert, 1971a). They form conspicuous masses visible with the light microcope, under which they appear rather compact with occasional alveolation. Ultrastructurally, the inclusions appear as accumulations of membranuos vesicles containing the usual network of fibrillar material, virus particles in various amounts (but increasing with the age of infection), ribosomes, and flaky electron-opaque material of unknown nature. The source of the vesicles is also unknown, but it was suggested that they originate *de novo* as receptacles of possible viral nucleic acid (Esau and Hoefert, 1971a). The developmental changes in the inclusions also suggest that virus synthesis is initiated in the vesicle aggregates, which would be formed first, while the increasing amount of virus later is converted into large fibrous or banded viral aggregates (Esau and Hoefert, 1971a).

Virtually identical cytoplasmic accumulations of membranous vesicles and virus particles have been reported for two additional members of the closterovirus group: beet yellow stunt virus (Hoefert *et al.*, 1970) and carnation necrotic fleck virus (Inouye and Mitsuhata, 1973). Beet yellow stunt virus infections, however, differ from those of beet yellows virus in that, instead of the flaky component, aggregates of granular material reminiscent of "viroplasm" are found interspersed with groups of virus particles.

Strikingly similar masses of fibril-containing vesicles and presumed virus particles have also been found associated with little-cherry–diseased sweet cherry plants (Raine *et al.*, 1975). Although it is not known whether the causal agent of little-cherry disease is a member of the closteroviruses, the similarities in outward aspect and localization (sieve tubes) of its cytopathological structures suggest an affinity with viruses of that group.

4. Other Anisometric Viruses

Some members of the carlavirus group induce amorphous cytoplasmic inclusions detectable with the light microscope after proper staining (e.g., Bos and Rubio-Huertos, 1972) but difficult to recognize with certainty at the ultrastructural level for lack of specific characterizing features.

Among viruses with anisometric particles that are not included in any of the above groups, soil-borne wheat mosaic virus (SBWMW) induces very interesting inclusions. These were first recorded by McKinney *et al.* (1923) and were later partially characterized cytochemically (Wada and Fukano, 1937). More recently, information on their ultrastructure has become available (Peterson, 1970; Tsuchizaki *et al.*, 1973; Hibino *et al.*, 1974a,b).

In a study of nine isolates of SBWMW, Hibino *et al.* (1974a) recognized three major groups of inclusions. Two of these consisted of accumulations of membranes in the form of more or less dilated endoplasmic reticulum cisternae, tangled tubules, vesicles, and concentrically arrayed membranes. Virions were also present, loosely scattered or in paracrystalline configurations, within or surrounding the inclusions. Inclusions of the third group were smaller and composed of scattered or tangled endoplasmic reticulum strands and ribosomes. They certainly lacked the massive membranous structures typical of the other two types, but virus particles were found at the periphery. The composition of these abnormalities changed as the infection aged, so that the masses of tangled tubules and endoplasmic reticulum found in younger leaves were no longer seen in older leaves, where myelin-like membranes predominated (Hibino *et al.*, 1974a). As reported by Hibino *et al.* (1974b) in a paper describing the sequential development of the inclusion bodies in rye leaves, membranous tubules and vesicles were the first to appear in infected cells, soon followed by concentric membranes, and finally by virus rods either associated with the myelin-like membranous structures or in bundles or paracrystalline aggregates in the cytoplasm and vacuole. The appearance of virus particles first in the inclusions was taken as an indication that such structures, and their membranous components in particular, may be sites for virus synthesis (Hibino *et al.*, 1974a,b). Hybridization experiments with two markedly different SBWMV strains have demonstrated that the inclusion type is a strain-dependent characteristic and that formation of such bodies is coded by the smaller RNA molecule present in the short particles (Tsuchizaki *et al.*, 1975). This seems to represent the first direct evidence confirming the long-standing concept that production of inclusion bodies is an expression of the viral genome rather than a mere reaction of the host to infection.

5. Nepo- and Comoviruses

Cytoplasmic inclusion bodies are a consistent ultrastructural feature of nepovirus infections. The inclusion bodies are sometimes as large as the nucleus next to which they lie. They appear in the early stages of infection and contain ribosomes, endoplasmic reticulum elements, and membranous vesicles with fine fibrils. Such inclusions have been found in cells infected with arabis mosaic (Gerola *et al.*, 1966a), cherry leafroll (Jones *et al.*, 1973), grapevine fanleaf (Gerola *et al.*, 1969b; Šarić and Wrischer, 1975), raspberry ringspot (Harrison *et al.*, 1974), and strawberry latent ringspot (Roberts and Harrison, 1970) viruses. Those of the last virus contain masses of hollow structures resembling empty viral capsids and tubular inclusions with virus particles at the periphery (Roberts and Harrison, 1970). Recent findings seem to indicate that these cellular modifications are not strain specific; at least with raspberry ringspot virus they were basically the same for different strains and their hybrids (Harrison *et al.*, 1974).

The function of these bodies is not clear, but their early appearance, before that of virus particles, and the presence of possible viral material within them, suggest that they may play a role in virus synthesis and/or assembly.

Similar cytological abnormalities have been found in plants infected with different members of the comovirus group (Honda and Matsui, 1972; Hooper *et al.*, 1972; Kim and Fulton, 1971, 1973a; Štefanac and Ljubešić, 1971; Van der Scheer and Groenewegen, 1971; De Zoeten *et al.*, 1974; Langenberg and Schroeder, 1975). As seen with the light microscope, the inclusions develop in the cytoplasm as several small granular bodies that soon aggregate into a single large entity, which, in the end, may exceed the size of the nucleus (Štefanac and Ljubešić, 1971).

The inclusions evoked by cowpea mosaic virus are perhaps the most studied at the utrastructural level. These consist of large areas of cytoplasm, often surrounded by mitochondria, in which normal cell constituents such as ribosomes and endoplasmic reticulum strands are intermingled with osmiophilic droplets, amorphous material with a finely granular texture containing a network of fibrils and virus particles. The origin of the vesicles is unknown, although there is the suggestion that they may derive from dictyosomes (Van der Scheer and Groenewegen, 1971) or from the endoplasmic reticulum (Langenberg and Schroeder, 1975). Similar vesicles associated with bean pod mottle virus infections were thought to originate from endoplasmic reticulum (Kim and Fulton, 1972).

Although variously named [i.e., "cytopathological structures" (De Zoeten *et al.*, 1974) or "viroplasms" (Langenberg and Schroeder, 1975)], the inclusions induced by different comoviruses have essentially the same constitution and aspect. Autoradiography of virus-infected cells labeled with tritiated uridine and biochemical studies following fractionation of CoMV-induced inclusions have provided good evidence that the replicative form of CoMV–RNA is associated with the membranous inclusions, and probably with their vesicular structures, which may therefore represent the site for viral RNA synthesis (De Zoeten *et al.*, 1974).

Virus-containing tubules are a consistent feature of nepo- and comovirus-infected cells and, as already mentioned, may be associated with cytoplasmic inclusions. When seen with the electron microscope as "squash homogenates" in negative-stain mounts, the tubules always appear single walled (Walkey and Webb, 1968, 1970), whereas in thin sections they are often double walled. The outer sheet is continuous with the endoplasmic reticulum (Roberts and Harrison, 1970) or with the plasma membrane (Kim and Fulton, 1971, 1973a), or derives from myelinic extracytoplasmic bodies thought to contain phospholipids and proteins (Kim *et al.*, 1974). Hence, it seems highly probable that the external envelope of the tubules has a lipoprotein composition.

The nature of the internal tubule, which tightly envelops virus particles, is less clear. It has been suggested recently that the virus-containing structures are normal cytoplasmic microtubules. The supporting micrographic evidence is intriguing (Kim and Fulton, 1975), although it remains to be seen whether microtubules are pliable enough to accommodate virus particles larger than their diameter. In this connection, Kim and Fulton (1975) suggest that virus particles do not migrate as such into microtubules or desmotubules, where they are also found, but enter these structures as protein and nucleic acid to be assembled there.

As to the function of the tubular inclusions, two hypotheses have been put forward: (1) they may favor intercellular spread of virions by mediating their cell-to-cell transport through plasmodesmata (Davison, 1969; Roberts and Harrison, 1970; Walkey and Webb, 1970); (2) they may be sites where virus nucleic acid is synthesized and assembled into virus particles, or where only virus assembly takes place (Roberts and Harrison, 1970; Murant *et al.*, 1973; Kim and Fulton, 1975). No biochemical evidence supporting the latter assumption is yet available.

6. *Bromoviruses*

Broad bean mottle virus is the only member of the bromovirus group known to induce formation of inclusion bodies whose structure and function have been studied in some detail. The inclusions are cytoplasmic, and they have a granular appearance when young and small but become vacuolated as the infection ages, enlarging to become several times the size of the nucleus (Rubio-Huertos and Van Slogteren, 1956). Ultrastructurally (Lastra and Schlegel, 1975), they consist of accumulations of intensely electron-opaque amorphous material, membranous vesicles, and virus-like particles, sometimes associated with mitochondria and endoplasmic reticulum. The initial stages of the inclusions are vesicles clustering around the nucleus. Significantly, the inclusion bodies incorporate tritiated uridine (De Zoeten and Schlegel, 1967b) and [^{125}I]-immunoglobin G (IgG) virus-specific antibodies (Lastra and Schlegel, 1975). Thus, they contain both viral nucleic acid and coat protein and may represent the site for RNA replication and particle assembly. In fact, Lastra and Schlegel's (1975) studies suggest that viral protein synthesis takes place in the nucleus and the final product accumulates in the cytoplasm, where it combines with the RNA moiety to produce complete virions.

Accumulations of elongated and vesicular membranous elements apparently connected with cowpea chlorotic virus formation have been observed in infected tobacco protoplasts (Motoyoshi *et al.*, 1973).

7. *Other Isometric Viruses*

A number of other isometric viruses not classified in the foregoing sections can form amorphous cytoplasmic inclusions of various types. Very few of them, however, have been investigated in detail, so that the following are examples selected to illustrate some of the best-known cases.

All broad bean wilt virus strains investigated so far produce highly complex inclusion bodies which may exhibit, according to the strain, an amorphous and a crystalline portion (Rubio-Huertos, 1962, 1968; Frowd and Tomlinson, 1972; Sahambi *et al.*, 1973; Russo *et al.*, 1973; Hull and Plaskitt, 1974; Miličić *et al.*, 1974; Russo and Martelli, 1975b; Weidemann *et al.*, 1975). The inclusions are large, lie next to the nucleus, and are often surrounded by mitochondria and dictyosomes along with rough and smooth endoplasmic reticulum elements. Internally two major areas, i.e., one membranous and one granular, can be visualized. The membranous area is made up of convoluted membranes and tubules plus a great number of small vesicles, small ring-shaped particles, and membrane-bounded enclaves. Osmophilic droplets are often present, but ordinarily cell organelles are rarely seen except for ribosomes and endoplasmic reticulum. Many of the vesicles contain fibrils similar to nucleic acid strands. The small ring-shaped particles are empty viral capsids which may be scattered or form discrete aggregates, sometimes with a circular profile. The granular area has a viroplasm-like aspect and consists of accumulations of electron-opaque bodies about 50 nm across with a denser central core about 18 nm in diameter. Membranous strands are interspersed and often connected these bodies. Membranous and granular areas are not sharply separated and merge with one another.

Crystalline arrays of virus particles in a tubular configuration, when present, are located at the periphery of the inclusion. Occasionally, sheet-like virus crystals may be present. The inclusion bodies contain lipids, proteins (which constitute the bulk of the body), and RNA but not DNA (Russo and Martelli, 1975b). The membranous vesicles seem to derive from the endoplasmic reticulum and, possibly from dictyosomes (Rubio-Huertos, 1968).

Undoubtedly, the inclusions are a primary source of intracellular virus and of noninfective viral capsids (top component), but whether they

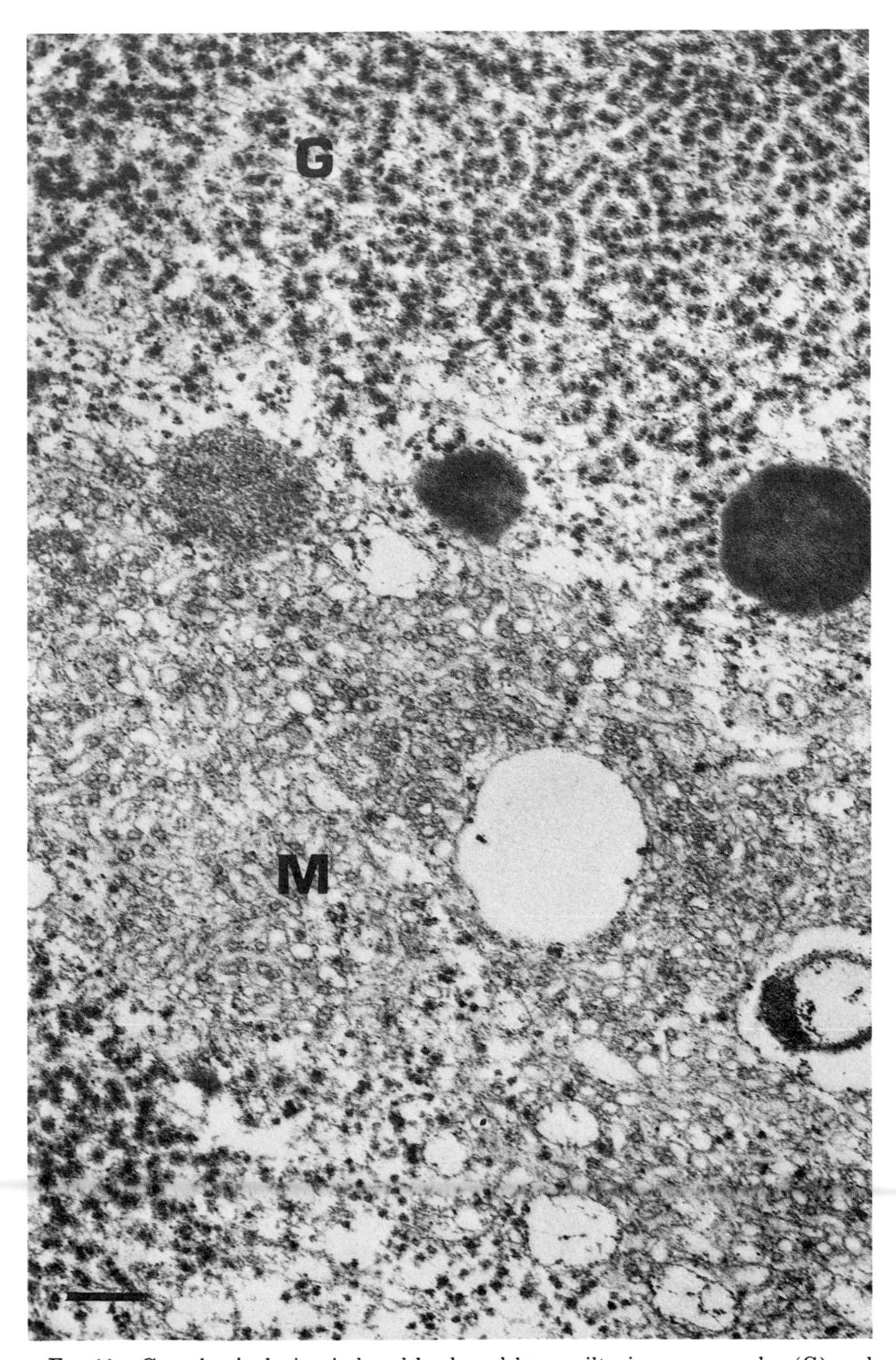

FIG. 11. Complex inclusion induced by broad bean wilt virus: a granular (G) and membranous (M) zone are clearly discernible. The granular zone may be a viroplasmic area. [From Russo and Martelli (1975b), courtesy of the copyright holder, *Journal of Submicroscopic Cytology*.] Magnification bar represents 250 nm.

represent sites of particle accumulation only is questionable. Their early appearance before virus particles are seen, and the presence of possible viral material within them, suggest that the inclusions have some bearing in virus synthesis and/or assembly (Sahambi *et al.*, 1973; Russo and Martelli, 1975b).

Inclusion bodies very similar to those of broad bean wilt virus have been recorded in cells infected by Brazilian eggplant mosaic virus (Kitajima and Costa, 1974b), a possible member of the comovirus group. The inclusions consist of membranous vesicles, tubular structures in a loose array, and osmiophilic droplets sometimes interspersed with normal cell components such as mitochondria, peroxisome-like bodies, and dictyosomes. Electron-dense granular areas are also encountered, made up of virus particles embedded in an amorphous matrix of medium density. The membranous complex seems to derive from endoplasmic reticulum, and some of the vesicular components contain fine threads. The inclusions follow this developmental sequence: in the early stages of infection they are recognizable as aggregates of convoluted tubules; these later give way to clusters of vesicles, flattened cisternae, and aggregates of virus particles, which finally occupy the whole area, while membranous material and other cell components disappear (Kitajima and Costa, 1974b).

The bulky cytopathological structures associated with parsnip yellow fleck virus infections are a further example of complex amorphous inclusions caused by small isometric viruses. Murant *et al.* (1975) described them as built up of accumulations of ribosome-studded vesicles originating, presumably, from endoplasmic reticulum and disappearing with time, clusters of smaller, less electron-opaque and smooth vesicles, straight tubules occurring singly or in groups of several elements, and densely staining granular material bounded by a single membrane. Mitochondria and Golgi bodies occur commonly at the periphery of the inclusions, which, although lacking a bounding membrane, are quite well separated from the surrounding cytoplasm.

The similarity in gross morphology and composition between inclusion bodies of the aforementioned viruses leaves little doubt that they are comparable structures with similar functions, which are likely to be related to virus multiplication.

G. Viroplasm Matrices

Viroplasms are cytoplasmic aggregates of finely textured, electron-dense material in which viral products are formed and/or virus assembly takes place. Therefore, strictly speaking, some of the membranous inclusions discussed earlier are also to be regarded as true viroplasms, their role as virus factories having been experimentally established. A separation between the two types of structures would, therefore, appear unjustified were it not for the differences in their general aspect and composition that are immediately perceivable at the fine-structure level. In fact, viroplasms constitute compact bodies which, especially in the advanced stages of infection, lack the membranous background characterizing other cytoplasmic inclusions.

"Classical" viroplasm matrices are consistently associated with infections by all members of the reolike virus and caulimovirus groups, but are also found in connection with several other widely different viruses.

1. Reolike Viruses

Reovirus-like plant viruses (or reolike viruses, for short) are not a homogeneous group. On the basis of RNA segmentation, vectors and capsid structure they can be subdivided into two subgroups represented by wound tumor and rice dwarf viruses on one side and, on the other, by maize rough dwarf and allied entities, or "acanthoviruses," as proposed by Milne and Lovisolo (1977). Despite these biological and structural differences, intracellularly they all seem to behave in the same way, eliciting comparable responses which have recently been reviewed (Milne and Lovisolo, 1977). The most striking intracellular feature of these viruses is, of course, the production of cytoplasmic inclusions which were first described in connection with Fiji disease of sugarcane, and considered to be protozoan, by Lyon (1921), McWhorter (1922) and Kunkel (1924). The inclusions of wound tumor virus were recorded and partially characterized chemically (presence of viral protein) in more recent times (Black, 1947; Littau and Black, 1952; Nagaraj and Black, 1961), and similar results were also obtained with light microscopic studies on rice dwarf [summarized by Iida *et al.* (1972)] and rice black-streaked dwarf (Kashiwagi, 1966) viruses.

Intensive investigations carried out in the last few years with the electron microscope have secured a conspicuous body of information on the ultrastructure and composition of these inclusions as well as on their function. In particular, inclusions of the following viruses have been studied electron microscopically: wound tumor (Shikata and Maramorosch, 1965, 1967a,b; Maramorosch *et al.*, 1968), rice dwarf (Shikata, 1966), rice black-streaked dwarf (Nasu, 1965; Shikata, 1969), maize rough dwarf (Gerola and Bassi, 1966; Gerola *et al.*, 1966c,d; Lovisolo and Conti, 1966; Kislev *et al.*, 1968; Vidano, 1966, 1970), oat sterile dwarf (Brčák *et al.*, 1972), cereal tillering disease (Lindsten *et al.*, 1973), *Arrhenatherum* blue dwarf (Milne *et al.*, 1974), *Lolium* enation dwarf (Lesemann and Huth, 1975), Fiji disease (Giannotti *et al.*, 1968; Teakle and Steindl, 1969; Francki and Grivell, 1972), and pangola stunt (Schank

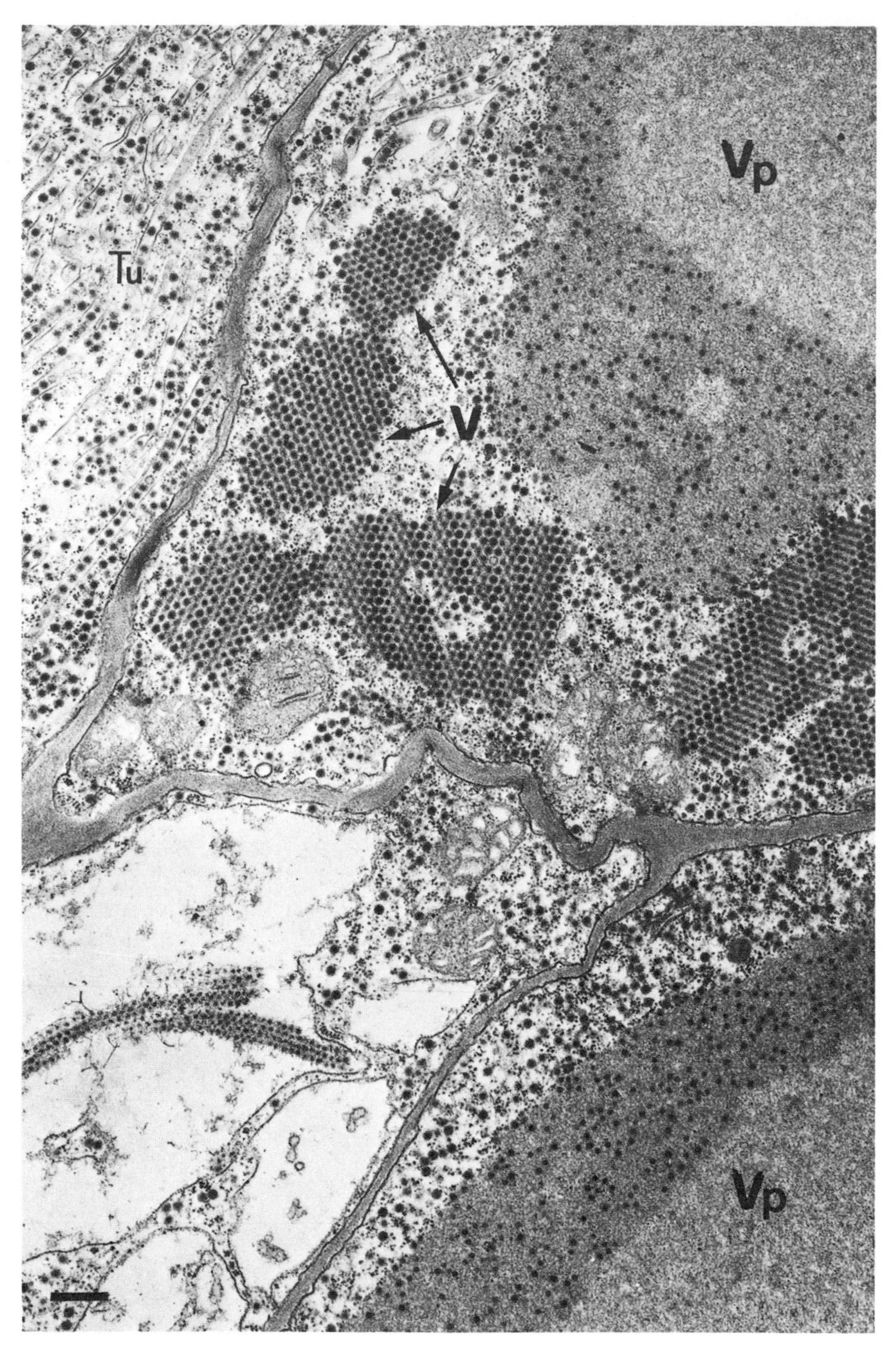

Fig. 12. Cells infected by maize rough dwarf virus, showing part of two large viroplasms (Vp) and mature virions in crystalline array (V) and within tubular structures (Tu). (Courtesy of O. Lovisolo). Magnification bar represents 250 nm.

et al., 1972; Kitajima and Costa, 1970). The inclusions all look alike, and do not differ basically from those of maize rough dwarf, which are by far the best known.

Viroplasms of maize rough dwarf virus are intensely electron-opaque aggregates of granular or finely fibrillar material mostly consisting of virus-coat protein (Bassi and Favali, 1972). They contain numerous dark, spherical bodies about 50 nm in diameter which, on a morphological basis, were formerly thought to be immature virus particles (Gerola and Bassi, 1966) and, later, were experimentally proven to consist of naked or incompletely coated RNA (Bassi and Favali, 1972).

Viroplasmic masses rich in immature virus particles incorporate tritiated uridine at a very high rate after 3-hour incubation and even incorporate it, though to a lesser extent, after a much shorter incubation period (20 minutes). This has been taken as evidence that viral RNA replicates in the viroplasm, where it becomes coated to form complete particles (Bassi and Favali, 1972; Bassi *et al.*, 1974c). Nuclei are relatively unaffected, are not labeled, and do not seem directly involved in virus synthesis. Data supporting the concept that these inclusions are virus factories have recently been obtained for Fiji disease virus, whose viroplasms were ascertained, by fluorescent microscopy, to contain both single- and double-stranded nucleic acid, presumably RNA (T. Hatta, personal communication). Within 48 hours mature virus particles move away from the viroplasm matrix into the surrounding cytoplasm (Bassi *et al.*, 1974c), where they remain scattered at random, or aggregate in crystals, or become aligned along or inside tubular structures.

Virus-containing tubules and virus crystals are two cytological features accompanying infections by all reolike viruses. Usually, these structures lie next to viroplasms. The tubules are about 100 nm in diameter, and in negative staining they have an overall aspect similar to that shown by the particle-containing tubules of smaller isometric viruses (Conti and Lovisolo, 1971). They have a wall of regularly repeating subunits about 4 nm in diameter (Lesemann, 1972) and are proteinaceous in nature (Bassi and Favali, 1972). As pointed out by Milne and Lovisolo (1977), the function of these tubules is unknown, and it also remains to be determined whether they are a gene product of the virus or the cell, and whether they differ chemically from one virus to another.

2. Caulimoviruses

Highly characteristic cytoplasmic inclusions are associated with infection by all recognized members of the caulimovirus group: cauliflower mosaic (Fujisawa *et al.*, 1967c; Rubio-Huertos *et al.*, 1968b; Martelli and Castellano, 1971), dahlia mosaic (Petzold, 1968; Kitajima *et al.*, 1969), carnation etched ring (Rubio-Huertos *et al.*, 1972; Lawson and Hearon, 1974), cole yellow vein banding (Kitajima *et al.*, 1965), strawberry vein banding (Kitajima *et al.*, 1973), and *Mirabilis* mosaic (Brunt and Kitajima, 1973). Perhaps the first ever to be recorded with the light microscope were those produced by dahlia mosaic virus (Goldstein, 1927). Their occurrence is consistent enough, even in symptomlessly infected plants, to be used as a valuable and reliable method for virus diagnosis (Robb, 1963). Additional light microscope studies on dahlia and cauliflower mosaic viruses have secured further information on the gross morphology and chemical constitution of the inclusions (Robb, 1964; Mamula and Miličić, 1968; Martelli and Castellano, 1971).

With the electron microscope they appear as rounded or elongated structures composed of virus particles embedded in a finely granular electron-dense matrix, broken by irregular, electron-lucent areas which may contain virions. Excellent micrographs illustrating ultrastructural details of dahlia mosaic virus viroplasms have been produced by Kitajima *et al.* (1969). Incorporation of tritiated thymidine, cytochemical staining reactions, and enzyme digestion tests have proven that the inclusions have protein, RNA, and DNA as major constitutents (Kamei *et al.*, 1969c; Martelli and Castellano, 1971; Conti *et al.*, 1972).

Golgi bodies are thought to be involved in the formation of inclusions of several caulimoviruses. The development of these structures for dahlia mosaic virus was described in detail by Petzold (1968), who postulated that they begin as accumulations of ribosomes that form small electron-opaque units, fuse together with strands of endoplasmic reticulum, and induce larger bodies wherein virus particles are produced.

Similar findings were reported by Kitajima *et al.* (1969), who emphasized the importance of ribosomes and dictyosomes in the possible genesis of the inclusions (see also Brunt and Kitajima, 1973; Lawson and Hearon, 1974). For cauliflower mosaic virus, a similar developmental sequence was observed, although without apparent involvement of Golgi bodies (Martelli and Castellano, 1971). Viroplasmic masses begin as small aggregates of electron-dense proteinaceous material surrounded by ribosomes. At this stage virus particles are few. As the infection ages, the inclusions become larger and the number of virus particles increases.

Although there is general agreement on the gross structure and constitution of viroplasms, the interpretation of their origin and significance is somewhat controversial. For instance, Rubio-Huertos *et al.* (1968b) for cauliflower mosaic virus and Rubio-Huertos *et al.* (1972) for carnation etched ring virus have proposed that the inclusion bodies represent only sites of particle accumulation, i.e., virus particles develop inde-

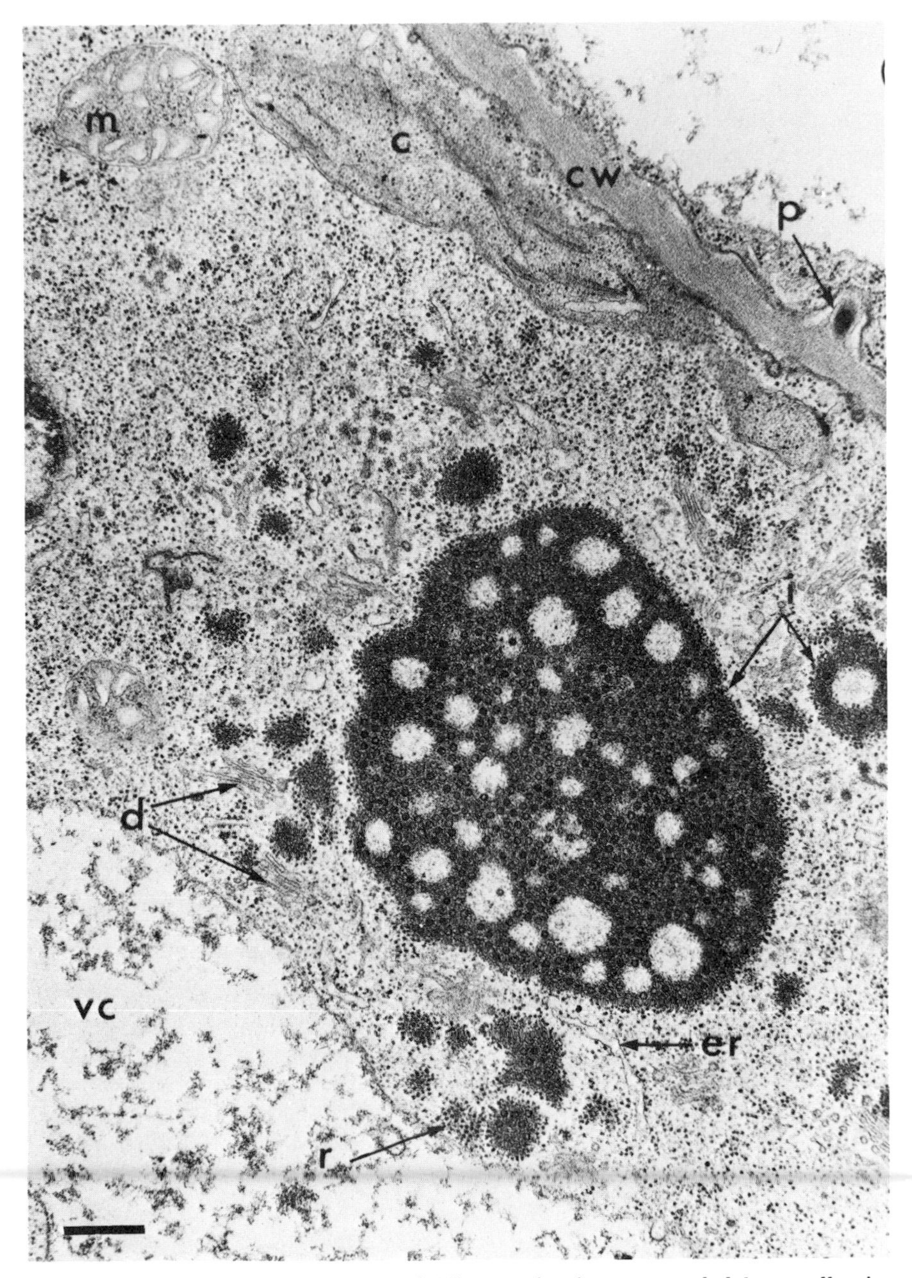

FIG. 13. A viroplasm induced by dahlia mosaic virus surrounded by smaller inclusions (i) and dictyosomes (d). Ribosomes (r) are associated with the surface of the inclusions; m, mitochondrion; c, chloroplast; cw, cell wall; p, plasmodesma; er, endoplasmic reticulum; vc, vacuole. [From Kitajima *et al.* (1969), courtesy of the copyright holder, Academic Press.] Magnification bar represents 500 nm.

pendently in the cytoplasm (cauliflower mosaic virus) or in the nucleus (carnation etched ring virus) and then assemble with the dark proteinaceous matrix to form the inclusion. The current trend, however, is to consider these inclusions as sites of virus synthesis and/or assembly. This has been postulated on the basis of circumstantial evidence for dahlia mosaic (Petzold, 1968; Kitajima *et al.*, 1969), cauliflower mosaic (Martelli and Castellano, 1971; Conti *et al.*, 1972), *Mirabilis* mosaic (Brunt and Kitajima, 1973), and carnation etched ring (Lawson and Hearon, 1974) viruses. Recent tritiated thymidine incorporation experiments (Favali *et al.*, 1973) have produced evidence that in the early stages of cauliflower mosaic virus replication DNA is synthesized in the nucleus, from which it is quickly transferred to the inclusion body, where its replication continues. Hence, both the nucleus and viroplasm are involved in viral DNA synthesis. Carnation etched ring virus represents an exception, as virus multiplication leading to formation of whole particles seems to take place both in the viroplasm and the nucleus. This organelle harbors a network of membranes which may be consequential in the synthesis of virus-coat protein (Lawson and Hearon, 1974).

Inclusion bodies rather similar to those of caulimoviruses have been reported for petunia vein clearing virus, which has isometric particles about 43 nm in diameter (Lesemann and Casper, 1973). However, a major difference is that the inclusions usually do not contain virus particles; these are scattered randomly in the cytoplasm or nucleus, or they are aligned along endoplasmic reticulum cisternae disposed in concentric layers, so as to give rise to a most unusual configuration. The origin, composition, and function of these bodies is not known, and likewise the inclusion of the causal virus in the caulimovirus group is not certain (Lesemann and Casper, 1973).

3. *Rhabdoviruses*

Besides *Gomphrena* virus, in which the greatly enlarged and vacuolated nucleolus containing, and probably manufacturing, viral nucleocapsids (Kitajima and Costa, 1966) may be regarded as a sort of viroplasm itself, in plant rhabdoviruses, viroplasm-like structures have been encountered in connection with infections by lettuce necrotic yellows (Wolanski and Chambers, 1971), barley yellow striate mosaic (Conti and Appiano, 1973), and melon variegation (Rubio-Huertos and Peña-Iglesias, 1973) viruses, plus two rhabdoviruses associated with diseases of *Colocasia esculenta* (James *et al.*, 1973) and orchids (Lawson and Ali, 1975). In lettuce necrotic yellows, presumptive viroplasms are small, dense, amorphous bodies with a variable texture, located at the cell periphery and arising early during the cytoplasmic phase of virus multiplication. Occasionally virus particles appear to be emerging from the inclusions, and rarely uncoated particles are seen within them (Wolanski and Chambers, 1971).

The cytoplasmic inclusions of barley yellow striate mosaic virus are more convincing as possible viroplasms. These are unique structures among plant rhabdoviruses, composed of massive accumulations of electron-dense material with a granular or finely fibrous texture. Virus particles are always found in abundance at their periphery, where they mature by acquiring the outer envelope from membranes deriving either from the endoplasmic reticulum or from the outer lamella of the nuclear envelope (Conti and Appiano, 1973). Similarly, naked nucleocapsids are found at the periphery of viroplasmic masses induced by the orchid rhabdovirus (Lawson and Ali, 1975), but unequivocal examples of membrane acquisition at those sites have not been demonstrated.

The identification of cytoplasmic inclusions evoked by lettuce necrotic yellows, barley yellow striate mosaic, and the orchid rhabdovirus as possible virus factories seems plausible, although based on circumstantial evidence only. A better characterization of inclusions found in melon and *C. esculenta* would be desirable.

4. Other Viruses

Several widely different viruses can occur in infected cells within or around inclusions that are often referred to as viroplasms. In general, little is known about these structures, for only short and poorly informative descriptions are available. Thus, for instance, rounded bodies with medium electron opacity and an apparently amorphous or finely granular texture have been found to contain particles of the presumed causal agent of the "waika" disease of rice (Doi *et al.*, 1975). The outward aspect of these bodies recalls that of the "pseudonucleolar regions" of sowbane mosaic virus–infected *C. amaranticolor* cells, consisting of a granular matrix interspersed with virus particles (Milne, 1967a). Likewise, clusters of rice tungro virions have been seen embedded in somewhat similar inclusions composed of filamentous or finely granular material (Favali *et al.*, 1975). Moreover, densely staining matrices with a granular texture have been found to contain particles of a rice strain of barley yellow dwarf virus (Amici *et al.*, 1975) and of the presumed agent of Eastern wheat striate disease (Nagaich and Sinha, 1974).

In no case has the origin, chemical composition, and significance of these intracellular aggregates of electron-opaque material and virus particles been ascertained. However, naming them viroplasms, as some authors did, suggests a possible implication in virus production.

Viroplasmic inclusions of tomato spotted wilt virus are certainly better known than any of the above, owing to the attention they have been paid in recent times (Martin, 1964; Kitajima, 1965; Milne, 1970; Francki and Grivell, 1970; Ie, 1971). The viroplasm proper, a dark staining material with a granular texture and striated dense spots, is part of a more complex inclusion whose components are membranous elements and vesicles, lipid droplets and clusters of virus particles. As reconstructed by Milne (1970), the sequence of events leading to virus production suggests an early appearance of the electron-dense viroplasmic matter, soon followed by the development of parallel paired membranes that bud to form enveloped particles. These incorporate viroplasmic material both in the center and between the two membranes. Then, the outermost envelopes of a group of particles fuse, giving rise to a cisterna into which singly enveloped mature virions are released. The source of the collective membranes enclosing groups of particles is somewhat controversial. Milne (1970) was not able to establish their origin in Golgi bodies, endoplasmic reticulum, or other usual cell membranes, thus implying a *de novo* synthesis. Other workers (Kitajima, 1965; Francki and Grivell, 1970; Ie, 1971), instead, reckon that such membranes derive from endoplasmic reticulum—the main argument in favor of this hypothesis being that they are often studded with ribosomes on the cytoplasmic side. According to Ie (1971), the enveloping of virus particles within endoplasmic reticulum represents a defense reaction of the host plant.

As for the viroplasm, the consensus is that it is directly involved in virus production. Milne (1970) postulated that the dark-staining matter is made up of ribonucleoprotein, and a partial experimental confirmation of his suggestion was provided by Ie (1971), who demonstrated occurrence of protein in the viroplasmic matrix.

VI. Summary and Concluding Remarks

In the last few years intracellular inclusions induced by plant viruses have been paid ever-increasing attention. The studies, which lately have been dominated by conventional electron microscopy, have provided an adequate body of information on the architecture and fine-structural organization of a number of such inclusions. Advances have also been made in our knowledge of their origin, chemical constitution, and role in the infection process, but in these fields the unknown still far exceeds the established facts. The available information on the relationship between source, structure, and function of these pathological entities is fragmentary and allows for little, if any, generalization. Therefore, an attempt to review inclusion bodies as a whole ends up in the presentation of a series of single-case histories rather than in a comprehensive picture in which biochemical, physiological, and cytological data are combined in order to locate and define the intracellular sites and pathways of virus multiplication.

A clear-cut dividing line can be drawn between those inclusion bodies which are relevant to virus multiplication and those which are not (e.g., various kinds of virus particle aggregates).

Both these types of inclusions can be virus specific. Sometimes, specificity and consistency in their outer aspect are amazing, even at the virus group level, thus providing the practicing plant virologist with rapid and reliable means for virus diagnosis. In this connection, it would be difficult to disagree with McWhorter (1965), who states that viruses "sign" the cells of the plants they inhabit and that learning how to read these signatures has many practical advantages.

For those who have an interest in more fundamental aspects of virology, inclusion bodies represent an even more challenging problem, for some of them have been recognized, beyond doubt, as sites of virus replication and/or assembly. The type and extent of their involvement in virus multiplication is just now being unraveled, and therefore little can be said at the moment about which steps of this process are accomplished by which kind of inclusions.

With cowpea mosaic virus (De Zoeten *et al.*, 1974) and turnip yellow mosaic virus [for review, see Matthews (1973)], viral RNA replication has been found associated with inclusion bodies, and there are clear indications that the same may apply to other viruses as well. On the other hand, with several animal viruses (e.g., Caliguiri and Tamm, 1970; Cardiff *et al.*, 1973; Lubiniecki and Henry, 1974; Boulton and Westway, 1976) it has been demonstrated that viral protein synthesis and viral RNA replication occur at the level of intracytoplasmic membranes. The Citrus exocortis viroid also appears to replicate in association with cellular membranes (Semancik *et al.*, 1976). Thus instances in which a membranous backbone seems essential for accomplishing some of the crucial steps leading to virus formation become increasingly frequent. Now, it is known that membrane proliferation is one of the early cytological signs of virus infection in plant cells. These membranes often become components of inclusion bodies, indeed, they may become a major part thereof. It is tempting, then, to correlate such intracellular accumulations of membranous material with transcription and/or translation activity. But it is also possible that membranes are implicated in the transport of viral RNA or RNA complements from the site of synthesis to the site of particle assembly (see also De Zoeten, 1976).

These ideas need better experimental support than is presently available. Therefore, it can be foreseen that in the years to come investigators studying plant virus–induced inclusions will evolve new methods that provide a deeper insight into the structure-function relationship.

Acknowledgments

We wish to thank Drs. J. Begtrup, E. Hiebert, G. A. De Zoeten, W. S. Gardner, T. Hatta, E. W. Kitajima, K. Maramorosch, R. E. F. Matthews, D. Miličić, and R. G. Milne for providing bibliographic material, unpublished data, or manuscripts prior to publication, and Drs. A. Appiano, L. Bos, R. G. Christie, K. Esau, E. W. Kitajima, O. Lovisolo, T. A. Shalla, and H. E. Warmke for kindly supplying photographs. We are much indebted to Dr. R. G. Milne for reviewing the manuscript at very short notice and making many helpful suggestions, and to Dr. A. Sarić for assistance in translations from Russian. We also express our appreciation to Mrs. N. Martelli for help with the language and for patient typing of an almost illegible manuscipt.

References*

Akimoto, T., Wagner, M. A., Johnson, J. E., and Rossman, M. G. (1975). *J. Ultrastruct. Res.* **53,** 306.
Allen, T. C. (1972). *Virology* **47,** 467.
Allison, A. V., and Shalla, T. A. (1974). *Phytopathology* **64,** 784.
Amelunxen, F. (1956). *Protoplasma* **45,** 164.
Amelunxen, F. (1958). *Protoplasma* **49,** 140.
Amelunxen, F., and Giele, T. (1968). *Z. Pflanzenphysiol.* **58,** 457.
Amelunxen, F., and Thaler, I. (1967). *Z. Pflanzenphysiol.* **57,** 269.
Amelunxen, F., Thaler, I., and Harsdorff, M. (1970). *Z. Pflanzenphysiol.* **63,** 199.
Amici, A., Faoro, F., and Tornaghi, R. (1975). *Riso* **24,** 353.
Andrews, J. H., and Shalla, T. A. (1974). *Phytopathology* **64,** 1234.
Appiano, A., and Pennazio, S. (1972). *J. Gen. Virol.* **14,** 273.
Arnott, H. J., and Smith, K. M. (1968). *Virology* **34,** 25.
Bald, J. G. (1949a). *Phytopathology* **39,** 395.
Bald, J. G. (1949b). *Am. J. Bot.* **36,** 335.
Bald, J. G. (1966). *Adv. Virus Res.* **12,** 103.
Bald, J. G., and Solberg, R. A. (1961). *Nature* (*London*) **190,** 651.
Barnett, O. W., De Zoeten, G. A., and Gaard, G. (1971). *Phytopathology* **61,** 926.
Barton, R. (1967). *Planta* **77,** 203.
Bassi, M., and Favali, M. A. (1972). *J. Gen. Virol.* **16,** 153.
Bassi, M., Favali, M. A., and Conti, G. G. (1974a). *Virology* **60,** 353.
Bassi, M., Favali, M. A. and Appiano, A. (1974b). *Riv. Patol. Veg.* **10,** 19.
Bassi, M., Favali, M. A., and Appiano, A. (1974c). *J. Gen. Virol.* **24,** 563.

* The survey of the literature pertaining to this review was concluded in March 1976.

Bawden, F. C. (1964). "Plant Virus and Virus Diseases," 4th Ed. Ronald Press, New York.
Bawden, F. C., and Sheffield, F. M. L. (1939). *Ann. Appl. Biol.* **26,** 102.
Bawden, F. C., and Sheffield, F. M. L. (1944). *Ann. Appl. Biol.* **31,** 33.
Beale, H. P. (1937). *Contrib. Boyce Thompson Inst.* **8,** 413.
Begtrup, J. (1976). *Phytopathol. Z.* **86,** 127.
Berkeley, G. H., and Weintraub, M. (1952). *Phytopathology* **42,** 258.
Bernal, J. D., and Fankuchen, I. (1937). *Nature (London)* **190,** 651.
Black, L. M. (1947). *Growth Symp.* **6,** 79.
Bock, K. R., Guthrie, E. -J., and Woods, R. D. (1974). *Ann. Appl. Biol.* **77,** 289.
Borges, M. L., and David-Ferreira, J. F. (1968). *Rev. Biol. (Lisbon)* **6,** 421.
Bos, L. (1969). *Neth. J. Plant Pathol.* **75,** 137.
Bos, L. (1970). "Symptoms of Virus Diseases in Plants," 2nd Ed PUDOC, Wageningen.
Bos, L., and Rubio-Huertos, M. (1969). *Virology* **37,** 377.
Bos, L., and Rubio-Huertos, M. (1971). *Neth. J. Plant Pathol.* **77,** 145.
Bos, L., and Rubio-Huertos, M. (1972). *Neth. J. Plant Pathol.* **78,** 247.
Boulton, R. W., and Westway, E. G. (1976). *Virology* **69,** 416.
Bové, J. M. (1975). *Int. Virol.* **3,** 85.
Bovey, R. (1971). *Ann. Phytopathol.* **3**(hors ser.), 225.
Brandes, J. (1956). *Phytopathol. Z.* **26,** 93.
Brčàk, J., Kràlìk, O., and Vacke, J. (1972). *Biol. Plant.* **14,** 302.
Brinton, C. C. (1965). *Trans. N.Y. Acad. Sci.* **27,** 1003.
Brunt, A. A., and Kitajima, E. W. (1973). *Phytopathol. Z.* **76,** 265.
Burgess, J., Motoyoshi, F., and Fleming, E. N. (1974). *Planta* **117,** 133.
Cadilhac, B., Marchoux, G., and Coulomb, P. (1972). *Ann. Phytopathol.* **4,** 345.
Caliguiri, L. A., and Tamm, I. (1970). *Virology* **42,** 100.
Camargo, I. J. B., Kitajima, E. W., and Costa, A. S. (1971). *Bragantia* **30,** 31.
Cardiff, R. D., Russ, S. B., Brandt, W. E., and Russell, P. K. (1973). *Infect. Immun.* **7,** 809.
Carlisle, C. H., and Donberger, K. (1948). *Acta Crystallogr.* **1,** 194.
Carroll, T. W. (1970). *Virology* **42,** 1015.
Caspar, D. L. D. (1956). *Nature (London)* **177,** 475.
Castro, S., Moreno, R., and Rubio-Huertos, M. (1971). *Microbiol. Esp.* **24,** 135.
Chalcroft, J. P., and Matthews, R. E. F. (1966). *Virology* **28,** 555.
Chalcroft, J. P., and Matthews, R. E. F. (1967a). *Virology* **33,** 167.
Chalcroft, J. P., and Matthews, R. E. F. (1967b). *Virology* **33,** 659.
Chamberlain, J. A. (1974). *J. Gen. Virol.* **23,** 201.
Chamberlain, J. A. (1975). *Ann. Appl. Biol.* **81,** 264.
Chamberlain, J. A., and Catherall, P. L. (1976). *J. Gen. Virol.* **30,** 41.
Chandra, H., and Hildebrandt, A. C. (1965). *Nature (London)* **206,** 325.
Chen, M.-H., Miyakawa, T., and Matsui, C. (1971). *Phytopathology* **61,** 279.
Chen, M.-H., Miyakawa, T., and Matsui, C. (1972). *Phytopathology* **62,** 663.
Chen, M.-J. (1974). *Plant Prot. Bull. (Taiwan)* **16,** 132.
Christie, R. G. (1967). *Virology* **31,** 268.
Christie, R. G. (1971). *Proc. Pest Control Conf., Univ. Fla.* p. 65.
Christie, S. R., Christie, R. G., and Edwardson, J. R. (1974). *Phytopathology* **64,** 840.
Conti, G. G., Vegetti, G., Bassi, M., and Favali, M. A. (1972). *Virology* **47,** 694.
Conti, M., and Appiano, A. (1973). *J. Gen. Virol.* **21,** 315.
Conti, M., and Lovisolo, O. (1971). *J. Gen. Virol.* **13,** 173.

Cooper, J. I., and Mayo, M. A. (1972). *J. Gen. Virol.* **16,** 285.
Cousin, R., Maillet, P. L., Allard, C., and Staron, T. (1969). *Ann. Phytopathol.* **1,** 195.
Craig, A. S., and Williamson, K. I. (1969). *Virology* **39,** 616.
Cremer, M. C., van Slogteren, D. H. M., and van der Veken, J. A. (1960). *Electron Microsc., Proc. Eur. Reg. Conf.* **2,** 974.
Cronshaw, J., Hoefert, L. L., and Esau, K. (1966). *J. Cell Biol.* **31,** 429.
Da Cruz, N. D., Medina, D. M., Kitajima, E. W., Costa, A. S., and Landin, C. C. (1972). *Bragantia* **31,** 217.
David-Ferreira, J. F., and Borges, M. L. (1958). *Bol. Soc. Broteriana* **32,** 329.
Davison, E. M. (1969). *Virology* **37,** 694.
De Bokx, J. A., and Waterreus, H. A. J. I. (1971). *Neth. J. Plant Pathol.* **77,** 106.
Deshpande, B. P., and Evert, R. F. (1970). *J. Ultrastruct. Res.* **33,** 483.
De Vecchi, L., and Conti, G. G. (1975). *Isr. J. Bot.* **24,** 71.
De Zoeten, G. A. (1975). *Int. Virol.* **3,** 84.
De Zoeten, G. A. (1976). *Handb. Pflanzenphysiol.* (in press).
De Zoeten, G. A., and Gaard, G. (1969a). *Virology* **39,** 768.
De Zoeten, G. A., and Gaard, G. (1969b). *J. Cell Biol.* **40,** 814.
De Zoeten, G. A., and Gaard, G. (1970). *Virology* **41,** 573.
De Zoeten, G. A., and Schlegel, D. E. (1967a). *Virology* **31,** 173.
De Zoeten, G. A., and Schlegel, D. E. (1967b). *Virology* **32,** 416.
De Zoeten, G. A., Gaard, G., and Diez, F. B. (1972). *Virology* **48,** 638.
De Zoeten, G. A., Assink, A. M., and van Kammen, A. (1974). *Virology* **59,** 341.
Doi, Y., Yamashita, S., Kusunoki, M., Arai, K., and Yora, K. (1975). *Nippon Shokubutsu Byori Gakkaiho* **41,** 228.
Doraiswamy, S., and Lesemann, D. (1974). *Phytopathol. Z.* **81,** 314.
Dudman, W. F. (1966). *Biochim. Biophys. Acta* **120,** 212.
Edwardson, J. R. (1966a). *Am. J. Bot.* **53,** 359.
Edwardson, J. R. (1966b). *Science* **153,** 883.
Edwardson, J. R. (1974). *Fla. Agric. Exp. Sta., Monogr.* **4,** 398 pp.
Edwardson, J. R., and Purcifull, D. E. (1974). *Phytopathology* **64,** 1322.
Edwardson, J. R., Purcifull, D. E., and Christie, R. G. (1966). *Can. J. Bot.* **44,** 821.
Edwardson, J. R., Purcifull, D. E., and Christie, R. G. (1968). *Virology* **34,** 250.
Edwardson, J. R., Zettler, F. W., Christie, R. G., and Evans, I. R. (1972). *J. Gen. Virol.* **15,** 113.
Esau, K. (1941). *Hilgardia* **13,** 427.
Esau, K. (1944). *J. Agric. Res.* **69,** 95.
Esau, K. (1960a). *Virology* **10,** 73.
Esau, K. (1960b). *Virology* **11,** 317.
Esau, K. (1967). *Ann. Rev. Phytopathol.* **5,** 45.
Esau, K. (1968). "Viruses in Plant Hosts. Form, Distribution and Pathogenic Effect." Univ. of Wisconsin Press, Madison.
Esau, K. (1975). *J. Ultrastruct. Res.* **53,** 235.
Esau, K., and Cronshaw, J. (1967). *J. Cell Biol.* **33,** 665.
Esau, K., and Gill, R. H. (1969). *Virology* **38,** 464.
Esau, K., and Hoefert, L. L. (1971a). *Protoplasma* **72,** 255.
Esau, K., and Hoefert, L. L. (1971b). *Protoplasma* **72,** 459.
Esau, K., and Hoefert, L. L. (1971c). *Protoplasma* **73,** 51.
Esau, K., and Hoefert, L. L. (1972a). *Virology* **48,** 724.
Esau, K., and Hoefert, L. L. (1972b). *J. Ultrastruct. Res.* **40,** 556.
Esau, K., and Hoefert, L. L. (1973). *Virology* **56,** 454.

Favali, M. A., and Conti, G. G. (1971). *Cytobiologie* **3,** 153.
Favali, M. A., Bassi, M., and Conti, G. G. (1973). *Virology* **53,** 115.
Favali, M. A., Pellegrini, S., and Bassi, M. (1975). *Virology* **66,** 502.
Francki, R. I. B. (1973). *Adv. Virus Res.* **18,** 257.
Francki, R. I. B., and Grivell, C. J. (1970). *Virology* **42,** 969.
Francki, R. I. B., and Grivell, C. J. (1972). *Virology* **48,** 305.
Frederick, S. E., Newcomb, E. H., Vigil, E. I., and Wergin, W. P. (1968). *Planta* **81,** 229.
Freeman, T. P., and Duysen, M. E. (1975). *Protoplasma* **83,** 131.
Frey-Wyssling, A., and Mühlethaler, K. (1965). "Ultrastructural Plant Cytology." Elsevier, Amsterdam.
Frowd, J. A., and Tomlinson, J. A. (1972). *Ann. Appl. Biol.* **72,** 189.
Fujisawa, I., Hayashi, T., and Matsui, C. (1967a). *Virology* **33,** 70.
Fujisawa, I., Matsui, C., and Yamaguchi, A. (1967b). *Phytopathology* **57,** 210.
Fujisawa, I., Rubio-Huertos, M., Matsui, C., and Yamaguchi, A. (1967c). *Phytopathology* **57,** 1130.
Gardner, W. S. (1967). *Phytopathology* **57,** 1315.
Gerola, F. M., and Bassi, M. (1966). *Caryologia* **19,** 13.
Gerola, F. M., Bassi, M., and Belli, G. (1965a). *Caryologia* **18,** 567.
Gerola, F. M., Bassi, M., and Betto, E. (1965b). *Caryologia* **18,** 353.
Gerola, F. M., Bassi, M., and Giussani Belli, G. (1966a). *Caryologia* **19,** 481.
Gerola, F. M., Bassi, M., and Giussani, G. (1966b). *Caryologia* **19,** 457.
Gerola, F. M., Bassi, M., Lovisolo, O., and Vidano, C. (1966c). *Phytopathol. Z.* **56,** 97.
Gerola, F. M., Bassi, M., Lovisolo, O., and Vidano, C. (1966d). *Caryologia* **19,** 493.
Gerola, F. M., Lombardo, G., and Catara, A. (1969a). *Protoplasma* **67,** 319.
Gerola, F. M., Bassi, M., and Belli, G. (1969b). *G. Bot. Ital.* **103,** 271.
Giannotti, J., Monsarrat, P., and Vago, C. (1968). *Ann. Epiphyt.* **19**(hors ser.), 31.
Gibbs, A. J., and Harrison, B. D. (1973). *CMI/AAB Descriptions Plant Viruses* **124,** 4 pp.
Gigot, C., Kopp, M., Schmitt, C., and Milne, R. G. (1975). *Protoplasma* **84,** 31.
Gill, C. C. (1974). *Can. J. Bot.* **52,** 621.
Giri, L., and Chessin, M. (1975). *Phytopathol. Z.* **83,** 40.
Goldin, M. I. (1938a). *Mikrobiologija* **7,** 353.
Goldin, M. I. (1938b). *Mikrobiologija* **7,** 1121.
Goldin, M. I. (1953). *C. R. Acad. Sci. URSS* **88,** 933.
Goldin, M. I. (1963). "Viral Inclusions in Plant Cells and the Nature of the Viruses," 2nd Ed. Akad. Nauk USSR, Moscow.
Goldstein, B. (1924). *Torrey Bot. Club Bull.* **51,** 261.
Goldstein, B. (1926). *Torrey Bot. Club Bull.* **53,** 499.
Goldstein, B. (1927). *Torrey Bot. Club Bull.* **54,** 285.
Granett, A. L. (1973). *Phytopathology* **63,** 1313.
Granett, A. L., and Shalla, T. A. (1970). *Phytopathology* **60,** 419.
Gunning, B. E. S., and Pate, J. S. (1969). *Protoplasma* **68,** 107.
Gunning, B. E. S., Pate, J. S., and Briarty, L. G. (1968). *J. Cell. Biol.* **37,** C7.
Halk, E. L., and McGuire, J. M. (1973). *Phytopathology* **63,** 1291.
Hampton, R. O., Phillips, S., Knesek, J. E., and Mink, G. I. (1973). *Arch. Virusforsch.* **42,** 242
Harrison, B. D., and Roberts, I. M. (1968). *J. Gen. Virol.* **3,** 121.
Harrison, B. D., and Roberts, I. M. (1970). *Rep. Scot. Hort. Res. Inst.* **1969,** 52.
Harrison, B. D., and Roberts, I. M. (1971). *J. Gen. Virol.* **10,** 71.

Harrison, B. D., Štefanac, Z., and Roberts, I. M. (1970). *J. Gen. Virol.* **6,** 127.
Harrison, B. D., Murant, A. F., Mayo, M. A., and Roberts, I. M. (1974). *J. Gen. Virol.* **22,** 233.
Hatta, T. (1976). *Virology* **69,** 237.
Hatta, T., and Matthews, R. E. F. (1974). *Virology* **59,** 383.
Hatta, T., and Matthews, R. E. F. (1976). *Virology* **73,** 1.
Hatta, T., and Ushiyama, R. (1973). *J. Gen. Virol.* **21,** 9.
Hatta, T., Nakamoto, T., Takagi, Y., and Ushiyama, R. (1971). *Virology* **45,** 292.
Hatta, T., Bullivant, S., and Matthews, R. E. F. (1973). *J. Gen. Virol.* **20,** 37.
Hayashi, T., and Matsui, C. (1967). *Virology* **33,** 47.
Herold, F., and Munz, K. (1967). *J. Gen. Virol.* **1,** 375.
Hibino, H., and Schneider, H. (1971).. *Arch. Virusforsch.* **33,** 347.
Hibino, H., Tsuchizaki, T., and Saito, Y. (1974a). *Virology* **57,** 510.
Hibino, H., Tsuchizaki, T., and Saito, Y. (1974b). *Virology* **57,** 522.
Hiebert, E. (1975). *Int. Virol.* **3,** 83.
Hiebert, E., and McDonald, J. G. (1973). *Virology* **56,** 349.
Hiebert, E., Purcifull, D. E., Christie, R. G., and Christie, S. R. (1971). *Virology* **43,** 638.
Hills, G. J., and Plaskitt, A. (1968). *J. Ultrastruct. Res.* **25,** 323.
Hiruki, C., and Shukla, P. (1973). *Can. J. Bot.* **51,** 1699.
Hiruki, C., and Tu, J. C. (1972). *Phytopathology* **62,** 77.
Hoefert, L. L. (1969). *Virology* **37,** 498.
Hoefert, L. L., Esau, K., and Duffus, J. E. (1970). *Virology* **42,** 814.
Hollings, M. (1957) *Ann. Appl. Biol.* **45,** 44.
Hollings, M., and Nariana, T. K. (1965). *Ann. Appl. Biol.* **56,** 99.
Honda, Y., and Matsui, C. (1968). *Phytopathology* **58,** 1230.
Honda, Y., and Matsui, C. (1972). *Phytopathology* **62,** 448.
Honda, Y., Matsui, C., Otsuki, Y., and Takebe, I. (1974). *Phytopathology* **64,** 30.
Honda, Y., Kajita, S., Matsui, C., Otsuki, Y., and Takebe, I. (1975). *Phytopathol. Z.* **84,** 66.
Hooker, W. J., and Summanwar, A. S. (1964). *Exp. Cell Res.* **33,** 609.
Hooper, G. R., and Wiese, M. W. (1972). *Virology* **47,** 644.
Hooper, G. R., Spink, G. C., and Myers, R. L. (1972). *Virology* **47,** 883.
Horne, R. W., and Pasquali Ronchetti, I. (1974). *J. Ultrastruct. Res.* **47,** 361.
Horne, R. W., Hobart, J. M., and Pasquali Ronchetti, I. (1975). *Micron* **5,** 233.
Horvat, F., and Verhoyen, M. (1975a). *Phytopathol. Z.* **83,** 328.
Horvat, F., and Verhoyen, M. (1975b). *Parasitica* **31,** 55.
Howatson, A. F. (1970). *Adv. Virus Res.* **16,** 196.
Hull, R., and Plaskitt, A. (1970). *Virology* **42,** 773.
Hull, R., and Plaskitt, A. (1973/1974). *Intervirology* **2,** 352.
Hull, R., Hills, G. J., and Plaskitt, A. (1969). *J. Ultrastruct. Res.* **26,** 465.
Hull, R., Hills, G. J., and Plaskitt, A. (1970). *Virology* **42,** 753.
Ie, T. S. (1971). *Virology* **43,** 468.
Iida, T. T., Shinkai, A., and Kimura, I. (1972). *CMI/AAB Descriptions Plant Viruses* **102,** 4 pp.
Inouye, T. (1970). *Nisshoku Byoho* **36,** 186.
Inouye, T. (1971). *Ber. Ohara Inst. Landwirtsch. Biol., Okayama Univ.* **15,** 69.
Inouye, T., and Mitsuhata, C. (1973). *Ber. Ohara Inst. Landwirtsch. Biol., Okayama Univ.* **15,** 195.
Israel, H. W., and Ross, A. F. (1967). *Virology* **33,** 272.

Iwanowski, D. (1903). *Z. Pflanzenkr.* **13,** 1.
Jacoli, G. C., and Ronald, W. P. (1974). *J. Ultrastruct. Res.* **46,** 34.
James, M., Kenten, R. H., and Woods, R. D. (1973). *J. Gen. Virol.* **21,** 145.
Johnson, J. E., Rossmann, M. G., Smiley, I. E., and Wagner, M. A. (1974). *J. Ultrastruct. Res.* **46,** 441.
Jones, A. T., Kinninmonth, A. M., and Roberts, I. M. (1973). *J. Gen. Virol.* **18,** 61.
Juretić, N., Miličić, D., and Schmelzer, K. (1970). *Acta Bot. Croat.* **29,** 17.
Kamei, T., Goto, T., and Matsui, C. (1969a). *Phytopathology* **59,** 1795.
Kamei, T., Honda, Y., and Matsui, C. (1969b). *Phytopathology* **59,** 139.
Kamei, T., Rubio-Huertos, M., and Matsui, C. (1969c). *Virology* **37,** 506.
Kapil, R. N., Pugh, T. D., and Newcomb, E. H. (1975). *Planta* **124,** 231.
Kashiwagi, I. (1966). *Nippon Shokubutsu Byori Gakkaiho* **32,** 168.
Kassanis, B. (1939). *Ann. Appl. Biol.* **26,** 705.
Kassanis, B., and Milne, R. G. (1971). *J. Gen. Virol.* **11,** 193.
Kassanis, B., and Sheffield, F. M. L. (1941). *Ann. Appl. Biol.* **28,** 360.
Kassanis, B., and Turner, R. H., (1972). *J. Gen. Virol.* **14,** 119.
Kassanis, B., Vince, D. A., and Woods, R. D. (1970). *J. Gen. Virol.* **7,** 143.
Kikumoto, T., and Matsui, C. (1961). *Virology* **13,** 294.
Kim, K. S., and Fulton, J. P. (1969). *Virology* **37,** 297.
Kim, K. S., and Fulton, J. P. (1971). *Virology* **43,** 329.
Kim, K. S., and Fulton, J. P. (1972). *Virology* **49,** 112.
Kim, K. S., and Fulton, J. P. (1973a). *J. Ultrastruct. Res.* **45,** 328.
Kim, K. S., and Fulton, J. P. (1973b). *Plant Dis. Rep.* **57,** 1029.
Kim, K. S., and Fulton, J. P. (1975). *Virology* **64,** 560.
Kim, K. S., Fulton, J. P., and Scott, H. A. (1974). *J. Gen. Virol.* **25,** 445.
Kislev, N., Harpaz, I., and Klein, M. (1968). *Acta Phytopathol. Acad. Sci. Hung.* **3,** 3.
Kiso, A., Yamamoto, T., and Kitani, K. (1974). *Shikoku Nogyo Shikenjo Hokoko* **27,** 1.
Kitajima, E. W. (1965) *Virology* **26,** 89.
Kitajima, E .W., and Costa, A. S. (1966). *Virology* **29,** 523.
Kitajima, E. W., and Costa, A. S. (1969). *J. Gen. Virol.* **4,** 177.
Kitajima, E. W., and Costa, A. S. (1970). *Proc. Int. Congr. Electron Microsc., 7th* **3,** 323.
Kitajima, E. W., and Costa, A. S. (1974a). *Bragantia* **33,** 45.
Kitajima, E .W., and Costa, A. S. (1974b). *Phytopathol. Z.* **79,** 289.
Kitajima, E. W., and Costa, A. S. (1976). *J. Electron Microsc.* **25** (in press).
Kitajima, E. W., and Galves, G. E. (1973). *Cienc. Cult. (Sao Paulo)* **25,** 979.
Kitajima, E. W., and Lauritis, J. A. (1969). *Virology* **37,** 681.
Kitajima, E. W., and Lovisolo, O. (1972). *J. Gen. Virol.* **16,** 265.
Kitajima, E. W., Oliveira, A. R., and Costa, A. S. (1965). *Bragantia* **24,** 219.
Kitajima, E. W., Camargo, I. J. B., and Costa, A. S. (1968). *J. Electron Microsc.* **17,** 144.
Kitajima, E. W., Lauritis, J. A., and Swift, H. (1969). *Virology* **39,** 240.
Kitajima, E. W., Caetano, V. R., and Costa, A. S. (1971). *Bragantia* **30,** 101.
Kitajima, E. W., Betti, J. A., and Costa, A. S. (1973). *J. Gen. Virol* **20,** 117.
Kitajima, E. W., Blumenschein, A., and Costa, A. S. (1974a). *Phytopathol. Z.* **81,** 280.
Kitajima, E. W., Tascon, A., Gamez, R., and Galvez, G. E. (1974b). *Turrialba* **24,** 393.
Kitajima, E. W., Giacomelli, E. J., Costa, A. S., Costa, C. L., and Cupertino, F. P. (1975). *Phytopathol. Z.* **82,** 83.
Klug, A., and Caspar, D. L. D. (1960). *Adv. Virus Res.* **7,** 225.

Klug, A., and Finch, J. T. (1960). *J. Mol. Biol.* **2,** 201.
Klug, A., Finch, J. T., and Franklin, R. E. (1957a). *Nature (London)* **179,** 683.
Klug, A., Finch, J. T., and Franklin, R. E. (1957b). *Biochim. Biophys. Acta* **25,** 242.
Knuhtsen, H., Hiebert, E., and Purcifull, D. E. (1974). *Virology* **61,** 200.
Koenig, R., and Lesemann, D. (1974). *Phytopathol. Z.* **80,** 136.
Kolehmainen, L., Zech, H., and van Wettstein, D. (1965). *J. Cell Biol.* **25,** 77.
Kojima, M., Shikata, E., Sugawara, M., and Murayama, D. (1969). *Virology* **39,** 162.
Kozar, F. E., and Sheludko, Y. M. (1969). *Virology* **38,** 220.
Krass, C. J., and Ford, R. E. (1969). *Phytopathology* **59,** 431.
Kubo, S., Harrison, B. D., Robinson, D. J., and Mayo, M. A. (1975). *J. Gen. Virol.* **27,** 293.
Kunkel, L. O. (1922). *Science* **55,** 73.
Kunkel, L. O. (1924). *Bull. Exp. Stn. Hawaii. Sugar Assoc.* **3,** 99.
Kuppers, P., Nienhaus, F., and Schinzer, U. (1975). *Z. Pflanzenkr. (Pflanzenpathol.) Pflanzenschutz* **82,** 183.
Labaw, L. W., and Wyckoff, W. G. (1957). *Arch. Biochem. Biophys.* **67,** 225.
Laflèche, D., and Bové, J. M. (1968). *C. R. Acad. Sci. Ser. D* **266,** 1839.
Langenberg, W. G., and Schlegel, D. E. (1967). *Virology* **32,** 167.
Langenberg, W. G., and Schlegel, D. E. (1969). *Virology* **37,** 86.
Langenberg, W. G., and Schroeder, H. F. (1972). *J. Ultrastruct. Res.* **40,** 513.
Langenberg, W. G., and Schroeder, H. F. (1973a). *Phytopathology* **63,** 1003.
Langenberg, W. G., and Schroeder, H. F. (1973b). *Phytopathology* **63,** 1006.
Langenberg, W. G., and Schroeder, H. F. (1973c). *Virology* **55,** 218.
Langenberg, W. G., and Schroeder, H. F. (1974a). *Phytopathology* **64,** 750.
Langenberg, W. G., and Schroeder, H. F. (1974b). *J. Gen. Virol.* **23,** 51.
Langenberg, W. G., and Schroeder, H. F. (1975). *J. Ultrastruct. Res.* **51,** 166.
Lastra, J. R., and Schlegel, D. E. (1975). *Virology* **65,** 16.
Lawson, R. H., and Ali, S. (1975). *J. Ultrastruct. Res.* **53,** 345.
Lawson, R. H., and Hearon, S. S. (1971). *Virology* **44,** 454.
Lawson, R. H., and Hearon, S. S. (1974). *J. Ultrastruct. Res.* **48,** 201.
Lawson, R. H., Hearon, S. S., and Smith, F. F. (1971). *Virology* **46,** 453.
Leak, L. V. (1968). *J. Ultrastruct. Res.* **24,** 102.
Lee, P. E. (1965). *J. Ultrastruct. Res.* **13,** 359.
Lee, Y.-S., and Ross, J. P. (1972). *Phytopathology* **62,** 839.
Lesemann, D. (1972). *J. Gen. Virol.* **16,** 273.
Lesemann, D., and Casper, R. (1973). *Phytopathology* **63,** 1118.
Lesemann, D., and Doraiswamy, S. (1975). *Phytopathol. Z.* **83,** 27.
Lesemann, D., and Huth, W. (1975). *Phytopathol. Z.* **82,** 246.
Lindsten, K., Gerhardson, B., and Petterson, J. (1973). *Natl. Swed. Inst. Plant Prot. Contrib.* **15,** 375.
Littau, V. C., and Black, L. M. (1952). *Am. J. Bot.* **39,** 87.
Liu, K.-C., and Boyle, J. S. (1972). *Phytopathology* **62,** 1303.
Lovisolo, O., and Conti, M. (1966). *Atti Accad. Sci. Torino* **100,** 63.
Lubiniecki, A. S., and Henry, C. J. (1974). *Proc. Soc. Exp. Biol. Med.* **145,** 1165.
Lyon, H. L. (1921). *Bull. Exp. Sta. Hawaii. Sugar Plant Assoc.* **3,** 17.
Lyons, A. R., and Allen, T. C. (1969). *J. Ultrastruct. Res.* **27,** 198.
McDonald, J. G., and Hiebert, E. (1974a). *J. Ultrastruct. Res.* **48,** 138.
McDonald, J. G., and Hiebert, E. (1974b). *Virology* **58,** 200.
McDonald, J. G., and Hiebert, E. (1975). *Virology* **63,** 295.

McKinney, H. H., Eckerson, S. H., and Webb, R. H. (1923). *J. Agric. Res. (Washington, D.C.)* **26,** 605.
Macovei, A. (1971). *Ann. Phytopathol.* **3**(hors ser.), 221.
McWhorter, F. P. (1922). *Philipp. Agric.* **11,** 103.
McWhorter, F. P. (1940). *Phytopathology* **30,** 788.
McWhorter, F. P. (1941). *Stain Technol.* **16,** 143.
McWhorter, F. P. (1951). *Stain Technol.* **26,** 177.
McWhorter, F. P. (1965). *Annu. Rev. Phytopathol.* **3,** 287.
McWhorter, F. P., and Price, W. C. (1949). *Science* **109,** 116.
Mamula, D., and Miličić, D. (1968). *Phytopathol. Z.* **61,** 232.
Maramorosch, K., Shikata, E., Hirumi, H., and Granados, R. R. (1968). *Natl. Cancer Inst., Monogr.* **31,** 439.
Marchant, R., and Robards, A. W. (1968). *Ann. Bot.* **32,** 457.
Markham, R., and Smith, K. M. (1949). *Parasitology* **39,** 330.
Martelli, G. P., and Castellano, M. A. (1969). *Virology* **39,** 610.
Martelli, G. P., and Castellano, M. A. (1971). *J. Gen. Virol.* **13,** 133.
Martelli, G. P., and Russo, M. (1969a). *Virology* **38,** 330.
Martelli, G. P., and Russo, M. (1969b). *Ann. Phytopathol.* **1**(hors ser.), 339.
Martelli, G. P., and Russo, M. (1972). *J. Gen. Virol.* **15,** 193.
Martelli, G. P., and Russo, M. (1973). *J. Ultrastruc. Res.* **42,** 13.
Martelli, G. P., and Russo, M. (1976a). *Virology* **72,** 352.
Martelli, G. P., and Russo, M. (1977). "Insect and Plant Viruses: An Atlas." Academic Press, New York. In press.
Martelli, G. P., Russo, M., and Castellano, M. A. (1969). *Phytopathol. Mediterr.* **8,** 175.
Martin, M .M. (1964). *Virology* **22,** 645.
Matsui, C., and Yamaguchi, A. (1964). *Virology* **22,** 40.
Matsui, C., and Yamaguchi, A. (1966). *Adv. Virus Res.* **12,** 127.
Matthews, R. E. F. (1970). "Principles of Plant Virology." Academic Press, New York.
Matthews, R. E. F. (1973). *Annu. Rev. Phytopathol.* **11,** 147.
Matthews, R. E. F. (1975). *Int. Virol.* **3,** 85.
Miličić, D. (1954). *Protoplasma* **43,** 228.
Miličić, D. (1968). *Naturwissenschaften* **55,** 90.
Miličić, D. (1969). *Zast. Bilja* **20,** 101.
Miličić, D. and Juretić, N. (1971). *Tagungser. Dtsch. Akad. Landwirtschaftswiss. Berlin* **115,** 141.
Miličić, D., and Štefanac, Z. (1967). *Phytopathol. Z.* **59,** 285.
Miličić, D., and Štefanac, Z. (1971). *Acta Bot. Croat.* **30,** 33.
Miličić, D., Štefanac, Z., Juretić, N., and Wrischer, M. (1968). *Virology,* **35,** 365.
Miličić, D., Wrischer, M., and Juretić, N. (1974). *Phytopathol. Z.* **80,** 127.
Milne, R. G. (1966a). *Virology* **28,** 79.
Milne, R. G. (1966b). *Virology* **28,** 520.
Milne, R. G. (1966c). *Virology* **28,** 527.
Milne, R. G. (1967a). *Virology* **32,** 589.
Milne, R. G. (1967b). *J. Gen. Virol.* **1,** 403.
Milne, R. G. (1967c). *Sci. Prog. (London)* **55,** 203.
Milne, R. G. (1968). *Rothamsted Exp. Stn. Rep.* p. 126.
Milne, R. G. (1970). *J. Gen. Virol* **6,** 267.
Milne, R .G. (1972). *Bot. Gaz.* **113,** 401.

Milne, R. G., and Lovisolo, O. (1977). *Adv. Virus Res.* **21,** 267.
Milne, R. G., Kempiak, G., Lovisolo, O., and Muehle, E. (1974). *Phytopathol. Z.* **79,** 315.
Moline, H. E., and Fries, R. E. (1974). *Phytopathology* **64,** 44.
Mollenhauer, H. H., and Morré, D. J. (1966). *Annu. Rev. Plant Physiol.* **17,** 27.
Motoyoshi, F., Bancroft, J. B., and Watts, J. W. (1973). *J. Gen. Virol.* **20,** 177.
Mueller, W. C., and Koenig, R. (1965). *Phytopathology* **55,** 242.
Murant, A. F., Roberts, I. M., and Goold, R. A. (1973). *J. Gen. Virol.* **21,** 269.
Murant, A. F., Roberts, I. M., and Hutcheson, A. M. (1975). *J. Gen. Virol.* **26,** 277.
Nagaich, B. B., and Sinha, R. C. (1974). *Plant Dis. Rep.* **58,** 968.
Nagaraj, A. N., and Black, L. M. (1961). *Virology* **15,** 289.
Nagata, T., and Yamaki, T. (1973). *Z. Pflanzenphysiol.* **70,** 472.
Nasu, S. (1965). *Nippon Shokubutsu Byori Gakkaiho* **30,** 265.
Newcomb, E. H. (1967). *J. Cell Biol.* **33,** 143.
Nilsson-Tillgren, T., and Bekke, B. (1975). *Int. Virol.* **3,** 85.
Nome, S. F., Shalla, T. A., and Petersen, L. J. (1974). *Phytopathol. Z.* **79,** 169.
Otsuki, Y., and Takebe, I. (1973). *Virology* **52,** 433.
Otsuki, Y., Takebe, I., Honda, Y., and Matsui, C. (1972). *Virology* **49,** 188.
Otsuki, Y., Takebe, I., Honda, Y., Kajita, S., and Matsui, C. (1974). *J. Gen. Virol.* **22,** 375.
Paliwal, Y. C. (1970). *J. Ultrastruct. Res.* **30,** 491.
Paliwal, Y. C. (1972). *J. Invert. Pathol.* **20,** 288.
Peña-Iglesias, A., and Rubio-Huertos, M. (1971). *Microbiol. Esp.* **24,** 183.
Peña-Iglesias, A., and Ayuso-Gonzales, P. (1972). *Ann. INIA, Madrid* **2,** 89.
Peña-Iglesias, A., Rubio-Huertos, M., and Moreno-San Martin, R. (1972). *Ann. INIA, Madrid,* **2,** 123.
Pennazio, S., and Appiano, A. (1975). *Phytopathol. Mediterr.* **14,** 12.
Peterson, J. F. (1970). *Virology* **42,** 304.
Petzold, H. (1968). *Phytopathol. Z.* **63,** 201.
Petzold, H. (1972). *Phytopathol. Z.* **74,** 249.
Porter, C. A., and McWhorter, F. P. (1952). *Phytopathology* **42,** 518.
Pratt, M. J. (1969). *Plant Dis. Rep.* **53,** 210.
Price, W. C. (1966). *Virology* **29,** 285.
Purcifull, D. E., and Edwardson, J. R. (1967). *Virology* **32,** 393.
Purcifull, D. E., and Shepard, J. F. (1967). *Plant Dis. Rep.* **51,** 502.
Purcifull, D. E., Edwardson, J. R., and Christie, R. G. (1966). *Virology* **29,** 276.
Purcifull, D. E., Edwardson, J. R., and Christie, S. R. (1970). *Phytopathology* **60,** 779.
Purcifull, D. E., Hiebert, E., and McDonald, J. G. (1973). *Virology* **55,** 275.
Pyliotis, N. A., and Gooldchild, D. J. (1975). *Protoplasma* **85,** 277.
Quacquarelli, A., Vovlas, C., Piazzolla, P., Russo, M., and Martelli, G. P. (1972). *Phytopathol. Mediterr.* **11,** 180.
Ragetli, H. W. J., Weintraub, M., and Lo, E. (1970). *Can. J. Bot.* **48,** 1913.
Raine, J., Weintraub, M., and Schroeder, B. (1975). *Phytopathology* **65,** 1181.
Ralph, R. K., and Clark, M. F. (1966). *Biochim. Biophys. Acta* **119,** 29.
Ralph, R. K., Bullivant, S., and Wojcik, S. J. (1971). *Virology* **43,** 713.
Rawlings, T. E. (1957). *Phytopathology* **47,** 307.
Reddi, K. K. (1972). *Adv. Virus Res.* **17,** 51.
Reitberger, A. (1956). *Züechter* **26,** 106.
Resconich, E. C. (1961). *Virology* **15,** 16.
Rich, S. (1948a). *Stain Technol.* **23,** 19.

Rich, S. (1948b). *Phytopathology* **38,** 221.
Rich, S. (1948c). *Science* **107,** 194.
Robb, S. M. (1963). *Ann. Appl. Biol.* **52,** 145.
Robb, S. M. (1964). *Virology* **23,** 141.
Roberts, D. A., Christie, R. G., and Archer, M. C. (1970). *Virology* **42,** 217.
Roberts, I. M., and Harrison, B. D. (1970). *J. Gen. Virol.* **7,** 47.
Ross, A. F., and Israel, H. W. (1970). *Phytopathology* **60,** 755.
Rubio-Huertos, M. (1950). *Microbiol. Esp.* **3,** 207.
Rubio-Huertos, M. (1956). *Phytopathology* **46,** 533.
Rubio-Huertos, M. (1962). *Virology* **18,** 337.
Rubio-Huertos, M. (1968). *Protoplasma* **65,** 465.
Rubio-Huertos, M. (1972). *In* "Principles and Techniques in Plant Virology" (C. I. Kado and H. O. Agrawal, eds.), pp. 62–75. Van Nostrand-Reinhold, New York.
Rubio-Huertos, M., and Bos, L. (1973). *Neth. J. Plant Pathol.* **79,** 94.
Rubio-Huertos, M., and Garcia-Hidalgo, F. (1964). *Virology* **24,** 84.
Rubio-Huertos, M., and Garcia-Hidalgo, F. (1971). *Protoplasma* **72,** 449.
Rubio-Huertos, M., and Lopez-Abella, D. (1966). *Microbiol. Esp.* **19,** 77.
Rubio-Huertos, M., and Peña-Iglesias, A. (1973). *Plant Dis. Rep.* **57,** 649.
Rubio-Huertos, M., and Van Slogteren, D. H. M. (1956). *Phytopathology* **46,** 401.
Rubio-Huertos, M., and Vela-Cornejo, A. (1966). *Protoplasma* **57,** 184.
Rubio-Huertos, M., Vela, A., and Lopez-Abella, D. (1967). *Virology* **32,** 438.
Rubio-Huertos, M., Castro, S., Moreno, R., and Lopez-Abella, D. (1968a). *Microbiol. Esp.* **21,** 1.
Rubio-Huertos, M., Matsui, C., Yamaguchi, A., and Kamei, T. (1968b). *Phytopathology* **58,** 548.
Rubio-Huertos, M., Castro, S., Fujisawa, I., and Matsui, C. (1972). *J. Gen. Virol.* **15,** 257.
Russo, M., and Martelli, G. P. (1969). *Phytopathol. Mediterr.* **8,** 65.
Russo, M., and Martelli, G. P. (1972). *Virology* **49,** 122.
Russo, M., and Martelli, G. P. (1973). *Phytopathol. Mediterr.* **12,** 54.
Russo, M., and Martelli, G. P. (1975a). *Phytopathol. Z.* **83,** 97.
Russo, M., and Martelli, G. P. (1975b). *J. Submicrosc. Cytol.* **7,** 335.
Russo, M., Martelli, G. P., and Quacquarelli, A. (1968). *Virology* **34,** 679.
Russo, M., Quacquarelli, A., Castellano, M. A., and Martelli, G. P. (1972). *Phytopathol. Mediterr.* **11,** 118.
Russo, M., Martelli, G. P., and Vovlas, C. (1973). *Phytopathol. Mediterr.* **12,** 61.
Russo, M., Martelli, G. P., and Rana, G. L. (1975). *Phytopathol. Z.* **83,** 223.
Sahambi, H. S., Milne, R. G., Cook, S. M., Gibbs, A. J., and Woods, R. D. (1973). *Phytopathol. Z.* **76,** 158.
Šarić, A., and Wrischer, M. (1975). *Phytopathol. Z.* **84,** 97.
Schank, S. C., Edwardson,, J. R., Christie, R. G. and Overman, R. A. (1972). *Euphytica* **21,** 344.
Schlegel, D. E., and DeLisle, D. E. (1971). *Virology* **45,** 747.
Schmelzer, K., and Miličić, D. (1966). *Phytopathol. Z.* **57,** 8.
Schnepf, E. (1964). *Z. Naturforsch. B* **19,** 344.
Semancik, J. S., and Vanderwoude, W. J. (1976). *Virology* **69,** 719.
Semancik, J. S., Tsuruda, D., Zaner, L., Geelen, J. L. M. C., and Weathers, L. G. (1976). *Virology* **69,** 669.
Shalla, T. A. (1964). *J. Cell Biol.* **21,** 253.
Shalla, T. A. (1966). "Viruses of Plants." North-Holland Publ., Amsterdam.

Shalla, T. A. (1968). *Virology* **35,** 194.
Shalla, T. A., and Amici, A. (1967). *Virology* **31,** 78.
Shalla, T. A., and Petersen, L. J., (1973). *Phytopathology* **63,** 1125.
Shalla, T. A., and Shepard, J. F. (1972). *Virology* **49,** 654.
Sheffield, F. M. L. (1931). *Ann. Appl. Biol.* **18,** 471.
Sheffield, F. M. L. (1934). *Ann. Appl. Biol.* **21,** 430.
Sheffield, F. M. L. (1939). *Proc. Roy. Soc., Ser.* **126,** 529.
Sheffield, F. M. L. (1941). *J. Roy. Microsc. Soc.* **61,** 30.
Sheffield, F. M. L. (1946). *J. Roy. Microsc. Soc.* **66,** 69.
Shepard, J. F. (1968). *Virology* **36,** 20.
Shepard, J. F., and Shalla, T. A. (1969). *Virology* **38,** 185.
Shepard, J. F., Gaard, G., and Purcifull, D. E. (1974). *Phytopathology* **64,** 418.
Shikata, E. (1966). *J. Fac. Agric., Hokkaido Univ.* **55,** 1.
Shikata, E. (1969). *In* "The Virus Diseases of Rice Plant." (Int. Rice Res. Inst.), pp. 223–240. Johns Hopkins Press, Baltimore, Maryland.
Shikata, E., and Maramorosch, K. (1965). *Virology* **27,** 461.
Shikata, E., and Maramorosch, K. (1966). *Virology* **30,** 439.
Shikata, E., and Maramorosch, K. (1967a). *J. Virol.* **1,** 1052.
Shikata, E., and Maramorosch, K. (1967b). *Virology* **32,** 363.
Shukla, P., and Hiruki, C. (1975). *Physiol. Plant Pathol.* **7,** 189.
Silberschmidt, K., Weigl, D. R., and Salomao, T. A. (1970). *Phytopathol. Z.* **68,** 210.
Singh, M., and Hildebrandt, A. C. (1966). *Phytopathology* **56,** 901.
Singh, R. P., and Lopez-Abella, D. (1971). *Phytopathology* **61,** 333.
Smith, K. M., (1956). *Virology* **2,** 706.
Smith, K. M. (1958). *Protoplasmatologia* **IV/4a,** 1.
Smith, S. H., and Schlegel, D. E. (1965). *Virology* **26,** 180.
Solberg, R. A., and Bald, J. G. (1962). *Am. J. Bot.* **49,** 149.
Solberg, R. A., and Bald, J. G. (1964). *Phytopathology* **54,** 802.
Spencer, D. F., and Kimmins, W. C. (1971). *Can. J. Bot.* **49,** 417.
Steere, R. L. (1957). *J. Biophys. Biochem. Cytol.* **3,** 45.
Steere, R. L., and Williams, R. C. (1953). *Am. J. Bot.* **40,** 81.
Štefanac, Z., and Ljubešić, N. (1971). *J. Gen. Virol.* **13,** 51.
Štefanac, Z., and Ljubešić, N. (1974). *Phytopathol. Z.* **80,** 148.
Štefanac, Z., and Miličić, D. (1965). *Phytopathol. Z.* **52,** 349.
Stols, A. L. M., Hill-van der Meulen, G. W., and Toen, M. K. I. (1970). *Virology* **40,** 168.
Sugimura, Y., and Ushiyama, R. (1975). *J. Gen. Virol.* **29,** 93.
Takahashi, W. N. (1962). *Phytopathology* **52,** 29.
Takahashi, W. N., and Ishii, M. (1953). *Am. J. Bot.* **40,** 85.
Takebe, I. (1975). *Annu. Rev. Phytopathol.* **13,** 105.
Tamada, T. (1975). *CMI/AAB Descriptions Plant Viruses.* **144,** 4 pp.
Teakle, D. S., and Steindl, D. R. L. (1969). *Virology* **37,** 139.
Thaler, I. (1956). *Protoplasma* **45,** 486.
Thaler, I. (1966). *Protoplasmatologia* **II/B/2bγ,** 1.
Thaler, I., and Amelunxen, F. (1975). *Protoplasma* **85,** 71.
Thaler, I., Amelunxen, F., and Sommer, I. (1970). *Protoplasma* **71,** 251.
Tsao, P. W. (1963). *Phytopathology* **53,** 243.
Tsuchizaki, T. (1975). *CMI/AAB Descriptions Plant Viruses* **142,** 3 pp.
Tsuchizaki, T., Hibino, H., and Saito, Y. (1973). *Phytopathology* **63,** 634.
Tsuchizaki, T., Hibino, H., and Saito, Y. (1975). *Phytopathology* **65,** 523.

Tu, J. C., and Hiruki, C. (1970). *Virology* **42,** 238.

Tu, J. C., and Hiruki, C. (1971). *Phytopathology* **61,** 862.

Turner, R. H. (1971). *J. Gen. Virol.* **13,** 177.

Unzelman, J. M., and Healey, P. L. (1972). *J. Ultrastruct. Res.* **39,** 301.

Ushiyama, R., and Matthews, R. E. F. (1970). *Virology* **42,** 293.

Van Bakel, C. H. J., and Van Oosten, H. J. (1972). *Neth. J. Plant Pathol.* **78,** 160.

Van der Scheer, C., and Groenewegen, J. (1971). *Virology* **46,** 493.

Vela, A., and Lee, P. E. (1974). *J. Ultrastruct. Res.* **47,** 169.

Vidano, C. (1966). *Atti Accad. Sci. Torino* **100,** 731.

Vidano, C. (1970). *Virology* **41,** 218.

Villiers, T. A. (1968). *Planta* **78,** 11.

Vovlas, C., Martelli, G. P., and Quacquarelli, A. (1971). *Phytopathol. Mediterr.* **10,** 244.

Vovlas, C., Russo, M., and Martelli, G. P. (1973). *Phytopathol. Mediterr.* **12,** 80.

Vovlas, C., Martelli, G. P., and Quacquarelli, A. (1974). *Phytopathol. Mediterr.* **13,** 179.

Wada, E., and Fukano, H. (1937). *J. Imp. Agr. Exp. Stn.* **3,** 93.

Walkey, D. G. A., and Webb, M. J. W. (1968). *J. Gen. Virol.* **3,** 311.

Walkey, D. G. A., and Webb, M. J. W. (1970). *J. Gen. Virol.* **7,** 159.

Warmke, H. E. (1967). *Science* **156,** 262.

Warmke, H. E. (1968). *Virology* **34,** 149.

Warmke, H. E. (1969). *Virology* **39,** 695.

Warmke, H. E. (1974). *Virology* **59,** 591.

Warmke, H. E., and Christie, R. G. (1967). *Virology* **32,** 534.

Warmke, H. E., and Edwardson, J. R. (1966a). *Virology* **28,** 693.

Warmke, H. E., and Edwardson, J. R. (1966b) *Virology* **30,** 45.

Weber, E., and Kenda, G. (1952). *Protoplasma* **41,** 378.

Wehrmeyer, W. (1960a). *Protoplasma* **51,** 165.

Wehrmeyer, W. (1960b). *Protoplasma* **51,** 242.

Weidemann, H. L., Lesemann, D., Paul, H. L., and Koenig, R. (1975). *Phytopathol. Z.* **84,** 215.

Weintraub, M., and Ragetli, H. W. J. (1966). *Virology* **28,** 290.

Weintraub, M., and Ragetli, H. W. J. (1968). *J. Cell Biol.* **38,** 316.

Weintraub, M., and Ragetli, H. W. J. (1970a). *J. Ultrastruct. Res.* **32,** 167.

Weintraub, M., and Ragetli, H. W. J. (1970b). *Virology* **41,** 729.

Weintraub, M., and Ragetli, H. W. J. (1970c). *Virology* **40,** 868.

Weintraub, M., and Ragetli, H. W. J. (1971). *J. Ultrastruct. Res.* **36,** 669.

Weintraub, M., Ragetli, H. W. J., and John, V. T. (1967). *J. Cell Biol.* **35,** 183.

Weintraub, M., Ragetli, H. W. J., and Veto, M. (1968). *Am. J. Bot.* **55,** 214.

Weintraub, M., Ragetli, H. W. J., and Veto, M. (1969). *J. Ultrastruct. Res.* **26,** 197.

Weintraub, M., Ragetli, H. W. J., and Schroeder, B. (1971). *Am. J. Bot.* **58,** 182.

Weintraub, M., Agrawal, H. O., and Ragetli, H. W. J. (1973). *Can. J. Bot.* **51,** 855.

Weintraub, M., Ragetli, H. W. J., and Lo, E. (1974). *J. Ultrastruct. Res.* **46,** 131.

Wergin, W. P., and Newcomb, E. H. (1970). *Protoplasma* **71,** 365.

Wergin, W. P., Gruber, P. J., and Newcomb, E. H. (1970). *J. Ultrastruct. Res.* **30,** 533.

Wilkins, M. H. F., Stokes, A. R., Seeds, W. E., and Oster, G. (1950). *Nature (London)* **166,** 127.

Willison, J. H. M. (1976). *J. Ultrastruct. Res.* **54,** 176.

Wilson, H. J., and Israel, H. W. (1972). *J. Cell Biol.* **55,** 284.

Wilson, H. J., Israel, H. W., and Steward, F. C. (1974a). *Phytopathology* **64,** 588.

Wilson, H. J., Israel, H. W., and Steward, F. C. (1974b). *J. Cell Sci.* **15,** 57.
Wilson, H. J., Goodman, R. M., and Israel, H. W. (1974c). *Proc. APS* **1,** 147.
Wolanski, B. S., and Chambers, T. C. (1971). *Virology* **44,** 582.
Woods, M. W., and Eck, R. V. (1948). *Phytopathology* **38,** 852.
Yamashita, S., Doi, Y., and Yora, K. (1973). *Virology* **55,** 445.
Yang, A. F., and Hamilton, R. I. (1974). *Virology* **62,** 26.
Zaitlin, M., and Beachey, R. N. (1974). *Adv. Virus Res.* **19,** 1.
Zaitlin, M., Duda, C. T., and Petti, M. A. (1973). *Virology* **53,** 300.
Zech, H. (1952). *Planta* **40,** 461.
Zech, H. (1954). *Exp. Cell Res.* **6,** 560.
Zech, H. (1960). *Virology* **11,** 499.
Zettler, F. W. (1969). *Phytopathology* **59,** 1109.
Zettler, F. W., Christie, R. G., and Edwardson, J. R. (1967). *Virology* **33,** 549.
Zettler, F. W., Edwardson, J. R., and Purcifull, D. E. (1968). *Phytopathology* **58,** 332.
Zimmermann, A. (1893). "Beiträge zur Morphologie und Physiologie der Pflanzenzelle." H. Laupp, Tubingen.

MAIZE ROUGH DWARF AND RELATED VIRUSES

Robert G. Milne and Osvaldo Lovisolo

Plant Virus Laboratory, National Research Council,
Turin, Italy

I. Introduction

When only a few reolike plant viruses were known, maize rough dwarf, rice black streaked dwarf, and Fiji disease viruses were considered to be very similar to wound-tumor virus and rice dwarf virus.* Now that more

* Abbreviations of virus names used herein: ABDV, Arrhenatherum blue dwarf virus; BTV, bluetongue virus; CPV, cytoplasmic polyhedrosis virus; CTDV, cereal

examples have come to light and more is known about them, it is clear that the MRDV-like viruses are widespread and are rather different from WTV and RDV, though, like WTV, they all cause tumors in plants. This therefore seems an appropriate time to review current knowledge of those viruses with close affinities to MRDV.

The name Reoviridae has been proposed for the family containing all reolike viruses (Fenner *et al.*, 1974). These all have (a) genomes consisting of several (generally 10–12) segments of double-stranded RNA (dsRNA) with molecular weights ranging from 0.3 to 3.0×10^6 (see Fig. 1), all segments being encapsidated within a single virus particle; and (b) a quasi-spherical capsid 60–80 nm in diameter, exhibiting icosahedral symmetry elements.

Within the Reoviridae there are the reoviruses proper, found in warm-blooded vertebrates but having no apparent vectors. A second group, the orbiviruses (causing, for example, bluetongue, Colorado tick fever, and African horse sickness), also have a wide host range among vertebrates including man, but they are morphologically somewhat different from reoviruses, and are carried by and multiply in arthropods (insects and ticks). A third group comprises the cytoplasmic polyhedrosis viruses of insects and a fourth group, the rotaviruses, causes gastroenteritis in the young of man, cattle, and mice. A fifth group is made up of the reolike viruses of plants.

The reolike plant viruses are rather similar to the reoviruses morphologically but, like the orbiviruses and CPV, they multiply in their associated insects. They appear at present to fall into two subgroups: those like WTV and RDV that have leafhopper (cicadellid) vectors and 12 genome segments; and those like MRDV that have planthopper (delphacid) vectors and 10 genome segments (see Tables I and II). However, many of the viruses concerned are very imperfectly characterized, and we may yet be in for some surprises. We shall here consider the MRDV-like viruses of the second subgroup, earlier reviewed in some aspects by Maramorosch (1969a,b), Granados (1969), Harpaz (1972), Milne and Lovisolo (1974), and Lovisolo *et al.* (1974). The members to date are MRDV, RBSDV, FDV, OSDV, CTDV, ABDV, PSV, and LEDV. We also include some discussion of MWEV as it has many similar properties, although it is transmitted by cicadellids, not delphacids. One

---

tillering disease virus; FDV, Fiji disease virus; LEDV, Lolium enation disease virus; MRDV, maize rough dwarf virus; MWEV, maize wallaby ear virus; OSDV, oat sterile dwarf virus; PSV, pangola stunt virus; RBSDV, rice black streaked dwarf virus; RDV, rice dwarf virus; RLCV, rugose leaf curl virus of clover; WTV, wound-tumor virus.

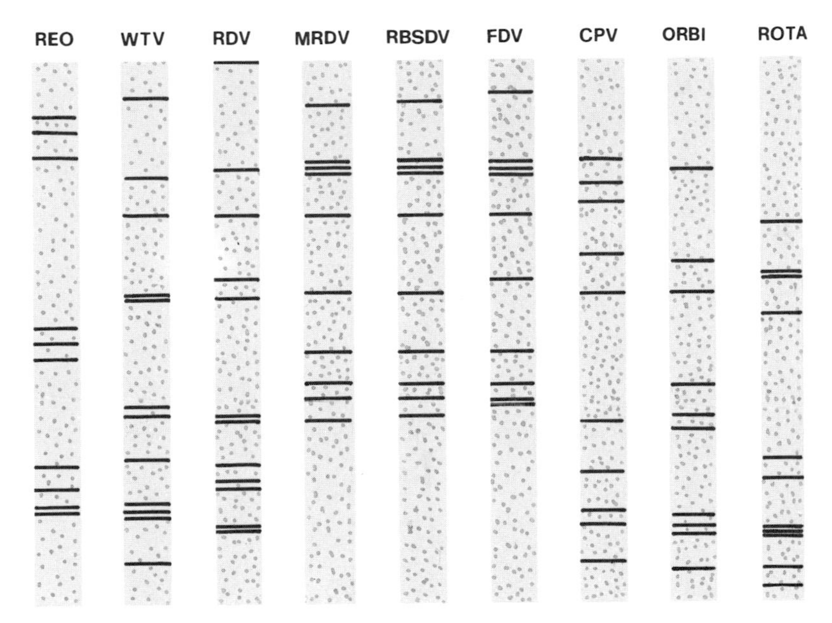

FIG. 1. Schematic representation of the mobilities of the genome segments of various members of the Reoviridae, when subjected to PAGE (polyacrylamide gel electrophoresis). The direction of movement is downwards. When two or more bands have the same or very similar mobility they are shown as separate but closely adjacent bands. Where both PTE and Loening's buffers were used, the data are those for Loening's buffer. See Section VII, D for discussion. [*References:* reovirus, Martin and Zweerink (1972); WTV and RDV, Reddy *et al.* (1974); MRDV and FDV, Reddy *et al.* (1975a); RBSDV, Reddy *et al.* (1975b); CPV, Fujii-Kawata *et al.* (1970); orbivirus, Verwoerd *et al.* (1972); rotavirus, Newman *et al.* (1975).]

or two other viruses of unclear status are discussed, and we have also briefly reviewed the virus-like effects of hopper salivary toxins as this helps to put the virus symptomatology in perspective.

Although some of the diseases caused by MRDV-like viruses have long been recognized, the viruses themselves are only now becoming known. There are several reasons for this. The viruses are not sap transmissible, and virions are only found in high concentration in the small tumors induced in plants, or in the vectors—which are also small. The tumors are not graft transmissible, perhaps primarily because of the lack of a wound-repair mechanism in the Gramineae. Laboratory propagation of adequate amounts of material is therefore difficult. Propagation of the viruses in cultured insect cells or in plant protoplasts has not yet been achieved. The virions themselves appear to be much less stable than those of WTV, RDV, or reoviruses, and this leads to problems with purification.

TABLE I

ESTIMATES OF THE MOLECULAR WEIGHTS[a] OF THE GENOME SEGMENTS OF WTV, RDV, MRDV, RBSDV, AND FDV COMPARED WITH THOSE OF REOVIRUS TYPE 3

Segment No.	Reo[b]	WTV[c]		RDV[c]		MRDV[d]		RBSDV[e]		FDV[d]	
		L[f]	P[g]	L	P	L	P	L	P	L	P
1	2.79	2.90	2.90	3.10	3.10	2.85	2.88	2.88	2.91	2.93	2.90
2	2.71	2.48	2.40	2.50	2.50	2.52	2.50	2.52	2.50	2.53	2.50
3	2.55	2.25	2.20	2.25	2.20	2.50	2.35	2.48	2.35	2.50	2.48
4	1.62	1.80	1.80	1.90	1.80	2.48	2.35	2.48	2.35	2.48	2.48
5	1.55	1.80	1.78	1.80	1.76	2.25	2.12	2.25	2.12	2.25	2.12
6	1.46	1.20	1.10	1.15	1.05	1.82	1.75	1.82	1.75	1.90	1.85
7	0.88	1.15	1.05	1.13	1.02	1.50	1.45	1.50	1.45	1.50	1.45
8	0.75	0.90	0.83	0.86	0.78	1.32	1.25	1.32	1.25	1.32	1.21
9	0.65	0.63	0.57	0.78	0.70	1.24	1.18	1.24	1.18	1.22	1.15
10	0.61	0.61	0.55	0.75	0.67	1.13	1.08	1.17	1.10	1.22	1.12
11	—	0.60	0.54	0.52	0.48	—	—	—	—	—	—
12	—	0.35	0.32	0.52	0.48	—	—	—	—	—	—
Total	15.6	16.7	16.0	17.3	16.6	19.6	18.9	19.7	19.0	19.9	19.3

[a] Daltons × 10^6.
[b] Martin and Zweerink (1972).
[c] Reddy *et al.* (1974).
[d] Reddy *et al.* (1975a).
[e] Reddy *et al.* (1975b).
[f] Estimates in Loening's buffer, taking reovirus RNA's (column 2) as standards.
[g] Estimaes in PTE buffer, taking reovirus RNA's (column 2) as standards.

Finally, with the traditional phosphotungstate negative staining, only the subviral particles may be observed in the electron microscope; this has given rise to a lot of confusion.

Nevertheless, there now exist fairly simple techniques and clear criteria for recognizing the MRDV-like viruses. They are the following:

(1) Gramineae only are infected, with symptoms of darker green color (except PSV), dwarfing, enation * production, increased tillering, and suppression of flowering on at least some hosts.

(2) Planthoppers (Delphacidae) propagatively transmit the viruses.

(3) Plant or hopper material, homogenized in small amounts of saline, clarified by low-speed centrifugation, and injected into the vectors, should render them inoculative in 1–2 weeks. (Mycoplasmas would fail to be transmitted in this way.)

* The term *enation* or *histoid enation* (Bos, 1970) is now generally accepted for neoplastic growths protruding from the veins. Some authors, especially when referring to FDV and MWEW, have used the term *gall* for the same entity, but this term properly refers to more complex growths induced by parasites.

TABLE II

SOME PROPERTIES OF THE PLANT REOLIKE VIRUSES

Properties	WTV	RLCV	RDV	MWEV	MRDV	RBSDV	FDV	OSDV	CTDV	ABDV	PSV	LEDV
Number of genome segments	12[a]	—	12[a]	—	10[b]	10[c]	10[b]	—	—	—	—	—
Total molecular weight of dsRNA $\times 10^6$	16.3[a,d]	—	16.9[a,d]	—	19[b,d]	19.3[c,d]	19.5[b,d]	—	—	—	—	—
Plant hosts Gramineae (G) or dicotyledons (D)	D	D	G	G	G	G	G	G	G	G	G	G
Insect hosts leafhoppers (L) or planthoppers (P)	L	L	L	L	P	P	P	P	P	P	P	—
Induce tumors	Yes	Yes	No	Yes	Yes	Yes	Yes	Yes	Yes	Yes	Yes	Yes
High percentage of egg transmission (more than 5%)	Yes[e]	Yes[f]	Yes[g]	Yes[h]	No[i]	No[i]	—	No[i]	No[i]	—	—	—

[a] Reddy *et al.* (1974).
[b] Reddy *et al.* (1975a).
[c] Reddy *et al.* (1975b).
[d] Means of values in Loening's and PTE buffers.
The symbol — indicates data not available.
[e] Nagaraj and Black (1962).
[f] Grylls (1954).
[g] Iida *et al.* (1972).
[h] Grylls (1975).
[i] See Section V.

(4) Thin sections of leaf enations and of infective hoppers reveal, in the cytoplasm, spherical virus particles about 70 nm in diameter containing dense cores of 50 nm diameter. Viroplasms and tubular structures also occur in infected cells. The viroplasms appear in the light microscope as X-bodies. In plants, the viruses are phloem associated.

(5) Simple dip or crush preparations of enations or swollen veins can be negatively stained for electron microscopy. Uranyl acetate reveals spherical particles of 65 nm diameter with double capsids and also spiked cores about 50 nm in diameter. Phosphotungstate reveals smooth spherical cores about 55 nm in diameter, though very occasionally the outer capsid may be retained.

(6) Antisera to three of the MRDV-like viruses are available. The viruses should also react with sera prepared against dsRNA or synthetic double-stranded polyribonucleotides such as poly I–poly C (polyinosinic–polycytidilic acid).

II. Virus-Like Effects of Hopper Salivary Toxins

The question of insect salivary toxins is introduced because in certain cases cicadellids and delphacids have been shown to cause swellings of the leaf veins, or dwarfing, in Gramineae. These symptoms mimic the effects of the viruses under discussion, and though some of the confusion generated has now been resolved, in other cases it still remains.

"Hopperburn" caused by delphacids has long been known (see Carter, 1973), but Storey was probably the first to note "enations" caused by hoppers (see Maramorosch *et al.*, 1961; Ruppel, 1965) when he found that the cicadellids *Cicadulina mbila* Naudé and *C. zeae* (China) caused leaf swellings on maize. These swellings were produced by a systemic factor, for insects feeding on the first leaf of a maize seedling would cause vein swelling on the third or fourth leaf. The symptoms were transitory and were followed by complete recovery. Maramorosch (1959) found that the cicadellid *Dalbulus elimatus* DeL. & W., in Mexico, could cause very convincing enations on maize leaves but that, again, the symptoms were transitory and not transmissible. Maramorosch *et al.* (1961) reported that the cicadellid *Cicadulina bipunctella* (Mats.), in the Philippines, caused similar symptoms of toxicity.

As species of *Cicadulina* and *Dalbulus* were known to induce vein swellings in the probable (but not proven) absence of viruses, Maramorosch (1959) and Maramorosch *et al.* (1961) received with caution the unpublished reports of N. E. Grylls and M. F. Day that the wallaby ear disease of maize in Australia was caused by a virus transmitted by *C. bipunctella bimaculata* (Evans). This was especially because Maramorosch understood that every individual of *C. bipunctella bimaculata* induced the disease. Following this, Granados (1969) and Harpaz (1972) both viewed the disease as an insect toxicosis, but Grylls (1975) has now shown that wallaby ear is indeed associated with a virus (see Section III, I, 3) and that a high proportion of, though not all, hoppers caught in the field can transmit the virus to plants. It remains possible that some wallaby ear symptoms, particularly the mild ones, might be caused by toxicosis, just as some of the "toxic" symptoms observed by H. H. Storey, K. Maramorosch, and others may have been due to transmitted viruses.

A disease which could have been caused either by a virus or by toxicosis was reported by Harder and Westdal (1971). It occurred on oats in Kenya, producing enations, some dwarfing, and abnormally dark green leaves which later became chlorotic. These symptoms were transmitted (or induced) by *C. mbila*. A second case which is so far undecided was reported by Bremer and Raatikainen (1975) from Turkey. Enations developed on barley, wheat, and oats after they were fed on by the cicadellid *Euscelis plebejus* ssp. *plebejus* (Fall.), but whether the symptoms persisted was not reported, and no attempt was made to observe virus particles. *E. plebejus* is already known to transmit a salivary toxin and a mycoplasma to clover (Maramorosch, 1953; Evanhuis, 1958; Carter, 1973).

In Finland, the delphacid *Javesella pellucida* (Fab.) causes toxic stunting of oats though enations are not induced and there are fewer rather than more tillers than normal on affected plants (see Nuorteva, 1962, 1965, and earlier papers; see also Mochida and Kisimoto, 1971; Carter, 1973). *J. pellucida* is common in Finland but causes this symptom only in one particular region. Although the situation is complicated by the presence of OSDV and of European wheat striate mosaic disease (EWSMD), the symptom is probably not caused by either of these, as male hoppers do not induce it, whereas both males and females transmit OSDV and EWSMD.

Sheffield (1968, 1969) reported from East Africa that enations not due to FDV were induced on sugarcane by an undetermined leafhopper. Enations were produced on leaves remote from the feeding site, but the symptoms were not transmissible and disappeared after the insect was removed.

Clearly, it has sometimes been difficult to decide between viruses and toxins. However, efforts should be made in future to apply some of the criteria outlined in Section I (and further discussed in the following sections) to help distinguish among the rival claims of viruses, toxins, and also mycoplasmas, which can cause confusingly similar symptoms and may have similar vectors.

III. Diseases Caused in Plants*

A. *Maize Rough Dwarf*

Much of the literature on natural and experimental infections caused by MRDV has been reviewed by Harpaz (1972) and summarized by Lovisolo (1971a,b). The disease was first described by Fenaroli (1949), Grancini (1949), and Biraghi (1949) from northern Italy, severe outbreaks in that year following closely the introduction in 1948 of higher-yielding but more susceptible North American maize hybrids. It seems clear that the virus was already present locally but that the old maize varieties were largely resistant or tolerant. A similar severe outbreak occurred in Israel in 1958, again following the introduction of American hybrids (Harpaz *et al.*, 1958; Harpaz, 1959). In both Italy and Israel, dent maize hybrids were susceptible but flint varieties were much more resistant.

MRDV has also been reported from France, Czechoslovakia, Switzerland, Yugoslavia, and Spain (see Lovisolo, 1971a; Harpaz, 1972), but little is known of its distribution in these countries. The Czechoslovakian report (Blattný *et al.*, 1965) probably refers to a mixed infection of maize by MRDV and a mycoplasma. In July 1975, O. Lovisolo, V. Valenta, and J. Matisová (unpublished data) did not succeed in finding MRDV in southern Czechoslovakia. The virus is apparently absent from the New World, where very large acreages of susceptible hybrids are grown and at least one vector is widely present (Harpaz, 1972; Lovisolo, 1971a). It is interesting that in northern Italy MRDV does not naturally infect rice, although maize and rice are often grown side by side and the vectors of the virus enter the rice crops. Experimentally, rice can be infected with difficulty and the symptoms are mild (M. Conti, personal communication, 1974).

A maize plant infected at an early stage in the field is easily recognized. The leaves become a darker green than normal, and the plant is dwarfed though the stem remains stout. The backs of the leaves are roughened by small enations developed on the veins. The enations may in extreme cases be over 10 mm long and more than 1 mm wide but are usually smaller. They are often prominent on leaf sheaths. Phloem swellings also occur in the roots but arise internally, causing longitudinal splitting, especially in adventitious roots. Later, apical necrosis may occur on the shoot, and secondary root rots set in and the plant dies; if infection occurs later, on older plants, symptoms may be milder and very late infections may produce enations, especially on the ears, but no dwarfing. Harpaz (1972) suggested that dwarfing was presumably caused by a

* For the diseases described here, the plant host ranges are given in Table III and the insect vectors in Table IV.

distinct, more severe strain of the virus, but experimentally the whole range from severe dwarfing to mild or no dwarfing can be produced by adjusting the age at which the plants are inoculated (M. Conti, unpublished data). While strains of differing severity may well exist, there is no strong evidence for them as yet.

Harpaz (1972) also suggested that maize rough dwarf symptoms might be caused by a double mycoplasma–virus infection, both agents being transmitted by the same vector, but we now know that the virus alone can produce the whole range of symptoms (Milne *et al.*, 1973).

Excess tillering, or the proliferation of shoots from the base of the plant, is not caused by natural MRDV infections in maize but is a strong symptom in experimentally infected Japanese varieties of maize (M. Conti and E. Shikata, unpublished data). Excess tillering is typical of infection in natural hosts such as *Digitaria sanguinalis* (Luisoni and Conti, 1970), *Cynodon dactylon* (Klein, 1967), and *Echinochloa crus-galli* (Vidano *et al.*, 1966a; Conti, 1972). It also occurs in experimentally infected wheat and barley (Vidano *et al.*, 1966b) and oats (Lovisolo *et al.*, 1966).

The hosts on which MRDV has so far been found naturally occurring are maize, *D. sanguinalis, E. crus-galli,* and *C. dactylon.* The virus has been transmitted experimentally to many of the Gramineae that have been tested (cf. Table III; see also Lovisolo, 1971a). Harpaz (1972) has reported failure to transmit MRDV to sugarcane and to any non-graminaceous host.

B. Rice Black Streaked Dwarf

RBSDV was first found in Japan, where it seems to have been present for many years but was only recognized as distinct from rice dwarf by Kuribayashi and Shinkai (1952). Work on the disease has been reviewed by Hashioka (1964), Iida (1969), Yasuo (1969), Ou (1972), Harpaz (1972), and Shikata (1974). RBSDV is also reported from mainland China (Anonymous, 1974), where outbreaks were first observed in 1963. The disease should be distinguished from rice leaf gall disease, in which rather similar symptoms are produced by toxins from the leafhopper *Cicadulina bipunctella* (Maramorosch *et al.*, 1961; see also Section II).

Rice plants are not usually killed by RBSDV but are stunted, dark green, and produce white waxy enations along the veins on the backs of the leaves, leaf sheaths, and culms. These enations later darken to give the typical black-streaked appearance [see the excellent photograph in Fukushi (1969)]. They are formed from phloem-derived neoplastic cells that contain cytoplasmic inclusion bodies, i.e., the viroplasms (Ogawa *et al.*, 1957; Ono, 1964; Kashiwagi, 1966). The leaves often become

TABLE III
PLANT HOST RANGES OF THE MRDV-LIKE VIRUSES

Host	Reaction to:							
	MRDV	RBSDV	FDV	OSDV	CTDV	ABDV	PSV	LEDV
Agropyron repens				−*a*	+*a*			
Agropyron tsukushiensis		−*b*						
Agrostis alba		−*b*						
Agrostis gigantea				−*a*	−*a*			
Agrostis verticillata	+*c*							
Alopecurus aequalis		+*b*						
Alopecurus japonicus		+*b*						
Alopecurus myosuroides				−*a*	+*a*			
Alopecurus pratensis		−*b*						
Andropogon distachyus	−*c*							
Anthoxanthum odoratum		−*b*						
Apera spica-venti				−*a*	+*a*			
Aristida caerulescens	−*c*							
Arrhenatherum elatius		−*b*		+*a*	−*a*	+*d*		
Avena fatua				+*a*	+*a*			
Avena orientalis				+*s*				
Avena sativa	+*e*	+*b*		+*a*	+*a*	+*d*		
Beckmannia syzigachne		+*b*						
Bromus arvensis				+?*a*	+*a*			
Bromus catharticus		−*b*						
Bromus inermis				−*a*	−*a*			
Bromus secalinus					+*a*			
Cenchrus echinatus	+*f*							
Chloris gayana	−*c*							
Coix lacryma-jobi		−*b*						

Cynodon dactylon	+*f*				
Cynosurus cristatus		+*b*	+*a*		
Dactylis glomerata			−?*a*	−?*a*	
Digitaria adscendens		+*b*			
Digitaria decumbens					±*k*,*l*
Digitaria diversinervis					−*l*
Digitaria eriantha					+*l*
Digitaria gazensis					+*l*
Digitaria longiflora					+*l*
Digitaria macroglossa					+*l*
Digitaria milanjiana					±*l*
Digitaria pentzii					+*l*
Digitaria sanguinalis	+*g*		−*a*	+*a*	+*t*
Digitaria setivalva					+*l*
Digitaria smutsii					+*l*
Digitaria swazilandensis					+*l*
Digitaria valida					±*l*
Digitaria violascens		+*b*			
Digitaria umfolozi					−*l*
Echinochloa crus-galli	+*h*	+*b*	−*a*	+*a*	
Echinochloa phyllopogon	+*h*				
Eleusine indica	−*c*	−*b*			
Eragrostis multicaulis		+*b*			
Festuca arundinacea		−*b*			
Festuca pratensis			−?*a*	+*a*	
Glyceria acutiflora		+*b*	+*a*	+*a*	
Holcus lanatus			+*a*	+*a*	
Hordeum astrachanense			+*s*		
Hordeum bulbosum	+*f*				
Hordeum murinum			+*s*		

(continued)

TABLE III (*continued*)

Host	Reaction to:							
	MRDV	RBSDV	FDV	OSDV	CTDV	ABDV	PSV	LEDV
Hordeum vulgare (= *sativum*)	+*h*	+*b*		+*a*	+*a*	+*d*		
Hordeum zeocriton				+*s*				
Lagurus ovatus				+*a*	+*a*	+?*r*		
Lolium multiflorum		+*b*		+*a*	+*a*	+*d*		+*m*
Lolium perenne	+*i*	+*b*		+*a*	+*a*	+*d*		+*m*
Lolium remotum				+*a*	+*a*			
Lolium temulentum				+*s*				
Oryza sativa	+*h,j*	+*b*						
Panicum miliaceum		+*b*						
Phalaris arundinacea		−*b*						
Phalaris canariensis				+*a*	+*a*			
Phleum bertolonii				−*r*				
Phleum pratense		+*b*		−?*a*	+*a*	+*d'*		
Poa annua		+*b*		+*a*	+*a*	+*d'*		
Poa pratensis				−?*a*	−?*a*			
Saccharum spp.	−*c*		+*n*					
Scleropoa rigida				+*s*				
Secale cereale		+*b*		+*a*	+*a*	+?*r*		
Setaria italica		+*b*						
Setaria verticillata	+*f*							
Setaria viridis		+*b*		+?*a*	+*a*			
Sorghum halepense	−*c*	−*b*						
Sorghum sudanense	+?*c*							
Sorghum vulgare	+?*h*							
Sorghum spp.		−*b*	+*o*					

Stipa bromoides	−*c*					
Trisetum bifidum		+*b*				
Trisetum flavescens						+?*d*
Triticum boeticum				+*s*		
Triticum macha				+*s*		
Triticum persicum				+*s*		
Triticum pyramidale				+*s*		
Triticum sphaerococcum				+*s*		
Triticum timofeevii				+*s*		
Triticum vulgare (= *aestivum*)	+*p*	+*b*		+*a*	+*a*	+*d*
Zea mays	+*q*	+*b*	+*o*	+?*a*	+*a*	+*d*

EXPLANATION OF SYMBOLS: +, susceptible; −, immune; ±, immune or susceptible according to clones; a blank indicates not tested. Italic letters refer to the following references: *a*, Lindsten *et al.* (1973); *b*, Shinkai (1957, 1962); *c*, Harpaz (1972); *d*, Mühle and Kempiak (1971); *e*, Lovisolo *et al.* (1966); *f*, Klein (1967); *g*, Luisoni and Conti (1970); *h*, Vidano *et al.* (1966b); *i*, Harpaz and Klein (1969); *j*, M. Conti and R. G. Milne (unpublished data); *k*, Dirven and van Hoof (1960); *l*, Hunkar *et al.* (1974); *m*, Lesemann and Huth (1975); *n*, Lyon (1910); *o*, Hutchinson *et al.* (1972); *p*, Vidano *et al.* (1966a); *q*, Biraghi (1949); *r*, Vacke and Vostřák (1974); *s*, Vacke (1974); *t*, Costa and Kitajima (1974).

TABLE IV
VECTORS OF THE MRDV-LIKE VIRUSES

Vector	Virus
Delphacodes propinqua (Fieber)	MRDV (not a natural vector but only when abdominally injected)[a]
Dicranotropis hamata (Boheman)	OSDV,[b,c] CTDV,[d] ABDV?[e]
Javesella discolor (Boheman)	OSDV,[b] ABDV?[e]
Javesella dubia (Kirsch.)	ABDV[f]
Javesella obscurella (Boheman)	OSDV[g]
Javesella pellucida (Fabricius)	ABDV,[f] MRDV,[h] OSDV,[i]
Laodelphax striatellus (Fallén)	CTDV,[d] MRDV,[j] RBSDV[k,s]
Perkinsiella saccharicida Kirkaldy	FDV[m,n,o]
Perkinsiella vastatrix (Breddin)	FDV[m,p]
Perkinsiella vitiensis Kirkaldy	FDV[m,l]
Sogatella furcifera (Horváth)	PSV[q]
Sogatella vibix (Haupt.)	MRDV[r]
Unkanodes albifascia (Matsumara)	RBSDV[s]
Unkanodes sapporona (Matsumara)	RBSDV[k]

[a] Harpaz and Klein (1969).
[b] Vacke (1964).
[c] Ikäheimo and Raatikainen (1963).
[d] Lindsten *et al.* (1973).
[e] Vacke and Vostřák (1975).
[f] Mühle and Kempiak (1971).
[g] Ikäheimo and Raatikainen (1961).
[h] Harpaz *et al.* (1965).
[i] Lindsten (1961b).
[j] Harpaz (1961).
[k] Shinkai (1966).
[l] Husain *et al.* (1965).
[m] Hughes and Robinson (1961).
[n] Mungomery and Bell (1933).
[o] North and Barber (1935).
[p] Ocfemia (1934a,b).
[q] Dirven and van Hoof (1960).
[r] Harpaz (1966).
[s] Shinkai (1967); Shikata (1974).

twisted and the roots stunted. The virus does not normally cause severe losses but may seriously attack maize, wheat, oats, and barley as well as rice (Ishii and Yasuo, 1967). The symptoms on barley, wheat, and maize are much the same as those of MRDV (Shikata and Kitagawa, 1976), and the only major difference in symptoms is that RBSDV readily attacks rice, whereas MRDV can hardly be persuaded to infect it.

The experimental host range of RBSDV is always within the Gramineae, but rather wide (Table III), reflecting perhaps the industry of the investigator (Shinkai, 1967) as much as the catholic tastes of the virus.

C. Fiji Disease

The disease of sugarcane caused by FDV probably first appeared in New Guinea and was described by Muir (1910) and Lyon (1910), who was sent samples from Fiji in 1908, following a severe outbreak of the disease. The virus is still found in New Guinea and Fiji, and has now spread to Australia, Madagascar, the New Hebrides, the Philippines, and Samoa (Edgerton, 1959; Hughes and Robinson, 1961; Hutchinson and Francki, 1973; Commonwealth Mycological Institute, 1975), where it causes major economic losses. It is still absent from several areas such as Hawaii where susceptible varieties of sugarcane are grown.

Infected plants are stunted and tufted, with shortened internodes and short, stiff, dark green leaves which usually have a deformed and "bitten" appearance. Enations up to 5 cm long appear on the leaf veins along the backs of the leaves and leaf sheaths. As with the other viruses in this group, the enations are derived from phloem proliferations (Kunkel, 1924; Giannotti and Monsarrat, 1968).

Only sugarcane *(Saccharum)* is naturally attacked, as far as is known, but the disease can be transmitted by planthoppers to other *Saccharum* species, *Sorghum* species, and maize (Hutchinson *et al.*, 1972; Hutchinson and Francki, 1973). Hayes (1974) has reported the probable occurrence of different strains of FDV in Australia, one strain producing consistently milder symptoms than normal.

For a further list of more than 250 references on FDV, see Hutchinson (1975).

D. Oat Sterile Dwarf (Oat Dwarf Tillering Disease)

OSDV was originally studied simultaneously in different parts of Europe—in Czechoslovakia (Průša, 1958; Průša *et al.*, 1959; Vacke, 1960; Vacke and Průša, 1962; Brčák *et al.*, 1972) and in Sweden (Lindsten, 1959, 1961a,b, 1966, 1970; Lindsten *et al.*, 1973). It is possible that the Czech and Swedish viruses differ slightly, but they have not been compared in the same laboratory; no sera are available, nor have the viruses yet been purified.

OSDV has also been found in Finland (Ikäheimo, 1960, 1961, 1964; Jamalainen, 1961), in Poland (Hoppe and Vacke, 1972), and in Britain (Catherall, 1970) and may be widespread in Europe. Recently, in southern England, Medaiyedu and Plumb (1976) found that 12.6% of all *Javesella pellucida* caught transmitted OSDV. Possibly the same virus was described from the Netherlands by Wit (1956) and a Czechoslovak disease first called dwarf of tall meadow oat grass, strongly attacking *Arrhenatherum elatius*, is now considered to be caused by OSDV (Vacke and Vostřák, 1975). Lindsten (1973a, 1976) suggests that in Sweden there is a mild strain of OSDV in addition to the normal one. ABDV (discussed later) is probably very similar to OSDV.

In oats, OSDV causes dwarfing, dark blue-green leaf color, and profuse tillering. Infected plants remain green and grass-like at harvest and are without heads. In field infections, the enations are very small and appear as gray streaks on the backs of the leaves. However, conspicuous enations appear on plants inoculated and reared in the glasshouse (Lindsten, 1961a; Catherall, 1970). The virus causes losses that are probably slight so far in Britain and not very severe in Sweden, but are said to be considerable in Czechoslovakia (Brčák *et al.,* 1972).

Symptoms in wheat and barley are less severe than in oats, but field infections of *Lolium perenne* and *L. multiflorum* cause stunting, tillering, a dark green color in the leaves, conspicuous enations, and premature death (Catherall, 1970).

E. Cereal Tillering Disease (Bestockningsjuka)

Cereal tillering disease was first noted in Sweden in 1971 as a severe disease of barley and oats that was distinct from OSDV (Lindsten and Gerhardson, 1971; Lindsten, 1973b). Natural infections on barley characteristically produce excess tillering and severe dwarfing with malformed leaves having serrated margins. The disease is similar but less severe on oats. No enations are produced on oats or barley, even in the glasshouse. Experimental infections in maize produce enations, dark green color, moderate dwarfing, and root splitting indistinguishable from those caused by MRDV.

The virus can be differentiated from OSDV, which occurs in the same regions in Sweden, by its host range, symptoms, and vector transmission, and on these criteria CTDV is closer to MRDV (Lindsten *et al.,* 1973; Milne *et al.,* 1975). For example, MRDV and CTDV easily infect maize, causing dwarfing and enations; OSDV infects maize with difficulty if at all, and enations do not form. MRDV and CTDV but not OSDV give clear symptoms on *Echinochloa crus-galli* and *Digitaria sanguinalis. Laodelphax striatellus* transmits MRDV and CTDV but not OSDV.

MRDV and CTDV are geographically separated but both are transmitted by *L. striatellus,* and there seem to be no firm biological differences between them. Nonetheless, Italian *J. pellucida* transmits MRDV, whereas Swedish *J. pellucida* does not transmit CTDV. But does the difference lie in the hoppers or the viruses? Apparently it is the hoppers, because M. Conti and K. Lindsten (unpublished data) have found that MRDV, like CTDV, is not transmitted by Swedish *J. pellucida.* The reciprocal test, to see if CTDV is transmitted by Italian *J. pellucida,* has not yet been made.

*F. Arrhenatherum Blue Dwarf (Blauverzwergung des Glatthafers)**

The disease caused by ABDV has been known in East Germany since 1941 (Mühle and Kempiak, 1971; Schumann, 1971; Kempiak, 1972; Milne *et al.*, 1974). It produces in *Arrhenatherum elatius* a dark blue-green coloring, excessive tillering, stunting, enations, suppression of flowering, sterility of such inflorescences as do form, and distortion and transverse splitting of the leaves.

Comparative tests in one and the same laboratory have not been done, but it appears on the basis of symptoms, host range, and vectors that ABDV is very similar to OSDV and is perhaps only a strain of this virus with a particular taste for *Arrhenatherum* (Milne *et al.*, 1974, 1975).

A disease similar to ABDV has been described in Czechoslovakia by Vacke and Vostřák (1975), and comparative tests suggest that it may be identical to OSDV. The relationships of the MRDV-like viruses occurring in Europe are not entirely clear, but Table V summarizes some of the differential properties of MRDV, CTDV, OSDV, and ABDV.

TABLE V

SOME DIFFERENTIAL PROPERTIES OF THE EUROPEAN MRDV-LIKE VIRUSES[a]

Properties	MRDV	CTDV	OSDV	ABDV
Naturally transmitted by *Laodelphax striatellus*	+	+	−	−
Egg transmitted by *L. striatellus*	+	−	NA	NA
Naturally transmitted by *Javesella pellucida*	−	−	+	+
Easily infects maize	+	+	−	−
Causes dwarfing	+	+	+	−
Causes enations	+	+	−	+
Easily infects *Arrhenatherum elatius*	NA	−	−	+
Causes dwarfing and enations	NA	−	+	+

[a] Modified from Milne *et al.* (1975). LEDV has not been included for lack of data. NA indicates data not available.

* It should be noted that a quite unrelated agent transmitted by the leafhopper *Macrosteles fascifrons* (Stål.) has been described under the name oat blue dwarf virus (see Long and Timian, 1975). The virus induces enations, dwarfing, and a blue-green color but has spherical particles 28–30 nm in diameter and also infects dicotyledons. Vacke (1970) described a similar agent under the name oat dwarf.

G. Pangola Stunt

Pangola grass is a sterile triploid of *Digitaria decumbens* grown as a pasture grass on millions of hectares throughout the world, particularly in Florida, the Caribbean, and Central and South America (see Schank *et al.,* 1972). As it is vegetatively propagated and genetically homogeneous, all plantings are in principle equally susceptible to viruses. Thus pangola stunt, first described as a devastating disease of pangola grass in Surinam (Dirven and van Hoof, 1960), is a serious threat to pangola grass cultivation. The disease is now reported from Guyana, Brazil, Peru, Fiji, and Taiwan (see Schank and Edwardson, 1968; Hunkar *et al.,* 1974), though perhaps not all examples of "pangola stunt" are the same (Hunkar *et al.,* 1974).

The symptoms include reduced growth and vigor, shortened internodes, increased tillering and branching, and dwarfing and distortion of leaves, all these producing a tufted appearance. The leaves become yellowish (a darker green coloration has not been reported), and enations have not been described, though small enations or vein swelling do in fact occur at least in the glasshouse (J. Giannotti and R. G. Milne, unpublished data). The incubation period in the plant, after insect inoculation, is reported as 2–3 months, not 10–20 days, as is typical of the group (Dirven and van Hoof, 1960). As with the other MRDV-like viruses, the virus particles are associated with the phloem. Neoplasia occurs but is limited, at least in pangola grass. Certain cells in the bundle sheaths develop abnormally thick walls, according to Schank *et al.* (1972), and this development is said to be a diagnostic feature of the disease. The host range and syptomatology outside the genus *Digitaria* have not been studied.

H. Lolium Enation Disease

Lesemann and Huth (1975) have described a disease in *Lolium perenne* and *L. multiflorum* in West Germany, that has symptoms strongly resembling those of the MRDV-like viruses. The enations contained MRDV-like virus particles, and the disease is almost certainly caused by the virus though it has not yet been shown to be the causal agent. No data on serology, host range, or vector specificity are yet available.

I. Diseases that Might Be Caused by Viruses Similar to MRDV

1. Rice Grassy Stunt

At present it is not clear whether the causal agent of rice grassy stunt is a virus or a mycoplasma. A preliminary and unpublished search for virus particles was reported briefly by Maramorosch (1969b); he and

Shikata found virus-like particles 70 nm in diameter in the cytoplasm of fat-body cells in infectious vectors, but no particles were found in infected rice plants. In a personal communication to Ou (1972), Shikata also reported the presence of mycoplasma-like bodies in diseased rice, but none of four tetracycline-type antibiotics has been successful in controlling or minimizing the symptoms (Ou, 1972; IRRI, 1968). The symptoms and vector tramsmission, described next, are consistent with both virus and mycoplasma etiology.

The disease was first recognized in the Philippines in 1962 (IRRI, 1964; Rivera *et al.*, 1966), and the symptoms have also been reported from Sri Lanka [formerly Ceylon] (Abeygunawardena, 1969), India, Malaysia, and Thailand (Ou, 1972; Ou and Rivera, 1969; Ling, 1969; Ling *et al.*, 1970). It may be the same disease as that called rice rosette (Ling, 1969; Bergonia *et al.*, 1966). The symptoms on rice are severe stunting, excess tillering, and a tufted appearance, with very erect leaves and stems. The leaves may be green or yellowish but are not dark green or blue-green in color. White or yellowish stripes and rusty spots can occur on the leaves, but enations have not been reported (though, as with pangola stunt, small enations might be present and have escaped notice). Panicle production is retarded and yield reduced, though the disease is not at present considered to be very important agriculturally. Ling *et al.* (1970) screened a number of rice varieties for resistance and found that most were susceptible, though particular lines showed some resistance.

Rice grassy stunt is transmitted in a persistent manner by the brown planthopper *Nilaparvata lugens* Stål. (see previous references). Up to 30% of the insects collected in nature were able to acquire and transmit the disease. Infectivity could be acquired in about 30 minutes and retained for up to 30 days, following a minimum incubation period of 6 days (average: 15 days). Retention of infectivity was not affected by vector molting. The minimum inoculation feeding period was 5 minutes. Tests for transmission through the egg were negative, as were attempts to transmit the agent through rice seed.

2. Oat Pseudorosette (Zakuklivanie)

Zakuklivanie occurs in oats, barley, wheat, rye, millet, and maize in Siberia and the Ukraine (Sukhov and Vovk, 1938; Protsenko, 1958; see also discussions by Lindsten, 1961a,b; Slykhuis, 1967; Harpaz, 1972). The agent is probably propagative in its vector, *Laodelphax striatellus*, and the symptoms are said to include dwarfing, excess tillering, reddening and mosaic on the leaves, and virescence and distortion of the panicles. Inclusion bodies that could have been viroplasms were described in infected plants (Sukhov, 1940a) and insects (Sukhov, 1940b). Protsenko (1958) found virus particles of dimensions 20 nm by 150 nm in infected plants, but it was not established that these virions were the cause of the disease. In sum, the status of zakuklivanie is not too clear, and indeed the symptoms may be caused by a number of different viruses or mycoplasmas.

3. Maize Wallaby Ear

Maize wallaby ear has been discussed in Section II in relation to insect toxicosis. We do not know whether the early reports of wallaby ear were due to toxins or to a virus or both, but Grylls (1975) has now shown that the classical disease is caused by a reolike virus transmitted by cicadellid hoppers.

The disease was first reported from Queensland, Australia, by Tryon (1910). Schindler (1942) showed that it was transmitted by *Cicadulina bimaculata* and suggested that it was of viral origin. A second cicadellid vector, *Nesoclutha pallida* (Evans), was found by Grylls (1975). The disease has been discussed by Granados (1969) and Harpaz (1972), who both, following Maramorosch *et al.* (1961), viewed it as an insect toxicosis.

MWEV produces symptoms very similar to those of MRDV in maize: dwarfing, a dark geen color, and numerous enations on the backs of the leaves. Leaves of infected plants stand up more stiffly than do those of normal plants, hence the name "wallaby ear." MWEV has been experimentally transmitted to rice, wheat, rye, sorghum, sugarcane, and nine species of grasses.

The transmission of the virus by leafhoppers, not planthoppers, should place it outside the group we are discussing, and the resemblance of the symptoms to those of MRDV may simply reflect the fact that both viruses inhabit the phloem. However, one further justification for considering MWEV here is that in phosphotungstate (PTA) negative stain, the virus closely resembles the PTA-generated smooth cores of MRDV-like viruses (see Fig. 7C in Section VII, A). This is interesting because when WTV or RDV (both transmitted by leafhoppers) are treated with PTA, the particles remain intact and no core is generated. Indeed, MRDV-like cores are not easily produced by any treatment of WTV or RDV (L. M. Black, personal communication, 1976; Iida *et al.*, 1972), though cores somewhat resembling those of MRDV can be produced from WTV by treatment with trypsin or urea (Streissle and Granados, 1968).

4. Colombian Enanismo

A disease called *enanismo,* causing stunting, rosetting, and enations in barley, wheat, and oats in Colombia, was reported by Gálvez *et al.* (1963). It was transmitted by the leafhopper *Cicadulina pastusae* Rup. &

Del. Virus particles were not looked for, but it seems probable that a virus similar to either MWEV or oat blue dwarf (see Section III, F) was involved.

5. *Reolike Viruses in Peregrinus maidis*

Recently, Trujillo *et al.* (1974) reported from Venezuela a virus in maize causing stunting (but not enations or dark green color) that was transmitted by the planthopper *Peregrinus maidis* (Ashm.). It gave, in PTA negative stain, particles typical of PTA-generated MRDV cores.

In a paper not referred to by Trujillo *et al.*, Herold and Munz (1967) had earlier found, also in Venezuela, a reolike virus in *P. maidis* while they were studying the quite different virus of maize mosaic, transmitted by the same insect. The virus had, in thin sections, a typically reolike appearance, with a dense core and a weakly contrasted envelope. The virus was said to be 54 nm in diameter in thin sections, but measurements taken from the figures presented indicate diameters varying from 50 to 75 nm for the whole virion. The authors noted the similarity of their virus to MRDV, but attempts to transmit it to maize, to six other species of Gramineae, and to various dicotyledons were unsuccessful. The authors also noted the presence of smaller particles generated by damaging the larger, and suggested that the damage was caused by PTA.

If these two viruses are the same or related, it is curious that the Herold and Munz virus could not be transmitted to plants. On the other hand, it is also curious that this virus remains, so far, the only reolike planthopper virus on record without a plant host.

A disease transmitted by *P. maidis* and thought to be caused by maize line virus (MLV) was reported in East Africa by Kulkarni (1973). The symptoms were reminiscent of those of MRDV, especially regarding the numerous small enations on the underside of the leaves. The isometric particles associated with the disease were 28 and 34 nm in diameter, but Kulkarni did not succeed in transmitting the disease with them. The only other viruses known to induce enations on Gramineae are oat blue dwarf (transmitted by cicadellids) and the MRDV-like viruses. It seems therefor that MLV is a third enation-inducing type of virus in Gramineae. Alternatively, it is possible that the small spherical virus was not responsible for the enations and that these were due to an MRDV-like virus which escaped notice.

6. *A Reolike Virus in Typhlocyba*

Maillet and Folliot (1967) reported a reolike virus in the leafhopper *Typhlocyba douglasi* Edw. No plant hosts for this virus have yet been reported.

IV. Transmission by Sap, Grafting, Dodder, and Seed

Classical sap transmission involving rubbing the leaves with infectious material plus an abrasive has never succeeded with the MRDV-like viruses. This is not surprising, as they must presumably enter viable phloem in order to propagate. Harpaz (1959) reported that MRDV could be inefficiently transmitted by injecting infectious material into the stems of young maize seedlings or pricking them with pins dipped in inoculum. At the time, this was useful, for no insect vector had yet been found.

Grafting is likewise a completely negative or very poor transmission technique, as the hosts are Gramineae. Harpaz (1972) reported no success with MRDV; Blattný *et al.* (1965) claimed two transmissions of MRDV out of 14, but the symptoms transmitted were leaf stripes not typical of MRDV (see Section III, A). Grylls (1975) had some success in transmitting MWEV by side-approach grafting of adventitious root buds.

There are no reports of the transmission of MRDV-like viruses by dodder *(Cuscuta)* or through seed, despite many attempts.

V. Virus–Vector Relationships

A. *Transmission*

All the plant reolike viruses are transmitted either by leafhoppers or planthoppers. These insects are classified, described, and discussed by Kisimoto (1973) and Ishihara (1969), and some planthopper virus vectors are described in more detail by Nasu (1969) and Harpaz (1972). Raatikainen (1967) and Mochida and Kisimoto (1971) have published reviews of the biology of *Javesella pellucida,* and the flight behavior of leaf- and planthoppers has been described by Raatikainen and Vasarainen (1973). The leafhoppers (Cicadellidae) belong to the superfamily Cicadelloidea, the planthoppers (Delphacidae) to the superfamily Fulgoroidea, and both superfamilies are classified as Auchenorrhyncha, Homoptera, Hemiptera.

Table IV lists the known vectors of the MRDV-like viruses. Some of these vectors are only known to transmit the viruses under experimental conditions but not in nature. All the vectors are planthoppers and in no case have leafhoppers been shown to transmit. Viruses like WTV and RDV, on the other hand, have leafhopper vectors exclusively. In view of this (so far) clear distinction, which parallels other virus properties (Table II), it is interesting that MWEV is transmitted by leafhoppers yet has symptoms and host range similar to those of MRDV. N. E. Grylls (personal communication, 1975) has tested the planthopper *Pere-*

grinus maidis and found that it does not transmit MWEV, but neither is this insect known to transmit any of the recognized MRDV-like viruses.* A critical test using locally available insects such as *Perkinsiella,* which are known vectors of MRDV-like viruses, has not yet been made with MWEV but would be interesting.

The general pattern of transmission with MRDV-like viruses is shown in Table VI. Hoppers of either sex and any instar or biotype can acquire virus by feeding on infected plants for periods of about 24 hours [though FDV can apparently not be acquired after the first nymphal instar (Hutchinson and Francki, 1973)]. Usually, young instars acquire virus more easily than do mature insects. After acquisition there follows a latent or incubation period of 1–3 weeks during which the virus multiplies in the hoppers and invades most tissues, including the salivary glands. For example, Shikata and Kitagawa (1976) reported that RBSDV in *L. striatellus* first multiplied in the gut, then the hemolymph, then the salivary gland and, last of all, the fat body. The insects began to transmit virus on the same day that infectivity was first detected in the salivary gland.

The incubation period completed, the insects can transmit virus for the rest of their lives by feeding on susceptible plants for minimum periods of about an hour. For RBSDV the minimum has been shown to be as little as 5 minutes (Shinkai, 1962; Shikata, 1974), and with MRDV the minimum so far observed is 15–30 minutes (M. Conti, personal communication, 1975). Details of times for acquisition, incubation, and transmission vary from virus to virus and in most cases have not been defined precisely. They may well reflect local and experimental differences rather than real divergences.

Usually about 30% of hoppers that have had access to virus can finally transmit, though why the other 70% do not is at present unclear. When hoppers are injected abdominally with virus preparations (to be discussed shortly), the percentage of survivors that transmit is often 50% (e.g., see Milne *et al.,* 1973). This may be because the injected preparations are particularly infectious, but the gut may normally form a partial obstacle to virus establishment, as has been found for other viruses and vectors (see Sinha, 1968, 1973). Sinha and Reddy (1964) found that individuals of the leafhopper *Agallia constricta* (V.D.) could contain WTV antigens irrespective of whether or not they could act as vectors. This indicates that viruses may have trouble not only getting into their vectors, but also getting out.

Lindsten (1962) suggested that different races of *J. pellucida* may have

* Kisimoto (1973) incorrectly lists *P. maidis* as a vector of MRDV.

TABLE VI

VIRUS–VECTOR RELATIONSHIPS OF THE MRDV-LIKE VIRUSES

Virus	Natural vector(s)	Minimum efficient times for:			Egg transmission	Ref.
		Acquisition	Incubation	Inoculation		
MRDV	*Laodelphax striatellus*	1 day	10 days	5 hours	Yes, max. 4%	*a,b*
RBSDV	*L. striatellus*	30 minutes	7 days	5 minutes	No	*c,d*
FDV	*Perkinsiella saccharicida* (in Queensland, Australia)	—	—	1 day	—	*e*
	P. vastatrix (in the Philippines)	—	—	1 day	—	*e*
OSDV	*Javesella pellucida*	—	7 days	—	Yes, 0.2%	*f,g*
CTDV	*L. striatellus*	1 day	14 days	1 day	No	*h,i*
ABDV	*J. pellucida*	1 hour	16 days	4 hours	—	*j*
PSV	*Sogatella furcifera*	—	2 months	—	—	*k*

[a] Harpaz (1972).
[b] Conti (1966).
[c] Shinkai (1962).
[d] Shikata (1974).
[e] Hutchinson and Francki (1973).
[f] Vacke (1966).
[g] Lindsten (1970).
[h] Lindsten (1973a).
[i] Lindsten *et al.* (1973).
[j] Mühle and Kempiak (1971).
[k] Dirven and van Hoof (1960).

differing success in transmitting OSDV. More recently, K. Lindsten and M. Conti (unpublished data) have found that *J. pellucida* from Sweden will transmit neither MRDV nor CTDV, though the morphologically identical *J. pellucida* from Italy is a vector of MRDV. *L. striatellus* has been shown to vary genetically in its ability to acquire and transmit rice stripe virus (Kisimoto, 1967) though similar work has not been done with the MRDV-like viruses.

In practical terms, infectivity tests are conducted by allowing hoppers to make an acquisition feed of 24 hours on the infected plants, then transferring them for the duration of the incubation period to seedlings of a suitable food plant (e.g., wheat or barley for *L. striatellus*). After, say, 1 week, individual hoppers are caged on indicator seedlings at the coleoptile stage (germinated for 48 hours in the dark) and again allowed to remain for at least 24 hours (see Harpaz *et al.*, 1965; Harpaz, 1972). With *L. striatellus* and MRDV, it happens that maize is a good indicator for the virus but a poor host for the hopper (Harpaz, 1961), so seedlings of wheat or barley are included in the cage with the maize (Milne *et al.*, 1973). With RBSDV (Shikata and Kitagawa, 1976), the hoppers, after virus acquisition, are caged on healthy rice seedlings for up to 15 days and then tested on 2- to 3-day-old maize, rice, or barley seedlings.

Purified preparations of RBSDV (Kitagawa and Shikata, 1969a,b; Shikata and Kitagawa, 1976) or MRDV (Milne *et al.*, 1973) have been injected into *L. striatellus* to check that infectivity is indeed associated with the virus particles. The technique is to anesthetize the hoppers with CO_2 in a cold room and inject them ventrally in the abdomen with about 0.1 μl of the extract, using a glass capillary (see Conti, 1969). Because of injury, about 25% of the hoppers die within 24 hours, and a further 25% do not survive the minimum of 7 days necessary to become inoculative. Of the rest, some 30–60% transmit virus if originally injected with infectious extracts. Ikegami (1975) did not investigate the infectivity of his purified FDV because there is a high mortality of the vector, *Perkinsiella saccharicida*, when it is anesthetized by such means as low temperature, CO_2, or nitrogen (Sinclair and Osborn, 1970). Further, R. I. B. Francki and C. J. Grivell (personal communication, cited in Ikegami, 1975) also found that injected insects survived poorly.

Because of these rather discouraging facts, attempts have been made (see Harpaz, 1972) to transmit MRDV to hoppers by letting them feed on virus preparations through stretched Parafilm M membranes. The hoppers fed and survived satisfactorily but did not in any instance become inoculative. It is not clear whether the concentration of infectious virus reaching the gut was simply too low or whether some additional acquisition factor was lacking (see Lung and Pirone, 1974). Perhaps the addition of poly-L-ornithine (Pirone and Shaw, 1973; Pirone and Kassanis, 1975) would be useful. For a review of methods used to rear leafhoppers and planthoppers (mostly *L. striatellus*) on artificial media, see Mitsuhashi (1974).

No attempts have so far been made to grow the MRDV-like viruses either in cultured insect cells or in plant protoplasts. The problem in each case is similar: both planthopper cells and graminaceous protoplasts have proved difficult to culture. However, the host range of a virus in cultured cells, either plant or animal, is sometimes wider than in intact organisms, so it might be possible to grow these viruses in leafhopper cells, for which the technology is well established (e.g., see Reddy and Black, 1972). The possibility of growing the MRDV-like viruses in plant protoplasts seems slim at present; an extra problem is that, at least in the intact plant, the viruses are phloem specific, and protoplasts come from leaf parenchyma.

The question of transovarial transmission is a difficult one because, when this occurs at a very low level, extreme precautions must be taken to make sure that the emerging nymphs do not accidentally acquire virus from infected plants and also that food plants have not become infected, though they may still be symptomless. Transovarial transmission was reported to occur with MRDV in *L. striatellus* (Harpaz and Klein, 1969; see also Harpaz, 1972) to a level of 4% at best. The authors started with about 1000 eggs laid by viruliferous females on wheat. Two days before hatching, the eggs were transferred to healthy wheat and allowed to develop. Only about 100 hoppers reached the fourth instar, and of these apparently four transmitted MRDV. However, the symptoms produced by these progeny in maize were atypical, mild, and nondwarfing; their relation to the normal MRDV symptoms is not at present clear. As wheat is susceptible to MRDV, Harpaz (1972) reported a second trial in which transmitting females of *L. striatellus* deposited eggs on sorghum, which is largely or completely immune to MRDV. None of 300 progeny from these eggs proved to be a transmitter.

M. Conti (unpublished data) obtained no transovarial transmission of MRDV in *L. striatellus* when testing 120 progeny of viruliferous mothers, and Lindsten (1974) reported finding no evidence for the transovarial passage of CTDV in *L. striatellus* or OSDV in *J. pellucida*. Vacke (1966) reported transovarial transmission of OSDV in *J. pellucida* at a level of 0.2%.

No transmission of RBSDV occurs through the eggs of the vectors (Shinkai, 1962; see also Shikata, 1974). The question has not been investigated for FDV (Hutchinson and Francki, 1973) or for the other MRDV-like viruses.

To sum up, it seems that the situation needs clarification, but that a very small percentage of egg transmission may occur with some of these viruses. This is probably not enough to affect the epidemiological balance. It should be borne in mind that hopper–virus relationships can be very plastic and that data from one experimental situation may not be valid elsewhere. For example, Nagaraj and Black (1962) showed that four races of *Agallia constricta* differed widely in their ability to acquire WTV and transmit it through the egg. The percentage of progeny transmitting the virus varied from 10 to 84%.

B. *Effect of the Virus on the Vector*

Although the effect of the virus on the vector is an interesting subject, it seems to have been studied only by Harpaz, who investigated the effects of MRDV in *L. striatellus*. His results appear in his book (Harpaz, 1972) but have yet to be repeated by independent workers.

Harpaz reported that inoculative females laid 30–50% fewer eggs than virus-free females and that the viability of these eggs was poor (14% as against the normal 99% hatch). In those cases where hatching did occur, the incubation period was up to 3 days (about 25%) longer than that of normal eggs, and the mortality of the resulting larvae was high. The cumulative mortality was about 95%. In contrast, Harpaz found little or no effect of the virus on the longevity of adult hoppers and concluded that, although the developing stages were affected severely, the adults were highly tolerant.

This situation is parallel to that of RDV infecting its vector, *Inazuma dorsalis* (Motsch.), where nymphs infected through the egg tended to die prematurely (Shinkai, 1962; see also Iida *et al.*, 1972). Also, Nasu (1963) reported that *L. striatellus* transmitting rice stripe virus laid eggs that failed to hatch. Similar results were reported for *J. pellucida* infected with European wheat striate mosaic disease, but Kisimoto and Watson (1965) pointed out that these may have been due to inbreeding of the hopper colonies. When this factor was taken into account, the evidence for a pathogenic effect of the agent became much less strong.

Klein and Harpaz (1969) and Harpaz (1972) reported that infection of certain plants by MRDV could modify the capacity of those plants to support MRDV vectors. Thus *L. striatellus* is unable to survive on healthy *Cynodon dactylon* (Bermuda grass) for longer than 4–7 days and does not molt or lay eggs on this plant, but it can survive and breed successfully if the *C. dactylon* is infected with MRDV or if the hopper is already MRDV inoculative when placed on the plant.

In a related but perhaps different phenomenon, Harpaz (1972) reported that virus-free *L. striatellus* did not breed well on sorghum but that if the hoppers carried MRDV they reproduced much more successfully. The effect of the virus was presumably on the hopper, not the plant, as sorghum is probably immune to MRDV. Harpaz (1972) also reported that the planthopper *Delphacodes propinqua* could only survive and develop on barley, wheat, and oats if it was carrying MRDV. For a more general review of the harmful and beneficial effects of plant viruses on insects, see Maramorosch and Jensen (1963) and Jensen (1969).

C. Epidemiology and Control

Egg transmission by hopper vectors seems to be of little or no importance for the spread of the MRDV-like viruses in the field. There is also no evidence so far that these viruses can spread directly from hopper to hopper in any other way, though this has not been thoroughly investigated. Likewise, transmission from infected plants to their seedlings seems never to occur, and no infection passes from plant to plant by contact or other direct means. In vegetatively propagated plants such as pangola grass or sugarcane, the viruses can of course be easily spread, but with annual cereals like rice, wheat, maize, oats, or barley, there has to be a cycle of infection involving both hoppers and plants.

This cycle must operate in one of three climatic and agricultural situations: (1) where a cold winter, such as occurs in parts of Europe or Japan, is followed by the growing of one main crop of cereals; (2) where, as in Israel, a cool winter is followed by an early growing season and this is succeeded by a hot dry summer; (3) where, in tropical or subtropical situations, growth of both plants and insects is continuous. In this last case, no weak links in the cycle are apparent, and this is especially so where, for example, two crops of rice per year or even five crops every 2 years are grown. How in fact do the viruses behave in these situations? What preventive measures can be taken?

(1) The known natural plant hosts of MRDV in northern Italy are all annuals that die off in the winter, so the virus probably does not survive the winter in plants (Conti, 1972, 1974, 1976). The perennial grass *C. dactylon* could constitute an overwinter reservoir but has not so far proved to be one. The only natural vector of importance is *Laodelphax striatellus,* which overwinters as a nymph in diapause. Conti (1972, 1976) has shown that it does harbor the virus over the winter and that in spring the MRDV can be transmitted to maize and other plants (see Fig. 2). Thus it seems that overwintering viruliferous hoppers are the main—perhaps the only—link between succeeding annual infections in plants. The hoppers can be attacked with a combination of weed control and insecticide applied to the borders of maize fields in early spring (Grancini and Corte, 1969; see also discussions in Conti, 1972; Harpaz,

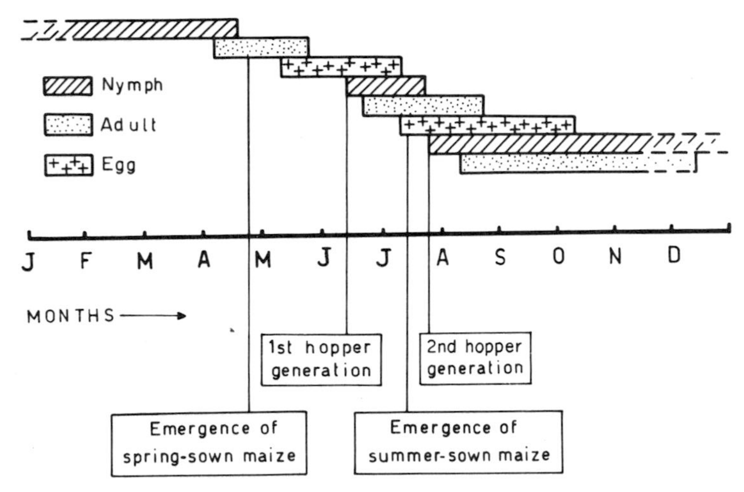

FIG 2. Representation of the life cycle of *Laodelphax striatellus* in northern Italy, in relation to the emergence of maize crops. (From Conti, 1975.)

1972). Conti (1972) also noted that *L. striatellus* is often parasitized by a dryinid hymenopteran which exercises a degree of biological control that might be further exploited.

In spring, both the numbers of *L. striatellus* and the proportion carrying MRDV increase to a peak. In summer both the numbers and the proportion infected fall off, perhaps because of high temperatures (as discussed later in this section), but in autumn both numbers and the proportion of virus carriers builds up again.

In Scandinavia, CTDV is transmitted by *L. striatellus,* though fortunately not by the Swedish biotype of *J. pellucida,* which is more common. The virus almost certainly overwinters in the diapausing hoppers and probably also survives in autumn-sown cereals in leys (Lindsten *et al.*, 1973; Lindsten, 1974). OSDV, in Europe and Scandinavia, overwinters in its vector *J. pellucida* (Lindsten, 1961a,b) and also in autumn-sown cereals, in leys, and perhaps among perennial weed grasses. Lindsten (1970, 1974) noted that the number of OSDV-carrying overwintering hoppers was particularly high in grass leys sown under oats (as opposed to barley or wheat), and even higher when these leys contained *Lolium multiflorum,* a good host for the virus. Adjustment of agricultural practices could therefore be of great benefit here.

RBSDV in Japan is transmitted mainly by *L. striatellus* and overwinters in the hoppers. It also overwinters in infected wheat, barley, and wild grasses and in secondary or ratoon growth from the rice stubbles (Shinkai, 1962; see also Harpaz, 1972; Iida and Shinkai, 1969).

(2) In Israel, as in Italy, the only known natural plant hosts of

MRDV are annual grasses that do not survive over the winter, and the vector *L. striatellus* carries the virus through the winter diapause. The hot summer, however, is a considerable obstacle for both the hoppers, which decline greatly in numbers (Harpaz, 1961; see also Harpaz, 1972), and the virus, which does not easily get established in the hopper if the mean daily temperature rises above 24°C (Klein and Harpaz, 1970). This happens (e.g., at Tel Aviv) during June, July, August, and September (Harpaz, 1972). It seems that high temperature does not adversely affect the virus itself but rather blocks some early step during incubation in the insect, as was earlier found for WTV by Sinha (1967). MRDV in Israel probably survives the summer mainly in weed grasses, particularly where there is irrigation, though Harpaz does not comment in detail on this situation. Infected maize crops are not a good source of virus for *L. striatellus*, and though they may constitute a considerable reservoir, they do not contribute much to virus survival (Harpaz, 1972).

In both situations (1) and (2), the diapausing hoppers mature into winged forms in spring and migrate over quite a short time period, perhaps a month (see Fig. 2). If this migration occurs when crops are at a susceptible stage, then damage may be severe; on the other hand, it may be negligible if the migration is over before the crops emerge or if the crop is already well established and somewhat resistant before the hoppers migrate. Thus, Lindsten (1966) noted that cereal crops in Sweden sown in autumn were essentially immune to OSDV brought in with the spring hopper migration, though spring-sown crops were susceptible. Maize also develops resistance to MRDV as it matures, and is most susceptible to infection when just emerging (Harpaz, 1972). When 2 weeks old, the maize is already largely immune, though it remains attractive to the hoppers when quite mature (M. Conti, personal communication, 1975). Thus, displacing the normal planting date either forward or backward to avoid the spring hopper migration is one of the most effective measures against MRDV, provided this displacement is compatible with other factors (Harpaz, 1961, 1972; Klein and Harpaz, 1969; Grancini, 1958; Blattný *et al.*, 1965).

Kisimoto (1969) noted that in Japan, the spring migration of *L. striatellus* in May and June coincides with the emergence of the rice seedlings. This is, however, important not so much on account of RBSDV, but because *L. striatellus* also carries the more damaging rice stripe virus into the crop. The leafhopper *Nephotettix cincticeps* Uhler likewise carries rice dwarf virus and rice yellow dwarf (a mycoplasma) into the crops in spring. The synchronized nature of the life cycle of the vectors and rice growing is considered to be the main cause of serious occurrences of these viruses, and attempts are being made in Japan to forecast vector density and virus occurrence in order to plan insecticide spraying programs (see Kisimoto, 1969; Sasaba, 1974).

(3) There is, unfortunately, little information available on the epidemiology of the MRDV-like viruses in tropical or subtropical countries (but see Anonymous, 1969). The best means of control is likely to be the development of resistant varieties wherever this is justified by the severity of the disease.

What is known about genetic sources of resistance is briefly given in the following paragraphs. Harpaz (1972) lists the maize varieties tested by himself and by Grancini (1962) and Scossirolli (1950, 1951) for susceptibility to MRDV. Some sources of resistance were found, but no breeding program has been developed because the virus is not generally severe and can be controlled by other means. Native flint varieties were more resistant than dent hybrids of American origin, and early-maturing varieties were more susceptible than those maturing later.

Similarly, Sakurai (1969) reported that little work had been done on the resistance of rice to RBSDV because the disease is not severe compared, for example, with rice dwarf and stripe. However, Ishii *et al.* (1966), Yasu *et al.* (1966), Morinaka and Sakurai (1965a,b, 1966, 1967) and Morinaka *et al.* (1969) noted examples of field resistance, tested many rice varieties, and investigated the genetics of resistance. All native *japonica* types of *Oryza sativa* were susceptible to RBSDV, but a few foreign *indica* varieties showed resistance. Considerable work has been done on the genetics of the resistance of rice to other virus diseases (see Toriyama, 1969).

Lindsten (1966) suggested there was not much possibility of developing resistance to OSDV in oats, whereas barley showed more promise.

Hutchinson and Francki (1973) noted that FDV in sugarcane cannot be controlled by heat treatments that are effective with some other sugarcane diseases. While general hygiene, roguing, and insecticides are partially successful, the only permanent control lies in the use of resistant varieties. Some of these have been developed to the commercial stage (Hutchinson and Francki, 1973; Hutchinson, 1969).

All pangola grass, *Digitaria decumbens,* is genetically the same, vegetatively propagated, and equally susceptible to PSV. However, Schank *et al.* (1972), and Hunkar *et al.* (1974) have reported on the susceptibility of some newly developed *Digitaria* hybrids. The genus and its hybrids vary widely in their response to PSV, and several new introductions promisingly showed no symptoms after 8 years of field exposure, though others showed extreme susceptibility.

VI. Structure of Infected Cells

Infected plant cells generally contain cytoplasmic X-bodies visible by light microscopy. The FDV bodies, now known to be viroplasms, were first described by McWhorter (1922) and Kunkel (1924), who suggested that they might be parasitic amoebas. (See note added in proof.)

The histopathology of the enations and root tumors induced by the MRDV-like viruses has been described by Kunkel (1924) and Giannotti and Monsarrat (1968) for FDV, by Biraghi (1952) for MRDV, by Kashiwagi (1966) for RBSDV, and by Schank *et al.* (1972) for PSV. The phloem of infected plants produces, in certain regions, an uncontrolled growth of undifferentiated or partly differentiated phloem parenchyma together with enlarged sieve tubes. Xylem or other tissues are not usually affected, though with PSV the bundle sheaths may be invaded (Schank *et al.,* 1972). The growths can be considered as neoplastic tumors similar to those produced by WTV (Black, 1972), but growth ceases after some time. As the tumors are produced in monocotyledonous plants, transplantation is difficult, and nothing has been published on the growth *in vitro* or the physiology of enation tissue. Ultrastructural studies on virus-infected tissues will be reviewed in this section.

A. *Maize Rough Dwarf in Plants*

All members of the Reoviridae (except the CPVs) so far examined in thin sections of infected tissue present a rather similar and characteristic appearance. Virus particles some 70 nm in diameter are found in the cytoplasm in association with viroplasms or virus factories, but nuclei and other parts of the cell are relatively unaffected and are not directly concerned in virus synthesis in any of the examples investigated. All the MRDV-like viruses follow this pattern; in plants, the phloem and phloem parenchyma are invaded throughout the vascular system of the plant, but other tissues are virus free. The viroplasms develop in the cytoplasm, and within these bodies immature virus particles, about 50 nm in diameter, appear (Figs. 3 and 4). Mature 70 nm in virions (Fig. 4) are later found outside the viroplasm, where they often form large crystalline arrays (Fig. 3). Frequently, the virions are enclosed in unbranched tubes (Figs. 4 and 5). Tissues invaded by MRDV often have cell walls that are distorted, thickened, or formed into finger-like projections.

This picture [summarized by Lovisolo (1971a)] has been built up by many workers, starting with Gerola *et al.* (1966a,b), Lovisolo and Conti (1966), and Gerola and Bassi (1966), who examined sections of MRDV-

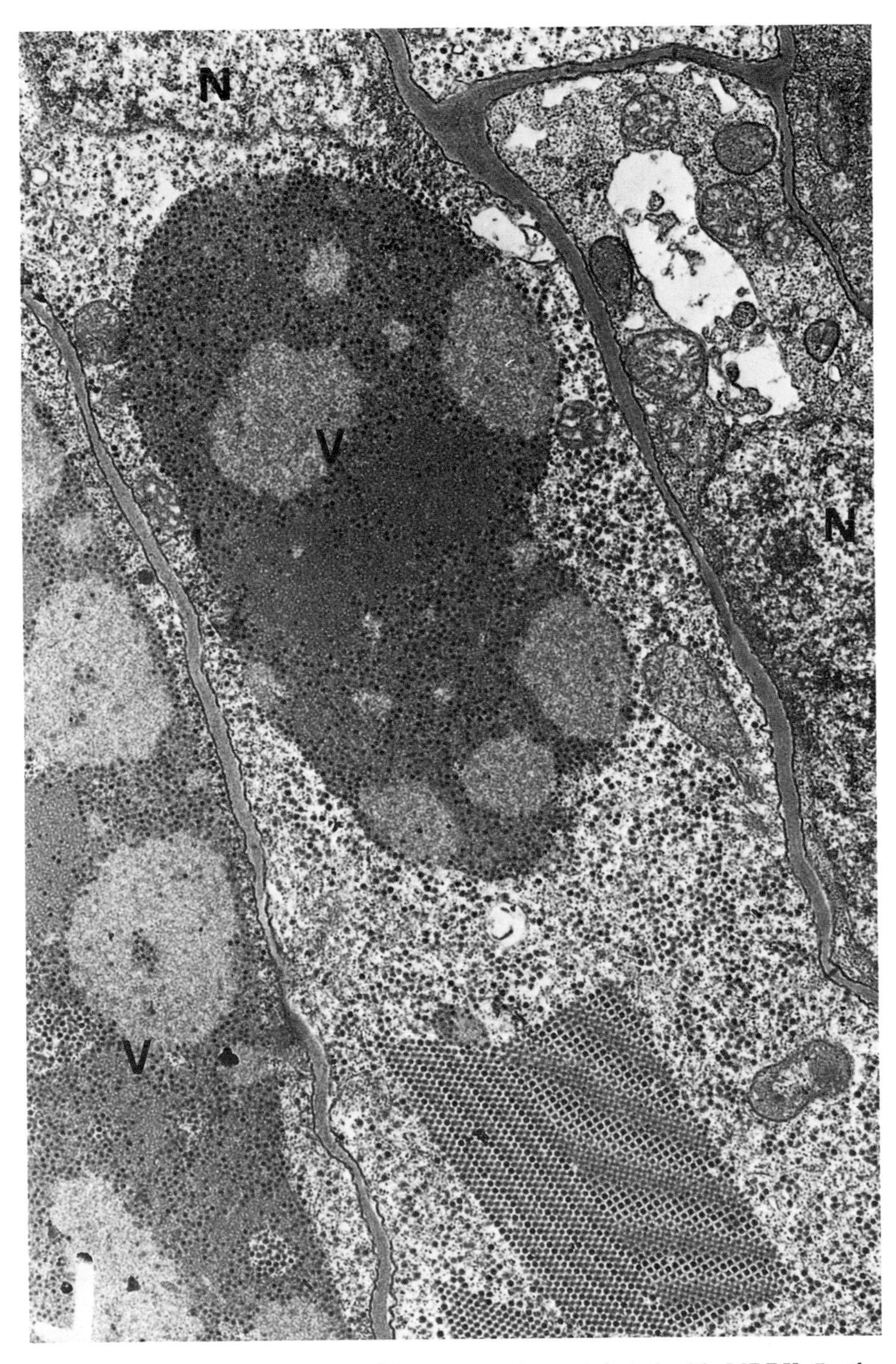

FIG. 3. Section of maize root phloem parenchyma infected with MRDV. In the central cell, part of the nucleus (N) is seen at the top. The viroplasm (V) is divided into distinct regions containing fibrillar material (pale), empty cores (darker), and full cores (black). Mature virions are seen as larger black spots free in the cytoplasm or in crystalline arrays (bottom). ×15,000. (Courtesy of A. Appiano, unpublished.)

induced enations in maize leaves. Confirmatory and additional results were obtained, with MRDV, by Kislev *et al.* (1968), Lesemann (1972), and Bassi *et al.* (1974). Bassi and Favali (1972) embedded MRDV leaf material in glycol methacrylate after glutaraldehyde fixation and found that pronase digested much of the viroplasm and the unbranched tubes, suggesting that they contained protein. The viral cores were digested by ribonuclease, but treatment with both pronase and ribonuclease was required to digest completed virions.

Favali *et al.* (1974) used [^{3}H]-uridine labeling and autoradiography to show that, as suspected, viral RNA is synthesized in the viroplasms and then incorporated into immature particles, which then mature and pass to other parts of the cytoplasm. There was no labeling of the nucleus, the chloroplasts, or the mitochondria. A sequence of events similar to that proposed by Favali *et al.* had already been suggested for WTV by Shikata and Maramorosch (1967) but without evidence from incorporation of label.

Bassi *et al.* (1974) also investigated by autoradiography the wall modifications induced by MRDV in plant using incorporation of [^{3}H]-glucose and [^{3}H]-phenylalanine.

The 50–55 nm cores of MRDV and related viruses that occur in the viroplasm are nascent virions. However, Lesemann (1972) reported "empty" cores free in the cytoplasm, and such cores have also been observed by A. Appiano and O. Lovisolo (unpublished data) and R. G. Milne (unpublished data). Because of the structure and relative instability of MRDV-like virions [see Section VII and the discussion of MWEV (Section III, I, 3)], it may be expected that free cores will be observed in sections, particularly in older infections, as breakdown products of the virion.

The tubes, some 100 nm in diameter, that often contain virus particles, are a curious feature of MRDV infections in both plants and insects (see Conti and Lovisolo, 1971). They are also found associated with the other MRDV-like viruses, and similar tubes are seen in cells infected with WTV (Shikata and Maramorosch, 1965; Hirumi *et al.*, 1967), with RDV (Fukushi *et al.*, 1962), and with BTV (Bowne and Jones, 1966). Lesemann (1972) showed that the MRDV-associated tubes have a wall of regularly repeating subunits. These subunits are about 4 nm in diameter, in both thin sections and negative stain [Milne (unpublished data); cf. Figs. 4 and 5B), and, as noted by Bassi and Favali (1972), they probably consist of protein. Questions of whether the tube is a gene product of the cell or the virus, what role it plays, and whether it differs chemically or antigenically from one virus to another are not yet answered.

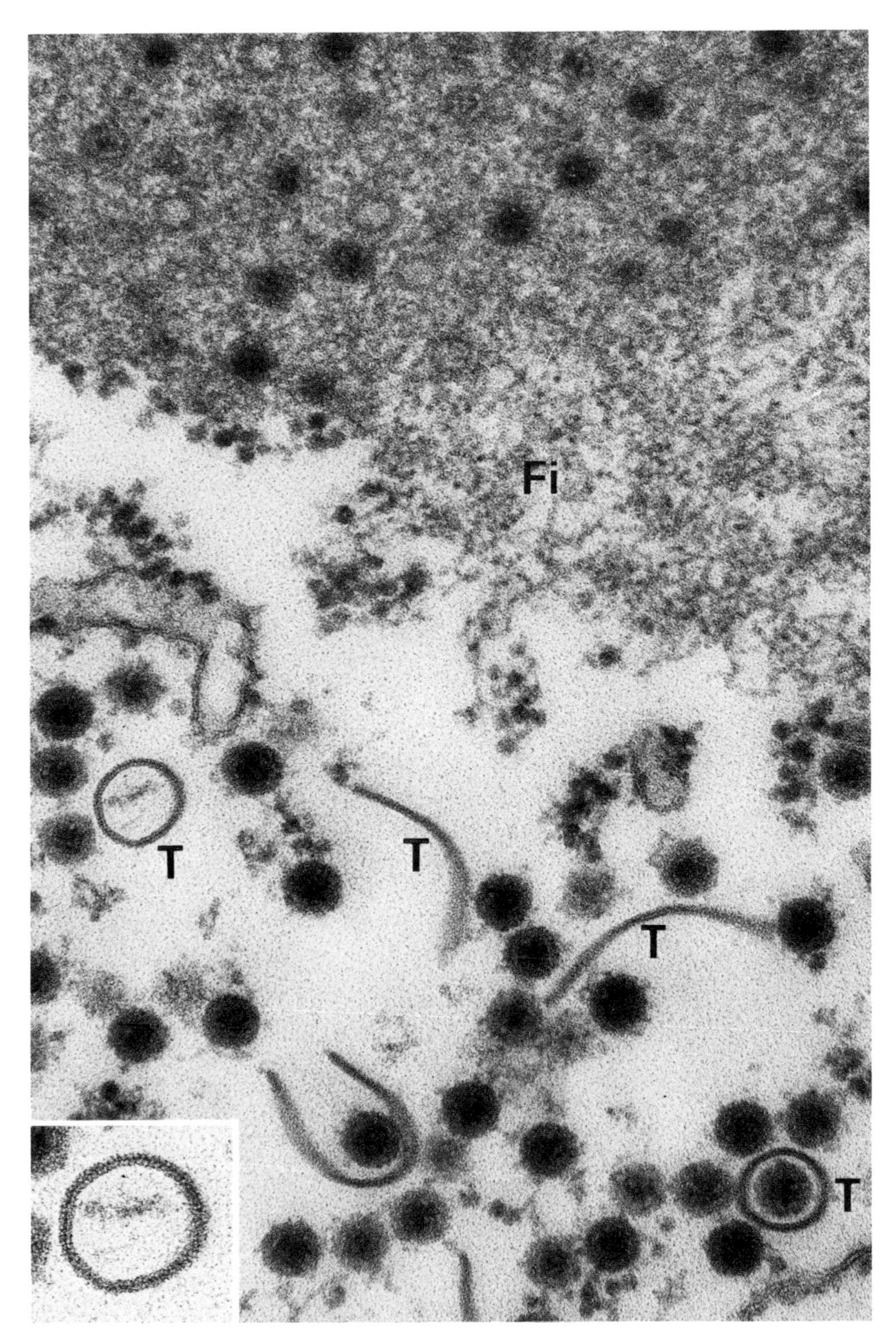

FIG. 4. Enlarged view of part of an MRDV-infected cell. At the top, part of a viroplasm is seen, containing empty cores, full cores, or fibrils only (Fi); below are mature virions. Tubes or tube-related structures are indicated by T. ×110,000. [One tube, in cross section, is enlarged to show its wall structure (*inset*). ×180,000.]

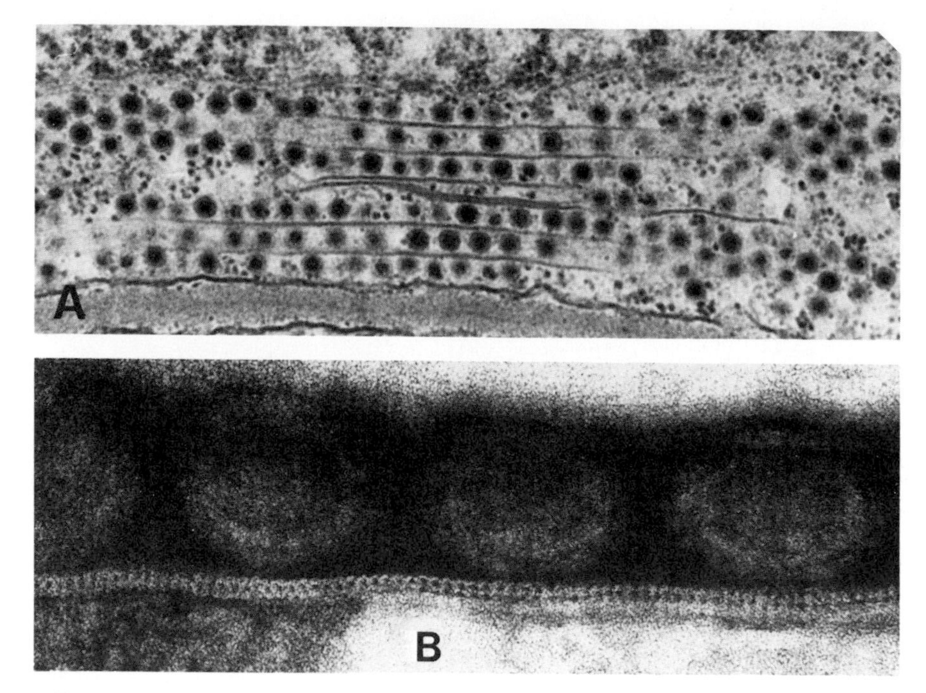

FIG. 5. (A) Part of an MRDV-infected cell showing tubular structures containing virus particles. ×40,000. (Courtesy of A. Appiano, unpublished.) (B) A uranyl acetate–contrasted crude preparation of MRDV showing a tube; note the structure of the wall. ×400,000.

B. Maize Rough Dwarf in Insects

While MRDV was being examined in thin sections of plants, parallel work with the virus in its vector *Laodelphax striatellus* was done by Vidano and his co-workers (Gerola *et al.*, 1966a; Vidano and Bassi, 1966; Vidano, 1966, 1967a,b, 1968, 1969a,b, 1970). It became clear that MRDV multiplied in the same manner as in plants, invading the cytoplasm, at first producing viroplasms, then immature particles, and finally mature virions and virus-containing tubes. The virus was found in nearly all organs of the insect, and most frequently in the mycetome and fat body.

During the course of his work, Vidano (e.g., see Vidano, 1970) elaborated a scheme for the morphogenesis of MRDV, involving lysosomes and tubules of various kinds that were found in association with the virus. This scheme was not based on kinetic studies and has not been confirmed by others who have worked on the development of MRDV-like viruses in insects or plants. It seems possible that some of the structures noted by Vidano may have been due to coinfection with a second insect virus unrelated to MRDV.

C. The Other MRDV-Like Viruses

All the other viruses here considered (Section III, A–H) have been shown to present an appearance similar to that of MRDV in thin sections of infected plants. Where the insect vectors have been examined, the result has been the same; in fact, in thin sections the MRDV-like viruses are indistinguishable from each other. The reported size of the virion has varied from 65–70 nm (Lesemann, 1972) to 80–90 nm (Shikata, 1969) but these differences are probably due to technical problems. A commonly reported diameter for the virion, in thin sections, is 70 nm, and for its densely staining core, 50 nm. Some details are given in the following paragraphs.

OSDV has been investigated in thin sections by Brčák *et al.* (1970) and by Brčák and Králík (1969a,b, 1971). Lindsten *et al.* (1973) compared OSDV and CTDV and found minor experimental but probably not real differences in the size and distribution of particles. ABDV has been examined in thin sections by Milne *et al.* (1974) and LEDV by Lesemann and Huth (1975).

The appearance of RBSDV in thin sections has been summarized by Shikata (1974) and described by Nasu (1965), Shikata (1969), Shikata and Kitagawa (1976), and the Virus Research Group, Shanghai Institute of Biochemistry (Anonymous, 1974). Shikata (1969) compared in some detail the appearance of cells infected with RBSDV, RDV, and WTV. He found little difference (apart from the lack of neoplasia with RDV), though he emphasized the presence of 50–55 nm viral cores in cells infected with RBSDV. It is not clear whether Nasu (1965) and Shikata (1969) observed free cores or only those associated with the viroplasm. The pictures of the Chinese workers are difficult to interpret as they used methacrylate embedding, but in addition to virions and cores, they found smaller spherical particles 32 nm in diameter in infected hoppers. It is unlikely that these particles were connected with RBSDV, and they may have been the particles of a virus native to the hopper. The problem was discussed by Wetter *et al.* (1969), who also found small spherical particles in preparations of both "healthy" and MRDV-infected hoppers (see Section VII, A).

FDV was first observed in thin sections by Giannotti *et al.* (1968) and has also been examined by Teakle and Steindl (1969) and Francki and Grivell (1972). The work was summarized by Hutchinson and Francki (1973). In all respects, FDV appears as a typical member of the MRDV group, except that virus-containing tubes have not been reported. This is a minor point, but it would be surprising if such tubes do not in fact occur.

PSV has been observed in thin sections by Schank and Edwardson (1968) and Schank *et al.* (1972), and was carefully described by Kitajima and Costa (1970). J. Giannotti (unpublished data) has confirmed and amplified their results. In both pangola grass and in the hopper vector, the virus conforms to the pattern of MRDV. Schank *et al.* (1972) suggested that the presence of thick-walled bundle sheath cells might be diagnostic for PSV.

VII. The Virions

A. Purification

To clarify the interpretation of the results summarized in this section, it may be useful to remind the reader of two problems. First, most if not all the MRDV-like viruses easily break down to give B-spiked cores (see Section VII, C and Fig. 6, E–G). Second, both these cores and complete virions are generally stripped down to smooth cores (Section VII, C and Fig. 7C) in the presence of PTA. Unfortunately, PTA has almost exclusively been the negative stain employed when assessing, by electron microscopy, the intermediate or final products of purification; this despite much evidence that PTA can be destructive. With many papers, it is hard to see whether the virus was reduced to cores during purification or only during the final negative staining step on the grid.

Only three of the MRDV-like viruses (MRDV, RBSDV, and FDV) have been purified or partially purified, and the starting points have generally been the methods used earlier for reoviruses, WTV, and RDV (Brakke, 1969; Suzuki, 1969). These methods have served as useful though not always appropriate models, so they are here briefly introduced.

Fig. 6. Images of purified MRDV in negative stain. (A) Virions contrasted with UA. Note the six knobs (A spikes) around the periphery. (B) The same, showing more clearly the "subunits" in the outer capsid. (C,D) The same, but with the outer capsid partially stripped to reveal the inner capsid and some B spikes; both A and B spikes are arrowed. (E,F,G) The same, but with the A spikes and labile parts of the outer capsid lost to reveal the B-spiked core. (H) A virion fixed in glutaraldehyde and contrasted with PTA; the virion has a larger diameter than in UA, and the A spikes are more acute. (I) An unfixed virion treated with PTA, showing the smooth inner core; some of the detached outer capsomeres are seen above the core. (J) Selected images of detached B spikes; first column, face view, second column, on edge. All ×400,000 except J, which is ×600,000.

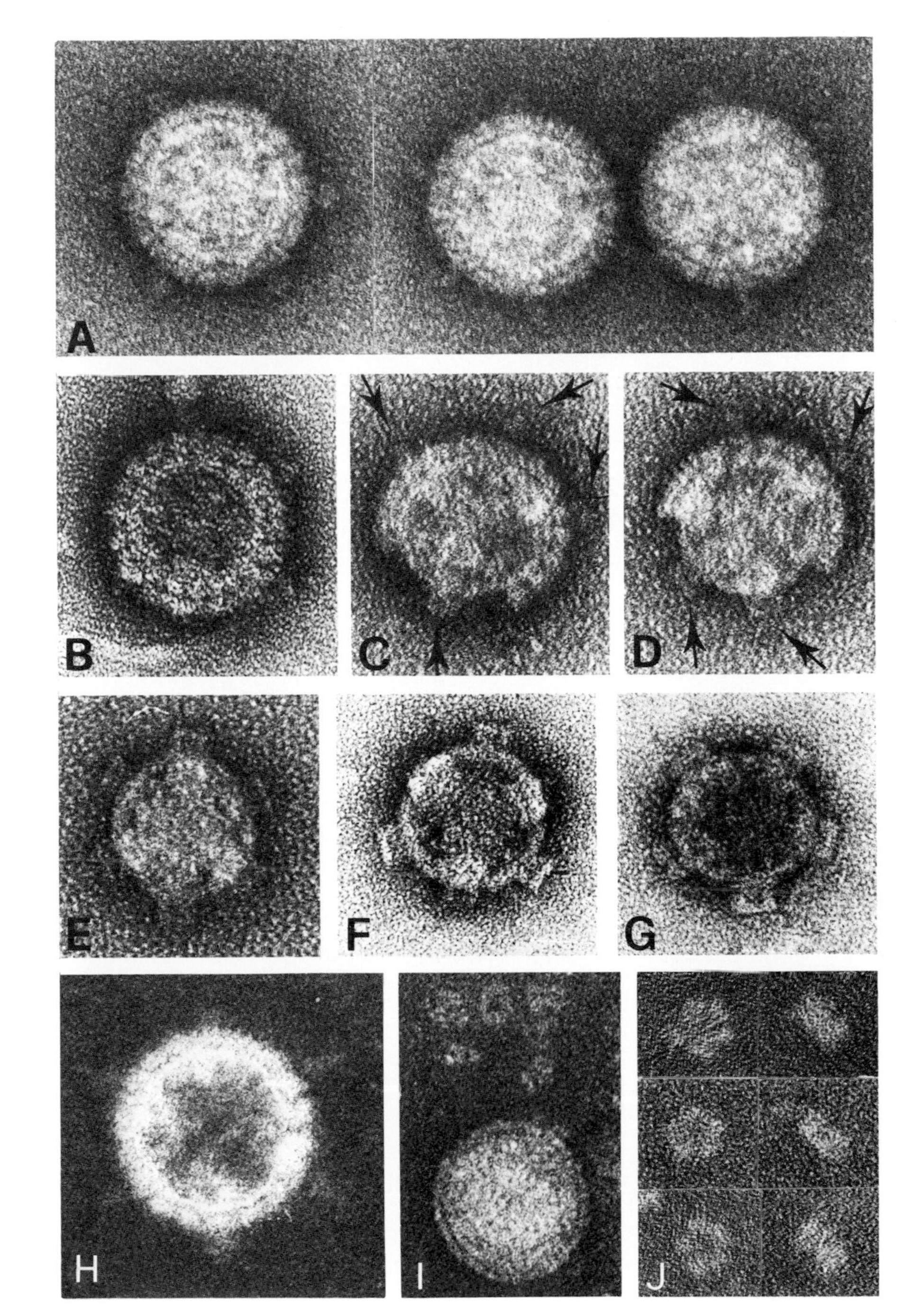
A
B
C
D
E
F
G
H
I
J

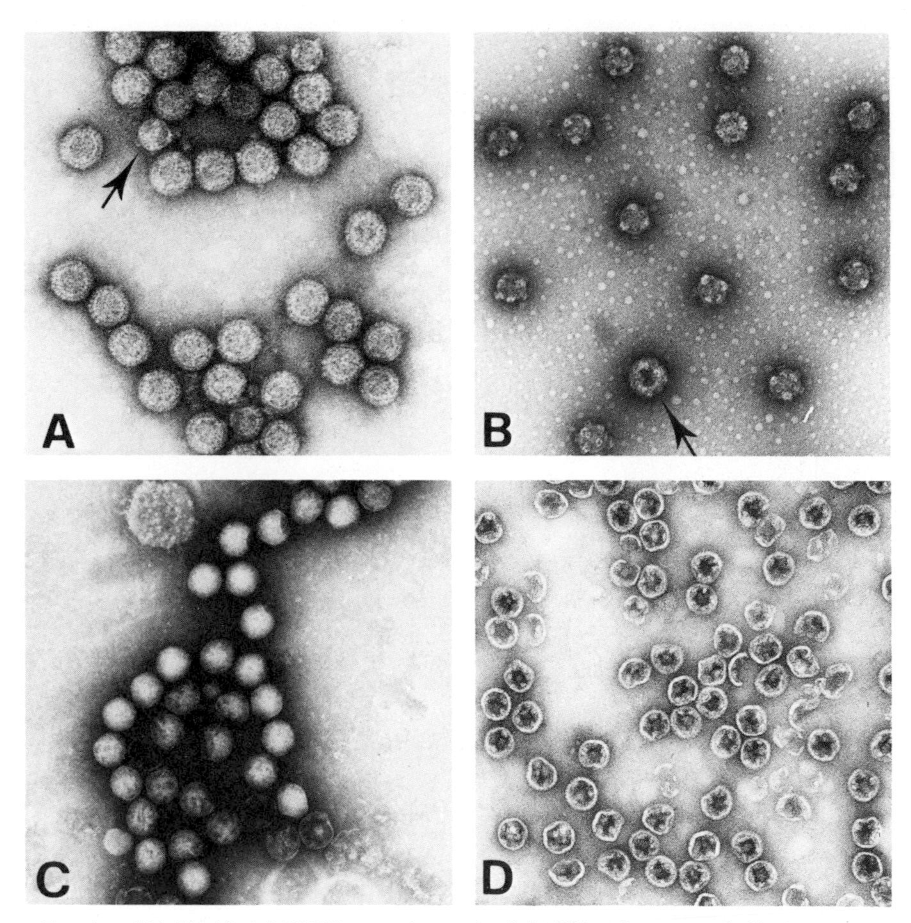

FIG. 7. (A) Purified MRDV negative-stained in UA; the arrow indicates a contaminating spiked core. (B) Purified spiked cores; the arrow indicates a contaminating virion. (C) Purified virions negative-stained in PTA, showing only smooth cores. (D) Purified smooth empty cores prepared by butanol treatment of virions, negative-stained in UA. All ×77,000.

Reoviruses are purified from cells, using clarification of the cell homogenate with Freon 113, differential centrifugation, sucrose density gradient banding, a further pelleting step, and then a density gradient separation on CsCl, followed finally by dialysis (e.g., see Smith *et al.*, 1969). The buffer used is generally dilute phosphate, pH 7.2, or SSC (i.e., 0.15 *M* sodium chloride, 0.015 *M* sodium citrate, pH 7.5). All operations are done in the cold.

WTV is currently purified from tumors (Reddy and Black, 1973a; Reddy and Lesnaw, 1971) using polyethylene glycol (PEG) 6000 (the same as Carbowax 6000) for clarification and concentration. The main advantage is that the specific infectivity of the resulting virus is much higher than with previous methods. After grinding up material, passing it through cheese cloth, and giving it a low-speed spin, Reddy and Black added solid NaCl to a level of 0.3 *M*, followed by PEG to a concentration of 4%. The mixture was left at 4°C for 2 hours to flocculate the virus, but if appreciably longer times were used, host materials also precipitated. The precipitate was suspended in a small volume of buffer and then banded on a sucrose density gradient. The buffer used for the PEG step contained 0.1 *M* histidine and 0.01 *M* $MgCl_2$, pH 5.7, whereas later steps took place in 0.1 *M* glycine, 0.01 *M* $MgCl_2$, pH 7 (Brakke, 1956). The authors noted that the conditions were rather critical and that those optimal for total virus yield were not always best for specific activity; for example, use of 0.2 *M* NaCl yielded one-third less virus, but this had a specific activity one-third higher compared with that prepared using 0.3 *M* NaCl. Reddy and Black also noted that when the expected virus yield is small, it may be useful to collect the purified virus using Millipore filters.

Streissle and Granados (1968) used Genetron 113 (the same as Freon 113, namely, 1,1,2-trifluoro-1,2,2-trichloroethane) in the purification of WTV, and Black and Knight (1970) used both Genesolv D (Freon 113) and CCl_4 for the same purpose. Neither paper gave data on the yield or specific activity of their virus; however, Reddy and Black (1973a) cited a personal communication from I. Kimura to the effect that WTV purified using Genetron had a reduced specific activity. As an alternative to a final density-gradient centrifugation, Suzuki (1969) described how WTV could be purified using density-gradient–stabilized electrophoresis, first described by Brakke (1953).

PEG has been used to clarify and concentrate RDV (Kimura, 1973), but RDV has also been purified by somewhat different methods. Toyoda *et al.* (1965) and Miura *et al.* (1966) ground up infected leaves in 0.033 *M* phosphate buffer, pH 6.8, passed the slurry through cheese cloth, and separated the virus by differential centrifugation. The pellet, resuspended in 0.025 *M* Tris, pH 7.2, was digested with crude pancreatic phosphatase for 48 hours at 5°C to remove an "outer membrane." The virus was then adsorbed to a DEAE cellulose column and eluted at concentrations of NaCl between 0.2 and 0.25 *M*, in the Tris buffer. The virus was pelleted once more and resuspended in water. Suzuki (1969; see also Suzuki and Kimura, 1969) used snake venom phospholipase as an alternative to pancreatic phosphatase.

Partial purification of MRDV was reported by Wetter *et al.* (1969), with preliminary reports of the work by Lovisolo *et al.* (1967), Luisoni *et al.* (1968), and Conti *et al.* (1968). Membranous material largely deriving from chloroplasts was a major contaminant of preparations made from leaves, and so roots were used. They contained relatively large amounts of virus, and the membrane problem was indeed reduced. Unfortunately, roots of plants raised and infected in the glasshouse are less satisfactory than those collected in the field.

Roots were ground in 0.1 *M* glycine and 0.01 *M* $MgCl_2$ (Brakke, 1956) or in an extraction solution containing 0.03 *M* Na_2HPO_4, 0.01 *M* Na_2SO_3, and 0.001 *M* EDTA (ethylenediaminetetraacetate); final pH 7.5–7.8. Clarification was attempted with Celite and with chloroform, and this was followed by differential centrifugation and banding on a sucrose density gradient. A light-scattering zone, occurring in infected but not in healthy extracts, contained smooth cores 55–60 nm in diameter when observed in PTA. Chloroform-treated particles appeared to be damaged, in comparison with those treated only with diatomaceous earth. Although ultraviolet spectra and infectivity data were not obtained, the preparations were found to be immunogenic. Sera raised from preparations treated only by differential and density-gradient centrifugation gave titers of 1:512 against the virus, and up to 1:16 against healthy plant material; sera obtained from preparations also clarified with chloroform had titers up to 1:2048 against virus and 0 or 1:2 against healthy plant extracts.

MRDV was also partially purified from viruliferous hoppers, though this was only possible using chloroform. In the hopper extracts, the usual 55–60 nm cores were observed, but in addition virus-like particles 35–40 nm in diameter appeared. Particles similar to these were also found in small amounts in extracts of nonviruliferous hoppers, so they were possibly the particles of a virus inhabiting only the hopper. Similar particles have been observed (Anonymous, 1974) in sections of hoppers carrying RBSDV, but the status of healthy hoppers was not reported in this case. Such particles have not been found associated with any other MRDV-like viruses; but while purifying RBSDV from hoppers, Shikata and Kitagawa (1976) found a 70–75 nm virus-like particle with prominent spikes in both healthy and RBSDV-carrying hoppers. This particle was quite different from RBSDV, which appeared as a 60 nm smooth sphere in their system.

Redolfi *et al.* (1973), using methods similar to those of Wetter *et al.* (1969), found that their product consisted of "full" 400-S particles containing RNA and "empty" 200-S particles containing little or no RNA. Redolfi and co-workers were the first to recognize these as subviral particles and not complete virions. Both types of particle appeared smooth in PTA, but when Lesemann (1972) examined the preparations in uranyl formate, or in PTA after fixation, he found the smooth cores adorned with 12 icosahedrally placed spikes. Complete virions were not present.

Milne *et al.* (1973) used Freon 113 followed by two sucrose density-gradient separations and a cycle of differential centrigugation to purify complete infectious virions of MRDV. The extraction solution of Wetter *et al.* (1969) was used, and the virus was observed in uranyl formate or uranyl acetate (UA) negative stains (Fig. 7A), or in PTA after fixation (Fig. 6H). Fixation of the virus in suspension did not succeed, but virus attached to support films on electron microscope (EM) grids could be fixed; virus also withstood PTA treatment after being washed with UA. The purified product corresponded in shape and size with the virus as seen in thin sections, and the yield was about 3 mg per 100 g of roots, judging from the fact that resuspending the final pellet in 2 ml of buffer gave 0.5 OD_{260} units per milliliter. Carbon tetrachloride was found to be less reliable than Freon 113 as a clarifying agent and gave a less infectious product; chloroform treatment stripped the virus down to spiked cores. Pelleting and resuspension was also found to degrade a proportion of the virus, and the final product always contained about 30% spiked cores.

Some methods by which subviral particles of MRDV can be obtained were described or confirmed by Milne *et al.* (1973), Redolfi and Boccardo (1974), and Boccardo and Milne (1975). By using GMT buffer (i.e., 0.3 *M* glycine, 0.03 *M* $MgCl_2$, and 0.05 *M* Tris–HCl, pH 7.5), 400-S cores were produced, having a full complement of RNA. Use of chloroform produced a mixture of 400-S (full) and 200-S (empty) spiked cores. Warming virions in the extraction buffer of Wetter *et al.* (1969) for 10 minutes at 50°C gave empty spiked cores (Fig. 7C), and further PTA treatment produced empty smooth cores. Shaking virus or spiked cores briefly with *n*-butanol yielded empty smooth cores (Fig. 7D). These various kinds of subviral particle have been purified and analyzed (Boccardo and Milne, 1975; Luisoni *et al.*, 1975), though the complementary missing parts of the virion were not recovered. Recently Milne and Boccardo (unpublished work) have found that treatment of virions at pH 7 with 2–10% CsCl or NaCl reduces them to smooth cores. This effect may be similar to that of PTA.

Milne *et al.* (1973) observed that the smooth cores produced by brief PTA treatment of virions were not infectious. Crude virus extracts clarified with $CHCl_3$ were very poorly infectious (2% transmission, compared with 46% in the non-$CHCl_3$-treated control) and purified virus shaken for 5 minutes with a one-third volume of $CHCl_3$ lost all detectable infectivity. This last result is in contrast, though not in direct conflict with, that obtained by Shikata and Kitagawa when using $CHCl_3$ on RBSDV (discussed next), and the question should be carefully reinvestigated for both viruses. In comparison, spiked cores of reoviruses, obtained by chymotrypsin treatment, generally have only 0.01–0.0001% the infectivity of whole virus (Joklik, 1972).

RBSDV was purified by Kitagawa and Shikata (1969a,b, 1971; see also Shikata, 1974). Leaves were ground with cold 0.1 *M* phosphate buffer, pH 7, and the homogenate, after passing through cheese cloth, was shaken with CCl_4 or a fluorocarbon (an unspecified trifluorotrichloroethane, probably the same as Freon 113) or with charcoal, as a clarifying step. Chloroform or chloroform–butanol mixtures were found to destroy infectivity almost entirely. The CCl_4 or fluorocarbon treatment was followed by one cycle of differential centrifugation, resuspension in 0.01 *M* phosphate buffer, pH 7, and then sucrose density gradient banding. [Shikata (1974) reports the buffer used as 0.001 *M* phosphate; this may be a misprint.] High infectivity, as judged by injection into hoppers, was associated with the visible virus band, and the ultraviolet absorption spectrum was typical of nucleoprotein. However, it is not clear whether complete virions or subviral particles were obtained, and the electron microscope revealed only the 60 nm smooth cores characteristic of PTA negative staining. Probably, the final product was a mixture of virions and B-spiked cores, familiar products with MRDV purified in a similar way. Shikata (1974) reported that the purification method used for plants was also successful in purifying the virus from insects.

The aforementioned authors also sometimes used a second treatment with CCl_4, after the first clarification and ultracentrifugation. R. I. B. Francki (personal communication, 1975) and Milne (unpublished data) have similarly subjected partially purified preparations of FDV and MRDV, respectively, to second treatments with Freon 113. In both cases this tended to disrupt the particles, though they had survived the first Freon treatment.

Shikata and Kitagawa (1976) have reported that, with RBSDV, $CHCl_3$-treated crude preparations are not infectious when injected into hoppers but that infectivity is regained as subsequent purification proceeds. This is interesting, if confirmed, because $CHCl_3$ invariably produces spiked cores with MRDV (as discussed earlier). If $CHCl_3$ also produces spiked cores with RBSDV (unfortunately, we do not know), then these cores are infectious. The same authors have also reported that purified virus treated with *n*-butanol retains some infectivity. As we have seen, with MRDV such treatment produces smooth empty cores whose infectivity has not been tested but which contain almost no nucleic acid.

The purification of FDV has been carefully investigated by Ikegami and Francki (1974) and Ikegami (1975), though, partly because of difficulties with the starting material, only subviral particles have so far been purified. These particles were not negative-stain artifacts, as they appeared the same in UA (which is not known to degrade any MRDV-like virus) or in freeze-dried and shadowed preparations. Unlike MRDV in maize, FDV appears not to be abundant in the roots of sugarcane, but only in the leaf galls or enations that must be painstakingly collected with a razor blade.

Ikegami and Francki ingeniously made use of the fact that sera prepared against poly I–poly C also react with any dsRNA—including that of FDV. Tissue fractions and preparations were first deproteinized and then assayed in agar double-diffusion plates for the presence of the viral dsRNA antigen. In this way, the following points were established. The glycine–$MgCl_2$ buffer used for WTV by Black (1965) and the SSC buffer used by Smith *et al.* (1969) to purify reovirus both led to severe losses of FDV in the low-speed pellets unless EDTA was included. The best resuspension buffer found contained 0.1 *M* NaCl, 0.05 *M* Tris–HCl and 0.005 *M* EDTA, pH 7.5 (STE buffer). Neutral or pH 7.5 buffers were suitable, but the use of pH 8 or 8.5 again led to loss of antigen to the low-speed pellets.

Use of chloroform, CCl_4, or Freon 113 resulted in serious losses, as did filtration through Celite (which in any case did not remove enough of the green membrane material). Attempted precipitation of the virus with 0.3 *M* NaCl and 4%, 6%, or 8% PEG was not successful, and with 10% PEG much of the host material was also precipitated. Both PEG 4000 and PEG 6000 were tried. [However, Reddy *et al.* (1975a) report the purification of FDV using PEG.] The nonionic detergent Nonidet P40 was successful in removing membrane contamination when used at 1% for 10 minutes at 4°C, and this step was incorporated in the purification.

The final purification schedule adopted involved extraction in 0.1 *M* glycine and 0.005 *M* EDTA, pH 8.5, adjustment to pH 7.5, differential centrifugation, and resuspension of the pellet in STE buffer. This suspension was treated with Nonidet P40, the virus was pelleted through a sucrose cushion, and the resuspended pellet banded in a sucrose density gradient, to be collected and dialyzed overnight against STE. The product was finally pelleted once more before resuspension in the required buffer. Preparations from about 10 g of galls finally suspended in 1 ml of water had an OD_{260} of between 0.15 and 0.25 units. The infectivity of the virus was not tested, but its antigenic properties have been extensively studied (see Section VIII). Sera obtained by injecting the purified virus into mice reacted up to 1:128 with the virus and did not react with healthy material.

In addition to virus cores, purified preparations of FDV generally contained very distinct flexible helical strands 15–20 nm in diameter and up to 1 μm in length. The nature of this material is at present not clear; morphologically, the strands could be some kind of viral nucleoprotein complex or "internal component," but this appears unlikely because such a component has never been found with any other member of the Reoviridae. Alternatively, the strands might be some kind of host component such as phloem P protein, which can assume helical forms (Cronshaw *et al.*, 1973). However, sera made by injecting strand-containing preparations into mice did not react with antigens prepared from healthy leaves. This suggests that the strands may not be a host component. Perhaps they are viroplasm fibrils (see Fig. 4).

It is not clear at which step the virus became degraded to subviral particles during the purification, as the process was not monitored by electron microscopy or infectivity tests. With MRDV, work in the Turin laboratory has shown that to obtain undamaged virions the purification should proceed very rapidly, taking only 7–8 hours, and that pelleting and resuspension should be avoided as much as possible. Ikegami and Francki took rather longer with their purification and pelleted the virus three times. Perhaps this led to the loss of the outer capsid. It is also conceivable that Nonidet P40 caused some damage.

Reddy *et al.* (1975a) and Outridge and Teakle (1976) purified FDV from gall tissue by repeated cycles of differential centrifugation or by precipitation with PEG, and finally collected the virus on a Millipore filter of 50 nm pore diameter. It is not clear what kind of particle was obtained, except that it possessed all ten genome segments (see Section VII, D).

In summarizing this section, it is worthwhile to consider whether there are any procedures that can be recommended for improving purifications already achieved or obtaining purified preparations of MRDV-like viruses that are still unpurified. One problem is that we do not know how closely MRDV, RBSDV, and FDV are related to other members of the group and to what extent their behavior on purification really differs, as they have been purified in different laboratories, starting from different parts of different hosts. However, some points do emerge:

(1) Purification should be monitored using negative staining with UA (not PTA) or with PTA after fixing the preparation on the grid. The fractions should be checked using antisera to poly I–poly C or to dsRNA. If possible, steps should be checked by an infectivity test.

(2) If the virus can be transmitted to maize and if it grows well in the roots, purification problems may be eased.

(3) The purification should proceed rapidly in the cold, not be left overnight, and not be pelleted more than necessary.

(4) Various buffers may be successful. They should have a pH around neutral and should contain EDTA.

(5) The situation concerning Freon 113, CCl_4, and even $CHCl_3$ is not clear, but they should be tried.

(6) A nonionic detergent such as Nonidet P40 or Triton X-100 may help to remove membrane contamination, but such agents could possibly have a stripping effect on the virion. Lipid-digesting enzymes should be tried.

(7) PEG may be useful to precipitate and concentrate the crude virus.

(8) Separation of the virus on a DEAE cellulose column eluted with NaCl sounds interesting and could be followed up. The problem with both this and the PEG procedure is that NaCl concentrations above 0.2 *M* may strip off the outer capsid (see p. 309).

(9) Millipore-type membranes may be useful in recovering the purified virus.

B. Properties in Vitro

Wetter *et al.* (1969) determined the survival of MRDV antigens (what we now know was a combination of core, B-spike, and dsRNA antigens; see Section VIII) under various treatments such as storage, heating, freezing, and pH changes. The antigens survived rather well, tempting these authors and Harpaz (1972) to remark that MRDV seemed a rather stable virus; but infectivity and morphological integrity were not followed, and these are in fact much more labile. M. Conti (unpublished data) found that MRDV infectivity survived almost indefinite freezing at −20°C in infected plants. The thermal inactivation point is between 55° and 60°C, for 10 minutes (M. Conti, unpublished data), and the pH optimum, at least in certain buffers, is around 6.5. (Klein, 1967; also in Harpaz, 1972). The survival of infectivity under other conditions has not been studied.

Various treatments that degrade the virion morphologically have already been described here. In addition, the purified virus survives a few hours at room temperature and a few days at 4°C before degradation to B-spiked cores (Milne, unpublished data). These cores are very stable, though gradually the B spikes are lost. Redolfi *et al.* (1973) found that the sedimentation coefficients of the B-spiked cores were about 400 S ("full" particles) and 200 S ("empty" particles). The S value of the virion has not been measured, though as Redolfi *et al.* suggested, it may lie between 500 and 600 S, by comparison with WTV [510 S (Black, 1970)] and reovirus [630 S (Shatkin, 1968)].

Kitagawa and Shikata (1973) and Shikata (1974) reported the following properties for RBSDV. The thermal inactivation point in rice leaf extracts was 60°C, the longevity *in vitro* in rice sap at 4°C was 7 days, and infectivity survived almost indefinitely in frozen leaves. The dilution end point of rice leaf extracts was 10^{-4}, and that of hopper extracts 10^{-5}. The virus in crude rice leaf extracts retained infectivity at pH values between 5 and 9. Shikata and Kitagawa (1976) add that three cycles of freeze-thawing of diseased leaf extracts did not reduce infectivity.

The behavior of the infectivity of FDV *in vitro* has not been investigated, nor has that of any other MRDV-like virus.

The MRDV-like viruses apparently consist only of dsRNA and protein; these are discussed in the next subsection. There is no evidence of any membrane or lipid component in the virion, though sensitivity to chloroform was originally taken as possible evidence for the presence of lipids in MRDV (Wetter *et al.*, 1969; Harpaz, 1972).

C. Morphology

Work with thin sections of embedded tissues (see Section VI) has shown that all the viruses we are concerned with here have spherical virions about 70 nm in diameter, containing a dense core some 45–50 nm in diameter. Likewise, for all these viruses, the complete virions have now been observed in negative stain: MRDV by Lovisolo (1967), Lesemann (1972), and Milne *et al.* (1973); OSDV and CTDV by Milne *et al.* (1975); ABDV by Milne *et al.* (1974); LEDV by Lesemann and Huth (1975); RBSDV by Shikata (1974); FDV by Teakle and Steindl (1969); and PSV by Kitajima and Costa (1970). (See note added in proof.)

In addition to the virions, and the smooth cores usually seen in PTA, the B-spiked cores (Fig. 6E–G) have been observed with MRDV, OSDV, CTDV, ABDV, and LEDV (references above) and with FDV (Ikegami, 1975) and PSV (Milne, unpublished data). The putative B-spiked cores of RBSDV have not yet been described, but only, we suspect, because the conditions appropriate to their observation have not been used. No difference has been detected between the virions, spiked cores, or smooth cores of any of these viruses except regarding the outer spikes (the A spikes, to be discussed shortly; see Fig. 6A,H). These spikes, projecting from the outer capsid, have so far been observed by Milne *et al.* (1973) for MRDV, by Shikata (1974), and Shikata and Kitagawa (1976) for RBSDV and by T. Hatta (see Ikegami, 1975) for FDV. The other MRDV-like viruses have perhaps not been observed sufficiently critically to see whether they possess A spikes or not.

The morphology of the MRDV virion has been studied most, so we shall take it as representative. In UA (or uranyl formate) negative stain, the virion appears spherical, about 65 nm in diameter, with a double capsid (Fig. 6A). A structure that is similar but a bit larger (70–75 nm in diameter) can be observed in neutral PTA if the virus is fixed first (Fig. 6H). Attempts to fix the virus in suspension have not been particularly successful (Luisoni *et al.*, 1975; for FDV, see Ikegami, 1975), but if the virus is first adsorbed to a support film, it can be fixed with glutaraldehyde or by a preliminary wash with UA, so that it subsequently remains intact in PTA (Milne *et al.*, 1973). Shikata and Kitagawa (1976) fixed RBSDV-infected leaves with formaldehyde, glutaraldehyde, or osmium tetroxide and were then able to observe intact virions in PTA dip preparations.

The A spikes look globular and 11 nm long in UA but acute and 16 nm long in fixed PTA preparations (Fig. 6A,H). Fixed and unfixed UA preparations look the same. Six or sometimes four spikes are seen in symmetrical positions that suggest an icosahedral structure resting on a 3-fold or a 2-fold symmetry axis (a model is shown in Fig. 8A). The spikes have been called A spikes to distinguish them from the B spikes on the inner capsid (discussed later).

The number of capsomeres in the outer capsid has not been clearly

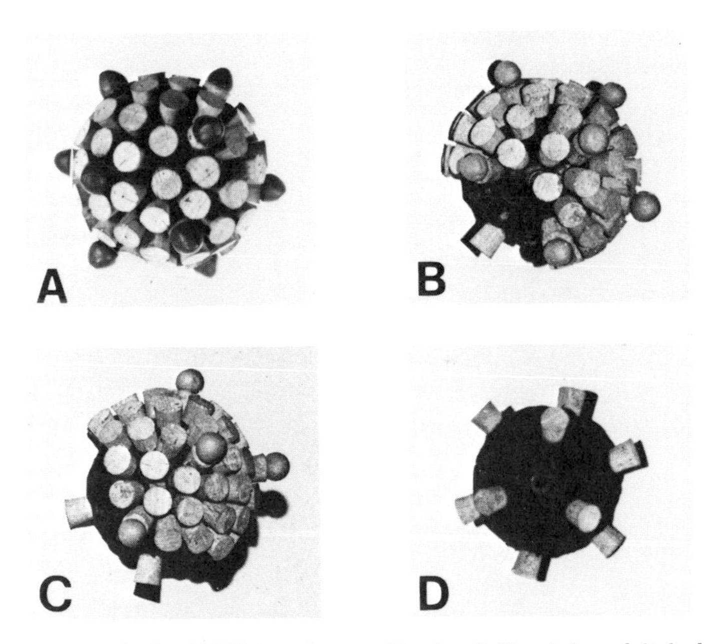

FIG. 8. A model of MRDV based on a 92-subunit T = 9 icosadeltahedron. (A) the complete virion; compare with Fig. 6A, B, and H. (B) the same, partially stripped as in Fig. 6C. (C) The model stripped as in Fig. 6D. (D) The model stripped to show the spiked core, as in Fig. 6E–G.

established for reovirus, WTV, RDV, or the MRDV-like viruses. The essential criterion of locating at least two 5-fold axes and observing the number of capsomeres in between them has not been rigorously met, though attempts have led to models, for reovirus, having 92 capsomeres (Vasquez and Tournier, 1962; Loh *et al.*, 1965), or 92 holes surrounded by indistinct capsomeres (Vasquez and Tournier, 1964; Jordan and Mayor, 1962; Amano *et al.*, 1971; see also Dales *et al.*, 1965). In all these, the basic structure would be a $T = 9$ lattice composed of 540 quasi-equivalent units (e.g., see Finch and Holmes, 1967).

A different and more hazardous approach is to derive the total number of capsomeres by counting the number visible around the edge of negatively stained particles. There is often, indeed, an appearance of "capsomeres" around the edges of reolike viruses, but, as can be appreciated if a suitable model is viewed from directly above (see Fig. 8A), these are almost certainly artifacts caused by partial superimposition of two or more real capsomeres. If that is true, it is misleading to count the peripheral "capsomeres" or attempt to establish, by photographic rotation, the number of peripheral "capsomeres," as did Luftig *et al.* (1972) for reovirus. They derived a figure of about 127 for the total capsomere number, an estimate that seems unlikely on geometrical grounds, as a capsid of quasi-equivalent units cannot be built from this number of capsomeres.

The structure of reovirus has been dwelt upon here at some length because the negatively stained images of the MRDV-like viruses are very similar to those of reovirus. The most likely figure for the number of capsomeres in MRDV is 92 (Milne *et al.*, 1973; see following page).

With WTV, the situation should be easier because, unlike reovirus or MRDV, it has a more angular capsid. A 92-subunit particle should have four subunits in an edge, and this is what Bils and Hall (1962) and Streissle and Granados (1968) observed. Black (1970) suggested that the number of capsomeres in WTV might be 32, but on the basis of new evidence (Reddy and McLeod, 1976; L. M. Black, personal communication, 1976) the earlier interpretations may be more correct.

Figures 6C and 6D show MRDV virions partially stripped, and Figs. 8B and 8C show a 92-capsomere model stripped in a similar way. Several points of interest emerge from such images. First, it is clear that the B spikes are parts of the outer capsid that resist stripping. Second, the A spikes are lost as the outer capsid strips, but the remaining A spikes, in partially stripped capsids, are in positions corresponding to the still unexposed B spikes; this indicates that one A spike is attached to the outer end of each B spike. Third, if the capsomeres that strip off are the same size as the B spikes (they probably are), then there is room for just two between adjacent B spikes. As is clear from Figs. 6E–G and 8D, the B spikes are arranged at the vertices of an icosahedron, and so the edge of a triangular facet would consist of four capsomeres. This is the T = 9 capsomere arrangement already suggested above.

The individual B spikes are hollow structures about 11 nm wide and 8 nm long in UA, and when stripped off in PTA (Fig. 6I, J) they appear to have clear 5-fold symmetry (Milne *et al.*, 1973). Ikegami (1975) has also observed this 5-fold symmetry in freeze-dried and shadowed preparations of FDV, but his measurements suggest that, when shadowed, the spike is 20.5 nm wide and 10.8 nm long. The B spikes resemble those found on reovirus type 3 by Smith *et al.* (1969) and by Luftig *et al.* (1972).

When the B-spiked core of MRDV is penetrated by UA negative stain, circular or possibly pentagonal structures are seen to lie under each B spike (Milne *et al.*, 1973).

The smooth cores obtained when the B spikes are stripped off are 50–55 nm in diameter in UA and 55–60 nm in PTA. No structures indicative of capsomeres have consistently been seen on these particles, though when penetrated by stain they are not uniform spheres, and they appear to have icosahedral symmetry. Sometimes, inside empty inner capsids (subviral ghosts, Fig. 7D) inward projecting structures can be seen (Milne *et al.*, 1973).

The morphology of the viral nucleic acid has not been examined, though other data indicate that there are 10 pieces of dsRNA within the inner capsid (see Section VII, D).

In conclusion, the structure of the MRDV-like viruses is complex and, judging from work on reoviruses (e.g., see Borsa *et al.*, 1973), more complexities are in store. However, the virion can be degraded in discrete steps, making it possible to begin to correlate morphological, biochemical, and immunological features. The capsid structure of MRDV and its relatives places them firmly in the Reoviridae and perhaps closest to the reoviruses. The ease with which the outer capsid strips down to spiked and smooth cores distinguishes it rather clearly from the capsids of WTV and RDV, though one does not yet know what to make of virus-like wallaby ear and rugose leaf curl (Grylls, 1954, 1975; Grylls *et al.*, 1974) which are apparently strippers but are vectored by leafhoppers, not planthoppers.

The orbiviruses differ from MRDV in apparently possessing a 32-capsomere T = 3 structure covered by an amorphous layer of protein (Els and Verwoerd, 1969; Verwoerd *et al.*, 1972) and the cytoplasmic polyhedrosis viruses (CPV's) differ in not having a true outer capsid (but these viruses have only been photographed in PTA). However, CPV's do have spikes (Lewandowski and Traynor, 1972; Asai *et al.*, 1972; Wood, 1973), and their particles somewhat resemble reovirus spiked cores (Lewandowski and Traynor, 1972; Payne and Tinsley, 1974) and especially MRDV spiked cores with the A spikes retained. Rotaviruses differ from the reoviruses, WTV, and the MRDV-like viruses in having clearly visible capsomeric structures around the periphery of a spherical particle. They do not have spikes (e.g., see Kapikian *et al.*, 1974; Flewett *et al.*, 1974).

Recently, in a very clear study of infantile gastroenteritis virus (a rotavirus), Martin *et al.* (1975) have suggested that the particle is composed of 32 large ring-shaped capsomeres in a T = 3 arrangement. However, each capsomere is in turn constructed of wedge-shaped subunits, and their arrangement follows the structure postulated for a T = 9 icosahedron. Thus on the level of structural units the particle conforms to a T = 9 lattice, whereas in the clustering of these units into 32 capsomeres it conforms to a T = 3 lattice. This finding may go far to reconcile the different structures that have been proposed for members of the Reoviridae, as it indicates that they may all be variants of the same basic architecture.

D. The RNA's

Earlier work on other members of the Reoviridae (e.g., see Wood, 1973; Ralph, 1969) made it likely that the MRDV-like viruses contain dsRNA; this was verified by Francki and Jackson (1972) for FDV and by Redolfi and Pennazio (1972) for MRDV. Independent evidence that FDV and MRDV contain dsRNA was obtained when Ikegami and Francki (1973) showed that antisera to these viruses and to RDV contain antibodies that react with FDV RNA and with poly I–poly C. Redolfi and Boccardo (1974) and Ikegami and Francki (1975) showed by PAGE (polyacrylamide gel electrophoresis) that MRDV and FDV RNA's had at least nine genome segments.

This was followed by a further resolution of the RNA of both viruses into 10 segments (Reddy *et al.*, 1975a), and a similar result was obtained with RBSDV (Reddy *et al.*, 1975b). The molecular weights of these RNA species were estimated with reference to those of reovirus type 3 (Martin and Zweerink, 1972) and of WTV and RDV (Reddy *et al.*, 1974), using both Loening's buffer (Loening, 1969) and PTE buffer (phosphate-Tris-EDTA, as described by Reddy and Black (1973b) or with slight modifications). The results are summarized in Table I. As absolute molecular weight values, the figures may not be very accurate but they were all obtained in the same laboratory against the same standards and therefore serve well to show up similarities and differences between the viruses. Figure 1 (see Section I) is a schematic representation of the values obtained in Loening's buffer, together with values for reovirus type 3 (Martin and Zweerink, 1972), a CPV (Fujii-Kawata *et al.*, 1970), an orbivirus (Verwoerd *et al.*, 1970), and a calf rotavirus (Newman *et al.*, 1975).

Inspection of Table I and Fig. 1 reveals several points. First, the three MRDV-like viruses form a homogeneous group, the overall pattern in both Loening's and PTE buffers being very similar, and quite different from that of the other viruses. Second, it seems that MRDV and RBSDV are very closely related, being indistinguishable except for small differences in segments 1, 3, and 10. FDV is perhaps less closely related, with only bands 5 and 7 occupying identical positions, though band 2 is close. Third, WTV and RDV form a related pair quite distinct from MRDV, FDV, and RBSDV and apparently no closer to them than to the other groups within the Reoviridae. Fourth, the general similarity of the genome patterns of all these viruses is striking and makes it virtually certain that they share the same genetic strategy.

One small but curious problem is that for RBSDV and FDV the 10 genome segments are present in equimolar amounts, but with MRDV material taken from both the glasshouse and the field (Reddy *et al.*, 1975a,b) contains segment 10 as a doublet, each sub-band being present in approximately half-molar amounts. It was suggested that segment 10A occurs in about half the virus population and 10B in the other half.

Reddy *et al.* (1975b) speculated that, because MRDV is transmitted transovarially, a part of the virus population might always be transmitted in this way from hopper to hopper without alternate passage through the plant, whereas the other part of the population went through the normal alternating cycle. Two virus populations could then be selected, differing, in this case, in segment 10. This would parallel the situation found with WTV (Martinez *et al.*, 1976), where prolonged passage in vector cells led to a difference from normal WTV in electrophoretic mobility.

This explanation seems unlikely for two reasons. First, both forms of segment 10 were isolated from virus grown in plants and transmitted by vectors, so there is no history of separation from the plant (or the insect) as there is with WTV. Second, if about half the virions are of a type maintained only by transovarial passage, a high rate of transovum transmission would be expected. In fact the rate is very low (see Section V, A). Nevertheless, some selection pressure must be at work to maintain the dimorphism, if that is what it is.

Once the RNA's of members of the Reoviridae have been isolated, it is clearly worthwhile to attempt hybridization between the segments of one virus and between the genomes of different viruses, to see whether there may be base sequences in common. No work of this kind has yet been done with the MRDV-like viruses, but it may be useful to summarize the results obtained with other Reoviridae (see also Shatkin, 1974). The question of intragenome homology was investigated by Bellamy and Joklik (1967) and Watanabe and Graham (1967), who detected no sequence homology when comparing the small, medium, and large RNA fractions of reovirus type 3. Martinson and Lewandowski (1974) studied sequence homology among the dsRNA's of CPV, WTV, and reoviruses types 1, 2, and 3. The genomes of reoviruses types 1 and 3 showed extensive though imperfect homologies, but reoviruses types 2 and 3 had little in common, and there was no detectable sequence homology among the genomes of reovirus type 3, CPV, and WTV. Likewise, there was little or no detected homology between the genome of RDV and single-stranded RNA (ssRNA) from reovirus type 3 (Shatkin and Rada, 1967) and between the genomes of reovirus type 1 and two orbiviruses (Verwoerd, 1970). Black and Knight (1970) could not demonstrate any homology between the ssRNA's of WTV and CPV produced by transcription *in vitro,* though they observed that the transcription might not have been complete.

These failures to detect sequence homologies among members of the Reoviridae are not, perhaps, surprising. As Martinson and Lewandowski (1974) pointed out, incomplete homologies would not be detectable if they led to more than about 20% mispairing. However, this degree of resolution may be just what is required to detect differences and similarities between the genomes of the MRDV-like viruses, which so far seem to form a remarkably homogeneous group. On the other hand, it seems highly unlikely that sequence homologies between the MRDV-like viruses and other members of the Reoviridae will be found, at least with techniques of the present sensitivity.

Finally, the MRDV-like viruses have not yet been examined to see if they contain substantial amounts of oligoadenylate in the virion, in addition to the dsRNA. In reovirus, about 20% of the total viral nucleotides are oligoadenylate of, so far, unknown function (e.g., see Joklik, 1972; Luftig *et al.,* 1972; Casjens and King, 1975).

E. The Proteins

Among the MRDV-like viruses, the capsid proteins of only MRDV have been examined, though a capsid-associated RNA-dependent RNA polymerase has been found with both FDV and MRDV. There are no data on other possible gene products.

Boccardo and Milne (1975) examined by PAGE the polypeptides

present in the virion, the spiked core, and the smooth core of MRDV. Six bands were detected in the virion fraction, probably three in the spiked core, and two in the smooth core. The polypeptides were labeled I–VI in order of decreasing molecular weight (see Table VII). It appears that peptides I and II come from the smooth core, that peptide III corresponds to the B spike, and that peptides IV, V, and VI are components of the outer capsid. The molar ratios of these peptides are not yet available, but peptides II, III, and VI are present in the largest amounts.

Table VII also compares the molecular weights of the peptides found with those of the hypothetical pepetides that could be encoded by all genome segments, assuming complete transcription and translation. The found values for the six peptides can be fitted rather closely (with differences of up to 6% from the closest hypothetical value), suggesting that the whole of each genome segment corresponding to peptides I–VI is translated into an intact peptide and that 60–70% of the genome is expressed as capsid proteins. Peptides II, III, and VI, present in the largest amounts, are apparently encoded by genome segments 3, 4, and 9.

How does this situation compare with that of other Reoviridae? The reoviruses have seven peptides in the capsid, four being in the spiked core (Loh and Shatkin, 1968; Smith *et al.*, 1969; Joklik, 1972). Verwoerd *et al.* (1972) found that the BTV capsid also contains seven peptides, five

TABLE VII

MOLECULAR WEIGHTS OF POLYPEPTIDES OF THE MRDV CAPSID, CORRELATED WITH THE MOLECULAR WEIGHTS OF THE GENOME SEGMENTS

Molecular weight of genome segments[a]	Molecular weight of peptides		Peptide number
	Expected[b]	Found[c]	
2.88	160,000	—	—
2.50	138,000	139,000	I
2.35	130,000	126,000	II
2.35	130,000	123,000	III
2.12	117,000	111,000	IV
1.75	97,000	97,000	V
1.45	80,000	—	—
1.25	68,000	—	—
1.18	65,000	64,000	VI
1.08	60,000	—	—

[a] Expressed in daltons × 10^6 (Reddy *et al.*, 1975a).
[b] Assuming the segment is fully transcribed, a conversion factor of 18 was used (Smith *et al.*, 1969).
[c] Data from Boccardo and Milne (1975).

being in an inner capsid and two constituting a diffuse outer layer (De Villiers, 1974). The capsid of the *Bombyx mori* CPV contains five peptides, but that of the *Nymphalis io* CPV apparently has three (Payne and Tinsley, 1974). WTV has seven peptides in its capsid, three or four of them major (Payne and Kalmakoff, 1974; Lewandowski and Traynor, 1972; Reddy, 1974; Reddy and MacLeod, 1976). The calf rotavirus capsid gives at least two major and three minor bands when analyzed by PAGE (Newmann *et al.*, 1975). Thus it appears that MRDV is generally similar to the other Reoviridae in capsid structure but that all member groups differ considerably from each other in detail (as they must, given their diverse genomes).

Verwoerd *et al.* (1972) pointed out that the four major components of the BTV capsid correlated best with genome segments 2, 3, 6, and 8, whereas the major reovirus polypeptides are considered to be derived from segments 1, 3, 4, and 9. The major capsid peptides of MRDV correlate best with segments 3, 4, and 9 as in reovirus, but segments 1 and 10 apparently do not code for capsid peptides, whereas in reovirus they do. Thus it is difficult at present to homologize the components of the MRDV capsid with those even of reovirus, which appears the most similar. It may not be useful to push comparisons further until more and firmer data are available for MRDV and its close relatives.

One thing that all these viruses have in common is a close correlation between the experimentally found molecular weights and those theoretically calculated assuming complete translation of the genome segments. Occasionally the correlation is less good, and it is interesting that Zweerink and Joklik (1970; see also Weber and Osborn, 1969; Verwoerd *et al.*, 1972) have found that the reovirus peptide with a molecular weight of 81,000 is derived not from segment 6, with which it correlates, but from segment 4, through cleavage after translation.

The only detailed work on the capsid-associated RNA-dependent RNA polymerase (or transcriptase) of the MRDV-like viruses has been done by Ikegami (1975) and Ikegami and Francki (1976), though such an enzyme has also been detected in MRDV (see Boccardo and Milne, 1975).

Ikegami and Francki found polymerase activity in concentrated extracts of FDV-diseased leaf galls but not in extracts of healthy tissue; the activity was correlated with FDV dsRNA antigen (which was presumably particle bound) in sedimentation studies, and this suggests that the polymerase is virion associated. However, when purified FDV and crude concentrated extracts were compared on the basis of antigen content, the purified virus (consisting of SVP's) was only about 10% as active in incorporating ^{32}P-guanosine triphosphate (GTP) as the crude extract. This probably merely reflects the difficulty of purifying undamaged FDV.

Optimal polymerase activity occurred at about 35°C and in the presence of 8 m*M* Mg^{2+} and 200 m*M* NH_4^+. It is interesting that the pH optimum was between 8.5 and 9 and that at pH 7 only some 13% of the optimal activity was expressed. It is also curious that incorporation of label proceeded rapidly for about 20 minutes but that after this, activity ceased. The polymerase product was a population of low molecular weight ssRNA's, over 80% of which annealed to FDV RNA, but this product was not released from the capsid and was separated from the dsRNA only after deproteinization.

The properties of RNA transcriptases associated with the virions of other viruses have been reviewed by Ikegami and Francki (1976) and Bukrinskaya (1973), so only certain points will be discussed. Such enzymes are an integral part of all other Reoviridae examined, including reovirus, BTV, CPV, WTV, and RDV (Borsa and Graham, 1968; Shatkin and Sipe, 1968; Lewandowski *et al.*, 1969; Black and Knight, 1970; Verwoerd and Huismans, 1972; Martin and Zweerink, 1972; Kodama and Suzuki, 1973). Furthermore, these transcriptases are all, except for that of BTV, similar in their requirements for 8 m*M* Mg^{2+}, though other virus groups such as rhabdoviruses and myxoviruses have different requirements, either for Mg^{2+} or Mn^{2+}. BTV apparently has a requirement for Mn^{2+} (Verwoerd *et al.*, 1972; Verwoerd and Huismans, 1972).

The short period of activity of the FDV polymerase is in contrast to the prolonged activity *in vitro* (over 10 hours) of the polymerase of reovirus (Levin *et al.*, 1970), and this short life is not yet explained. Like the polymerases of CPV and WTV (Lewandowski *et al.*, 1969; Black and Knight, 1970), that of FDV is already active in crude extracts or purified preparations and does not require activation by removing parts of the outer capsid, as is necessary for reovirus (see Shatkin and Lafiandra, 1972; Astell *et al.*, 1972). This may indicate that the polymerase of reovirus is in some way different from those of FDV, CPV, and WTV, but it may only mean that the latter three viruses have been partially stripped during preparation, so that their polymerases are already activated.

VIII. Serology of Viral Components

A. *The RNA's*

Nucleic acids, particularly double-stranded ones, and also synthetic polynucleotides, are antigenic, and their antigenicity is enhanced if they are combined, as haptens, with proteins. Though the antibodies they provoke are largely nonspecific, some specificities do exist, and though the titer of the antiserum is generally much lower than that obtainable with proteins, it is high enough to work with, and to be confused with reactions due to proteins (Stollar, 1973; see also, e.g., Moffitt and Lister, 1975). This was not generally known to plant virologists until Francki and Jackson (1972) detected FDV RNA immunologically and Ikegami and Francki (1973) found a positive serological reaction among FDV, RDV, and MRDV owing to their common possession of dsRNA.

Francki and Jackson (1972) used an antiserum prepared against the synthetic double-stranded polyribonucleotide poly I–poly C conjugated with methylated bovine serum albumin. This reacted with dsRNA's but not with ssRNA or DNA. With this serum they were able to detect FDV dsRNA in infected plants, and they concluded (correctly) that FDV is a dsRNA-containing virus. The end point of the serum, against poly I-poly C was 1:16. Ikegami and Francki (1973) found that the positive reactions in immunodiffusion tests between FDV antigen and either RDV or MRDV antisera could be abolished by intragel absorption with poly I-poly C, suggesting that the viral capsid proteins played no part in the reaction, which was due to dsRNA. The titers of the RDV, MRDV, and poly I-poly C antisera against FDV-infected sap, FDV nucleic acid, or poly I–poly C were generally 1:2 or 1:4 and (in one case) 1:8, but not higher (see Table VIII). The authors concluded by noting two precautions that should be taken when investigating the serological relationships of dsRNA-containing viruses: either (1) use antisera absorbed with RNA from the virus concerned, or (2) use viral protein preparations devoid of all RNA.

Ikegami and Francki (1974) purified the cores of FDV, as described in Section VII, A, using their antiserum to poly I–poly C as an assay for the

TABLE VIII
REACTIONS BETWEEN SOME DSRNA'S AND ANTISERA TO FDV, MRDV, RDV, AND POLY I–POLY C[a]

Antigen	Titer[b] of antiserum to: FDV	MRDV	RDV	Poly I–poly C
FDV RNA	16	2	8	16
Reo RNA	1	8–16	8	8
ϕ6 RNA	2	4–8	1	16
Poly I–poly C	1	4	8	16–32

[a] Data of Ikegami and Francki (1976).
[b] Reciprocals of the highest dilution reacting in gel diffusion tests.

virus. When making antisera against FDV, they used mice both in the normal way and by inducing antibody production in ascites fluid; mice, not rabbits, were used because of the small amount of antigen available for immunization. Interestingly, they found that antibodies directed against FDV RNA reacted relatively poorly with poly I–poly C, but antibodies directed against poly I–poly C had the same titer against FDV RNA as against the homologous antigen.

Ikegami and Fracki (1976) pursued this last point and examined the reactions of FDV RNA, bacteriophage $\phi 6$ dsRNA, reovirus RNA, and poly I–poly C to sera prepared against FDV, RDV, MRDV, and poly I–poly C. The result is illustrated by Table VIII, modified from their paper. Two conclusions are clear. First, all sera reacted with all antigens, confirming previous reports (Stollar, 1973) that antibodies to dsRNA react with a wide range of double-stranded polyribonucleotides. There was negligible or no reaction with the RNA's when they were heat-denatured, i.e., rendered single stranded. Second, some variation occurred; for example, anti-FDV serum had a higher titer against homologous antigen than against the heterologous antigens, whereas anti–poly I–poly C serum reacted more or less equally with all the antigens. However, some of the results are at present not easy to explain: for instance, the fact that MRDV antiserum reacted with reovirus RNA to 1:16 but with FDV RNA only to 1:2, although the RNA of MRDV would appear to resemble FDV RNA the more closely of the two.

Luisoni *et al.* (1975) also investigated the reactions of dsRNA and poly I–poly C to various antisera, as a background to studying the serology of MRDV capsid proteins. The results were in general agreement with those of Francki and his co-workers. One serum prepared by inoculating rabbits with MRDV virions reacted up to 1:16 with MRDV RNA and up to 1:8 with poly I–poly C, but when the serum was absorbed in tubes or intragel with excess poly I–poly C, it still reacted slightly with the homologous RNA. To remove all RNA-based reactions, therefore, it was necessary to absorb the serum with homologous RNA.

One interesting difficulty encountered by Luisoni *et al.* (1975) was that sera absorbed with a large excess of MRDV RNA had strongly reduced titers against capsid (protein) antigens, though sera absorbed with the same amounts of poly I–poly C had titers unchanged. For example, one serum had a titer against MRDV proteins of 1:2048 both before and after absorption with poly I–poly C, but after absorption with an equal amount of MRDV RNA the titer against proteins was reduced to 1:128. This effect was repeatable with sera totally unrelated to MRDV or dsRNA, was quite nonspecific, and was almost certainly due to binding of serum proteins to the added RNA [but not to the added poly I–poly C (see Fig. 9C)]. Stollar (1973) has pointed out that nucleic acids can combine nonspecifically with basic serum proteins, one class of which are antibodies.

The practical outcome of these findings is that a minimum of dsRNA should be used when absorbing sera. Poly I–poly C cannot generally be used for total absorption of antibodies to dsRNA, but it can be used for a preliminary absorption, followed by absorption with the minimum of homologous RNA.

B. The Proteins

As we have seen (Section VII, A), Wetter *et al.* (1969) partially purified MRDV and obtained antisera reacting up to 1:2048 in gel diffusion tests with MRDV antigens. Luisoni *et al.* (1973) tested two such MRDV sera and two similar sera prepared in Japan against RBSDV to see if MRDV and RBSDV were serologically related (see Table IX). There was no appreciable difference between homologous and heterologous titers with any of the four unabsorbed sera, suggesting that the antigens were closely related. The titers ranged from 1:512 to 1:8192, according to different sera and different kinds of assay. It should be noted that the precipitin ring test generally gave higher titers than the gel diffusion test (though the former is more wasteful of material).

The two anti-RBSDV sera were absorbed with MRDV in Italy and returned to Japan, where they were tested with RBSDV. One gave a titer of 1:256 and the other a titer of 1:512, whereas both the unabsorbed sera, also returned to Japan and retested, gave titers of 1:2048. Thus, although related, the viruses appeared to be not identical, as absorption with the heterologous virus reduced but did not abolish the homologous titer. Unfortunately it did not prove possible to make the reciprocal test, absorbing MRDV sera with RBSDV.

Luisoni and co-workers (1973) were not then aware of the possibility that the common dsRNA antigens might have influenced the result, as noted in section VIII A. However, with hindsight we know that the titers

---

FIG. 9. Immune electron microscopy of various kinds of MRDV particle. (A) Smooth cores reacted with serum A68 diluted 1:100. (B) Similar, but with normal serum. (C) A spiked core (center) and two virions fixed on the grid and treated with a serum (undiluted) raised against fixed virions. The spiked core has reacted strongly, but the virions have not reacted. The arrows indicate where we think serum proteins have adsorbed to extruded viral RNA. (D–G) Spiked cores treated with undiluted serum; D, E, and G show a serum raised against spiked cores and then absorbed with smooth cores to leave only the antibodies specific for the B spikes. (F) Normal serum. All approximately ×230,000.

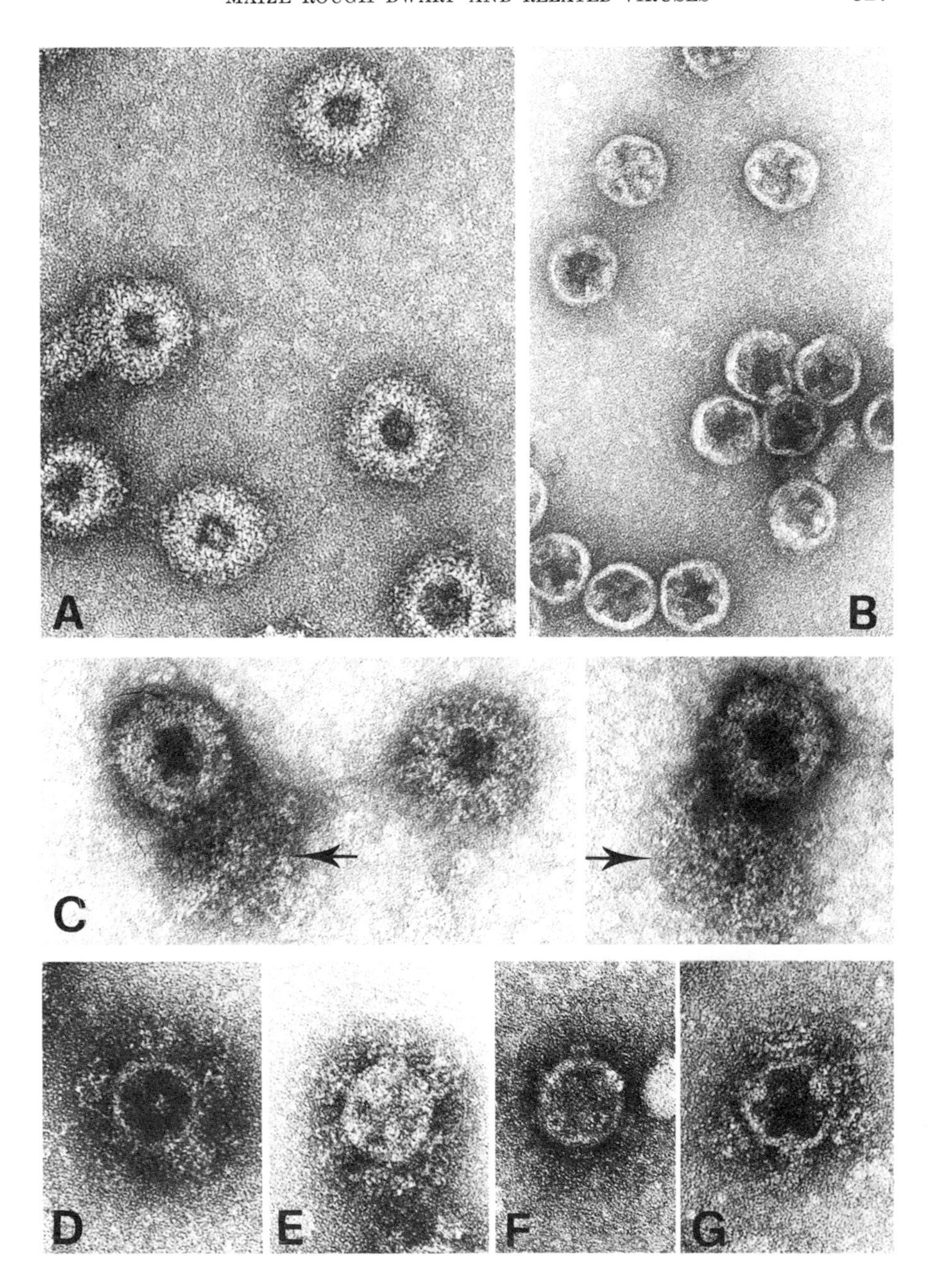
A
B
C
D
E
F
G

TABLE IX

REACTIONS OF TWO ANTI-MRDV SERA AND TWO ANTI-RBSDV SERA WITH HEALTHY PLANT PROTEINS AND WITH MRDV AND RBSDV ANTIGENS[a]

Antiserum	Antigen titers[b]				
	Healthy plant proteins		MRDV		RBSDV
	Agar gel diffusion test	Ring interface precipitin test	Agar gel diffusion test	Ring interface precipitin test	Ring interface precipitin test
MRDV No. 14	8	32	1024	2048	2048
MRDV No. 20	8	64	2048	4096	4096
RBSDV No. 1	0	2	512	2048	4096[c] (2048)[d]
RBSDV No. 3	4	2	1024	4096	8192[c] (2048)[d]
RBSDV No. 1 Abs.[e]	—	<8[f]	0	—	256
RBSDV No. 3 Abs.[e]	—	<4[f]	0	—	512

[a] Data of Luisoni *et al.* (1973).

[b] Reciprocals of the highest reacting dilution.

[c] Original values.

[d] Values in parentheses were obtained when sera were returned to Japan and retested.

[e] Absorbed with MRDV. Dashes indicate not done.

[f] Not tested at higher concentration.

due to dsRNA have never proved higher than 1:32 in gel diffusion tests, so that the observed heterologous titers of up to 1:1024 in gel diffusion and 1:4096 in the ring interface precipitin test are likely to have been produced in large part by common protein antigens.

To check this conclusion, Luisoni *et al.* (1975) tested the two original RBSDV sera against MRDV RNA. The titers were 0 and 1:2, respectively, that is, undetectable or negligible. Like the MRDV sera, the two RBSDV sera still reacted strongly with MRDV spiked cores after intragel absorption with MRDV RNA or with poly I–poly C. It seems clear, therefore, that the proteins of the MRDV and RBSDV capsids are related.

A further interesting point arose during the testing of the two RBSDV antisera and also several anti-MRDV sera against purified MRDV intact virions and against MRDV spiked cores. The results were observed in gel diffusion plates, in slide precipitation tests, and by immune electron microscopy (see Milne and Luisoni, 1975). First, it was apparent that contact and incubation with serum quickly stripped virions down to spiked cores unless the virions were previously fixed with glutaraldehyde. Second, both fixed and unfixed spiked cores reacted strongly with all the sera. Third, none of the sera reacted with intact fixed virions (Fig. 9C), except one anti-MRDV serum which reacted undiluted but not at 1:10.

Luisoni *et al.* (1975) concluded that when either RBSDV or MRDV preparations were injected into rabbits, the labile outer capsomeres were probably quickly lost and provoked little or no antigenic response, whereas the spiked cores were highly antigenic. A second conclusion was that the demonstrated serological relationship between the MRDV and RBSDV capsid proteins concerned the spiked cores, though whether the core proper, the B spike, or both were involved was not established. Whether the labile outer capsomeres are also serologically related is not known. Unfortunately injection of glutaraldehyde-"fixed" virions into rabbits resulted, as before, only in a serum reacting with spiked cores but not with fixed virions.

There are at least six polypepetides, and so there must be several antigen groups in the MRDV capsid. It should be possible to isolate sera specific to at least some of these. Although Luisoni *et al.* (1975) failed to raise an antiserum that reacted with the MRDV outer capsid, they did obtain a serum specific for smooth cores (Fig. 9A,B). This had a titer, in agar gel diffusion, of 1:1024 after absorption with MRDV RNA. The serum did not react with the B spikes. A second serum, against spiked cores, with a titer of 1:256, reacted with the core proper and the B spikes (Fig. 10) but not with MRDV RNA. When this serum was absorbed with purified smooth cores, it reacted only against the B spikes (Fig. 9D–G), with a titer of 1:64.

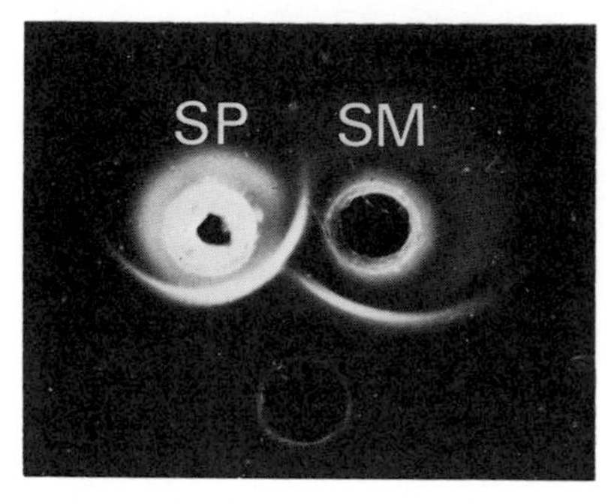

FIG. 10. Antiserum against spiked cores of MRDV (lower well) reacted with spiked cores (SP) and smooth cores (SM). The spur indicates the presence of antibodies against spiked cores that do not react with smooth cores.

Milne and Luisoni (unpublished data) are using these sera to investigate other MRDV-like viruses, utilizing the immune electron microscopic methods of Milne and Luisoni (1975, 1976) and Derrick (1973) and, where possible, agar gel diffusion tests. It is so far established that the spiked cores of both CTDV and PSV react strongly with anti-MRDV spiked core serum previously absorbed with MRDV RNA and healthy proteins. These three viruses are therefore serologically related in their spiked cores. The relationship between MRDV and CTDV confirms earlier unpublished work by K. Lindsten, M. Conti, and E. Luisoni. Further unpublished work of Milne and Luisoni shows that OSDV and ABDV spiked cores do not react with sera raised against MRDV, RBSDV, or FDV (all sera absorbed with dsRNA); similarly, anti-FDV serum does not react with the spiked cores of any other group member tested, namely MRDV, RBSDV, CTDV, PSV, OSDV, and ABDV. This nonreaction of FDV antiserum with heterologous viruses confirms the finding of Ikegami and Francki (1973) that MRDV antiserum does not react with FDV capsid proteins. In conclusion, it seems at present that the MRDV-like viruses fall into three serological groups: (a) MRDV, RBSDV, CTDV and PSV; (b) FDV, and (c) OSDV and ABDV.

Work of a similar kind is being done by Mayor (1975). Using three human reovirus serotypes, she found that there are common antigens on all three SVP's, and these are different from the antigens of the outer capsid. Kapikian *et al.* (1974) have used immune EM methods to good effect in studying the viruses causing infantile diarrhea. A summary of work on rotaviruses (Thornton and Zuckerman, 1975) suggests that with these the inner capsid may be the site of common group antigens, whereas the outer capsid antigens are probably type specific.

Finally, the work of Hiebert and his colleagues on the inclusion proteins of potyviruses (e.g., see McDonald and Hiebert, 1975; Hiebert, 1975) suggests that the tubes found in cells infected with MRDV-like viruses may consist of protein and may be viral gene products against which specific sera could be raised. All we know at present is that, with MRDV, the tubes do not react with antisera directed towards capsid proteins (Milne, unpublished data).

IX. Conclusions

We hope it may be clear from this review that the time is ripe for solution of many of the remaining problems concerning MRDV-like viruses. Recognition of these viruses should no longer be difficult, and confusion with other agents is now avoidable. Purification of most members will still not be easy, but at least the objectives and possible methods are reasonably well defined. Biochemical work should also be able to progress rapidly, at least with MRDV, FDV, and RBSDV.

The MRDV-like viruses are only just becoming recognized as typical members of the Reoviridae that are however quite distinct from WTV and RDV in a number of characteristics, notably their RNA segmentation, vectors, and capsid structure. A very interesting result to be expected shortly is firm information on the size distribution of the RNA segments of MWEV. This should indicate at once whether MWEV resembles WTV or MRDV, or perhaps neither, and may clarify the taxonomy of these viruses. (See note added in proof.)

Although this uncertainty remains at present, it seems appropriate to look for a suitable name to indicate "MRDV-like virus." We would like to propose the name ACANTHOVIRUS because of the spikes on both the outer and inner capsids, this combination being so far unique to these viruses.

The serology of the acanthoviruses is now at a tantalizing stage. First, any antiserum raised against dsRNA, poly I–poly C, or perhaps other synthetic polyribonucleotides should react with the dsRNA of these viruses, and this is a useful check for their possible presence. The variation in serological specificity among natural and synthetic dsRNA's is, needless to say, a very interesting problem in itself. Second, we suppose that smooth cores and spiked cores of several of these viruses will soon be purified and antisera raised against them, as has already been done with MRDV. Third, the elusive outer capsid antigen(s) must soon be brought to book, even if only after fixation, and this will add a fourth (and perhaps more) antigenic group(s) to the list of these known. At least the protein antigens may make useful markers for genetic studies as well as for classification and diagnosis. As many acanthoviruses are difficult to purify, it is interesting that, using carefully prepared sera and proper controls, immune electron microscopy can be used on crude unpurified virus preparations.

A curious feature of the acanthoviruses is that they all cause tumors in plants, and only a few other viruses are known to do so [see the discussions by Bos and Grancini (1968); Bos (1970); and Black (1972)]. Among these other viruses are WTV, MWEV, oat blue dwarf (Long and Timian, 1975), clover enation disease (Rubio-Huertos and Bos, 1969), and citrus vein enation virus (Hooper and Schneider, 1969). Reddy *et al.* (1974) noted that WTV is tumorigenic, whereas RDV is not, and suggested that a tumorigenic gene might be located on segment, 9, 10, 11, or 12 of the WTV genome that would be missing from the genome of RDV. Later studies by these authors have not supported this interesting idea, and an alternative approach may be to consider the site of multiplication of the virus. We suggest this because the small but diverse collection of tumorigenic plant viruses all seem to be phloem inhabiting (RDV, of course, is not), and it may be that this specific invasion of cells close to the meristem is an important factor in neoplasia. While all tumor-inducing plant viruses seem to be phloem specific, the reverse is not true; an example is the nontumorigenic but phloem-inhabiting potato leafroll virus (see Peters, 1970). A predilection for the phloem, then, may be necessary but is not a sufficient condition for tumor induction, and we are still far from knowing what the sufficient condition is.

We venture to conclude by airing one major problem in plant virology, well illustrated by the MRDV group. This is the unfortunate separation that exists between the majority of poorly known but important viruses and the bulk of well-qualified virologists. This separation is exacerbated by the difficulty of interchanging viruses and plant stocks between countries because of considerations of quarantine. Ever since the proposal of Maramorosch (1957, 1960), it has seemed to many people a good idea to set up one or more international quarantine and research centers to deal with these problems, but regrettably nothing concrete has ever emerged. Plant virus research in tropical and subtropical countries continues understaffed, undertrained, and underequipped, and it is hard to see how this situation will improve unless some permanent international laboratory can be set up that will attract the highest talent.

Acknowledgments

We want to thank Drs. A. Appiano, L. M. Black, F. Brown, M. Conti, R. I. B. Francki, J. Giannotti, N. E. Grylls, P. B. Hutchinson, M. Ikegami, K. Lindsten, E. Luisoni, R. T. Plumb, E. Shikata, J. Vacke, and their associates for providing unpublished data and manuscripts prior to publication. We wish also to thank members of the Laboratorio di Fitovirologia applicata, Turin, for their many helpful suggestions, Miss T. Sanna for typing the manuscript, and Mr. E. Piccolini for drawing the figures.

References

Abeygunawardena, D. V. W. (1969). *In* "The Virus Diseases of the Rice Plant," p. 53. Johns Hopkins Press, Baltimore, Maryland.
Amano, Y., Katagiri, S., Ishida, N., and Watanabe, Y. (1971). *J. Virol.* **8,** 805.
Anonymous (1969). "The Virus Diseases of the Rice Plant," Report of an IRRI Symposium, 1967. Johns Hopkins Press, Baltimore, Maryland.
Anonymous (1974). *Sci. Sin.* **17,** 273.
Asai, J., Kawamoto, F., and Kawase, S. (1972). *J. Invertebr. Pathol.* **19,** 279.
Astell, C., Silverstein, S. C., Levin, D. H., and Acs, G. (1972). *Virology* **48,** 648.
Bassi, M., and Favali, M. A. (1972). *J. Gen. Virol.* **16,** 153.
Bassi, M., Favali, M. A., and Appiano, A. (1974). *Riv. Patol. Veg.* **10,** 19.
Bellamy, A. R., and Joklik, W. K. (1967). *J. Mol. Biol.* **29,** 19.
Bergonia, H. T., Capule, N. M., Novero, E. P., and Calica, C. A. (1966). *Philipp. J. Plant Ind.* **31,** 47.
Bils, R. F., and Hall, C. E. (1962). *Virology* **17,** 123.
Biraghi, A. (1949). *Not. Mal. Piante* **7,** 1.
Biraghi, A. (1952). *Ann. Sper. Agrar.* **6,** 1043.
Black, D. R., and Knight, C. A. (1970). *J. Virol.* **6.** 194.
Black, L. M. (1965). "Encyclopedia of Plant Physiology," p. 236. Springer-Verlag, Berlin and New York.
Black, L. M. (1970). "CMI/AAB Descriptions of Plant Viruses," No. 34.
Black, L. M. (1972). *Progr. Exp. Tumor Res.* **15,** 110.
Blattný, C., Pozdena, J., and Prochazkova, Z. (1965). *Phytopathol. Z.* **52,** 105.
Boccardo, G., and Milne, R. G. (1975). *Virology* **68,** 79.
Borsa, J., and Graham, A. F. (1968). *Biochem. Biophys. Res. Commun.* **33,** 895.
Borsa, J., Copps, T. P., Sargent, M. D., Long, D. G., and Chapman, J. D. (1973). *J. Virol.* **11,** 552.
Bos, L. (1970). "Symptoms of Virus Diseases in Plants," 2nd Ed., p. 87. Pudoc, Wageningen.
Bos, L., and Grancini, P. (1968). *Phytopathol. Z.* **61,** 253.
Bowne, J. G., and Jones, R. H. (1966). *Virology* **30,** 127.
Brakke, M. K. (1953). *Arch. Biochem. Biophys.* **45,** 275.
Brakke, M. K. (1956). *Virology* **2,** 463.
Brakke, M. K. (1969). *In* "Viruses, Vectors and Vegetation" (K. Maramorosch, ed.), p. 527. Wiley (Interscience), New York.
Brčák, J., and Králík, O. (1969a). *Plant Virol., Proc. Conf. Czech. Plant Virol. 6th, Olomouc 1967* p. 138.
Brčák, J., and Králík, O. (1969b). *Biol. Plant.* **11,** 95.
Brčák, J., and Králík, O. (1973). *Proc. Conf. Czech. Plant Virol., 7th, High Tatras, 1971*, p. 253.
Brčák, J., Králík, O., and Vacke, J. (1970). *Int. Symp. Plant Pathol., New Delhi, 1966–1967*, p. 93.
Brčák, J., Králík, O., and Vacke, J. (1972). *Biol. Plant.* **14,** 302.

Bremer, K., and Raatikainen, M. (1975). *Ann. Acad. Sci. Fenn., Ser. A4* **203,** 1.
Bukrinskaya, A. G. (1973). *Adv. Virus Res.* **18,** 195.
Carter, W. (1973). "Insects in Relation to Plant Disease," 2nd Ed. Wiley, New York.
Casjens, S., and King, J. (1975). *Annu. Rev. Biochem.* **44,** 555.
Catherall, P. L. (1970). *Plant Pathol.* **19,** 75.
Commonwealth Mycological Institute (1975). "Distribution Maps of Plant Disease," 3rd Ed., No. 17.
Conti, M. (1966). *Ann. Fac. Sci. Agrar. Univ. Torino* **3,** 337.
Conti, M. (1969). *Ric. Sci.* **39,** 701.
Conti, M. (1972). *Actas III, Congr. Union Fitopatol. Mediterr., Oeiras, 1972,* p. 11.
Conti, M. (1974). *Mikrobiologija* **11,** 49.
Conti, M. (1976). *Mikrobiologija,* in press.
Conti, M., and Lovisolo, O. (1971). *J. Gen. Virol.* **13,** 173.
Conti, M., Wetter, C., Luisoni, E., and Lovisolo, O. (1968). *Atti Accad. Sci. Torino, Cl. Sci. Fis., Mat. Nat.* **102,** 563.
Costa, A. S., and Kitajima, E. W. (1974). *Fitopatologia* **9,** 48.
Cronshaw, J., Gilder, J., and Stone, D. (1973). *J. Ultrastruct. Res.* **45,** 192.
Dales, S. H., Gomatos, P. J., and Hsu, K. C. (1965). *Virology* **25,** 193.
Derrick, K. S. (1973). *Virology* **56,** 652.
De Villiers, E.-M. (1974). *Intervirology* **3,** 47.
Dirven, J. G. P., and van Hoof, H. A. (1960). *Tijdschr. Plantenziekten* **66,** 344.
Edgerton, C. W. (1959). "Sugarcane and its Diseases," p. 215. Louisiana State Univ. Press, Baton Rouge.
Els, H. J., and Verwoerd, D. W. (1969). *Virology* **38,** 213.
Evanhius, H. H. (1958). *Proc. Conf. Potato Virus Dis., 3rd, Lisse-Wageningen, 1957* p. 251.
Favali, M. A., Bassi, M., and Appiano, A. (1974). *J. Gen. Virol.* **24,** 563.
Fenaroli, L. (1949). *Not. Mal. Piante* **3,** 38.
Fenner, F., Pereira, H. G., Porterfield, J. S., Joklik, W. K., and Downie, A. W. (1974). *Intervirology* **3,** 193.
Finch, J. T., and Holmes, K. C. (1967). *Methods Virol.* **3,** 351.
Flewett, T. H., Bryden, A. S., Davies, H., Woode, G. N., Bridger, J. C., and Derrick, J. M. (1974). *Lancet* **ii,** 61.
Francki, R. I. B., and Grivell, C. J. (1972). *Virology* **48,** 305.
Francki, R. I. B., and Jackson, A. O. (1972). *Virology* **48,** 275.
Fujii-Kawata, I., Miura, K.-I., and Fuke, M. (1970). *J. Mol. Biol.* **51,** 247.
Fukushi, T. (1969). *In* "Viruses, Vectors and Vegetation" (K. Maramorosch, ed.), p. 279. Wiley (Interscience), New York.
Fukushi, T., Shikata, E., and Kimura, I. (1962). *Virology* **18,** 192.
Gálvez, G. E., Thurston, H. D., and Bravo, G. (1963). *Phytopathology* **53,** 106.
Gerola, F. M., and Bassi, M. (1966). *Caryologia* **19,** 13.
Gerola, F. M., Bassi, M., Lovisolo, O., and Vidano, C. (1966a). *Phytopathol. Z.* **56,** 97.
Gerola, F. M., Bassi, M., Lovisolo, O., and Vidano, C. (1966b). *Caryologia* **19,** 493.
Giannotti, J., and Monsarrat, P. (1968). *Ann. Epiphyt.* **19,** 707.
Giannotti, J., Monsarrat, P., and Vago, C. (1968). *Ann. Epiphyt.* **19,** 31.
Granados, R. R. (1969). *In* "Viruses, Vectors and Vegetation," (K. Maramorosch, ed.), p. 327. Wiley (Interscience), New York.
Grancini, P. (1949). *Not. Mal. Piante* **3,** 39.
Grancini, P. (1958). *Maydica* **3,** 67.
Grancini, P. (1962). *Maydica* **7,** 17.

Grancini, P., and Corte, A. (1969). *Maydica* **14,** 79.
Grylls, N. E. (1954). *Aust. J. Biol. Sci.* **7,** 47.
Grylls, N. E. (1975). *Ann. Appl. Biol.* **79,** 283.
Grylls, N. E., Waterford, C. J., Filshie, B. K., and Beaton, C. D. (1974). *J. Gen. Virol.* **23,** 179.
Harder, D. E., and Westdal, P. H. (1971). *Plant Dis. Rep.* **55,** 802.
Harpaz, I. (1959). *Nature (London)* **184,** 77.
Harpaz, I. (1961). *FAO Plant Prot. Bull.* **9,** 144.
Harpaz, I. (1966). *Maydica* **11,** 18.
Harpaz, I. (1972). "Maize Rough Dwarf," Israel Univ. Press, Jerusalem.
Harpaz, I., and Klein, M. (1969). *Entomol. Exp. Appl.* **12,** 99.
Harpaz, I., Minz, G., and Nitzany, F. (1958). *FAO Plant Prot. Bull.* **7,** 43.
Harpaz, I., Vidano, C., Lovisolo, O., and Conti, M. (1965). *Atti Accad. Sci. Torino, Cl. Sci. Fis., Mat. Nat.* **99,** 885.
Hashioka, Y. (1964). *Riso* **13,** 295.
Hayes, A. G. (1974). *Proc. Queensland Soc. Sugar Cane Technol., 41st Conf.* p. 105.
Herold, F., and Munz, K. (1967). *J. Virol.* **1,** 1028.
Hiebert, E. (1975). *Abstr., Int. Congr. Virol. 3rd, Madrid* p. 83.
Hirumi, H., Granados, R. R., and Maramorosch, K. (1967). *J. Virol.* **1,** 430.
Hooper, G. R., and Schneider, H. (1969). *Am. J. Bot.* **56,** 238.
Hoppe, W., and Vacke, J. (1972). *Pr. Nauk. JOR, Poznan* **14,** 145.
Hughes, C. G., and Robinson, P. E. (1961). *In* "Sugar-Cane Diseases of the World" (J. P. Martin, E. V. Abbott, and C. G. Hughes, eds.), Vol. 1, p. 389. Elsevier, Amsterdam.
Hunkar, A. E. S., Hung, A. T. A., Dulder, I. G., Schank, S. C., Holder, N., and Edwards, C. (1974). *Trop. Agric. (Trinidad)* **52,** 75.
Husain, A. A., Brown, A. H. D., Hutchinson, P. B., and Wismer, C. A. (1965). *Proc. Int. Soc. Sugar-Cane Technol.* **12,** 1154.
Hutchinson, P. B. (1969). *Sugarcane Pathol. Newsl.* **3,** 7.
Hutchinson, P. B. (1975). INCANDEX print-out of references to Fiji disease. See Hutchinson and Francki (1973).
Hutchinson, P. B., and Francki, R. I. B. (1973). "CMI/AAB Descriptions of Plant Viruses," No. 119.
Hutchinson, P. B., Forteath, G. N. R., and Osborn, A. W. (1972). *Sugarcane Pathol. Newsl.* **9,** 12.
Iida, T. T. (1969). *In* "The Virus Diseases of the Rice Plant," p. 3. Johns Hopkins Press, Baltimore, Maryland.
Iida, T. T., and Shinkai, A. (1969). *In* "The Virus Diseases of the Rice Plant," p. 125. Johns Hopkins Press, Baltimore, Maryland.
Iida, T. T., Shinkai, A., and Kimura, I. (1972). "CMI/AAB Descriptions of Plant Viruses," No. 102.
Ikäheimo, K. (1960). *J. Sci. Agric. Soc. Finl.* **32,** 62.
Ikäheimo, K. (1961), *J. Sci. Agric. Soc. Finl.* **33,** 81.
Ikäheimo, K. (1964). *Ann. Agric. Fenn.* **3,** 133.
Ikäheimo, K., and Raatikainen, M. (1961). *J. Sci. Agric. Soc. Finl.* **33,** 146.
Ikäheimo, K., and Raatikainen, M. (1963). *Ann. Agric. Fenn.* **2,** 153.
Ikegami, M. (1975). PhD. Thesis, Univ. of Adelaide, South Australia.
Ikegami, M., and Francki, R. I. B. (1973). *Virology* **56,** 404.
Ikegami, M., and Francki, R. I. B. (1974). *Virology* **61,** 327.
Ikegami, M., and Francki, R. I. B. (1975). *Virology* **64,** 464.

Ikegami, M., and Francki, R. I. B. (1976). *Virology* **70,** 292.
IRRI (1964). "Annual Report." Int. Rice Res. Inst., Los Baños, Philippines.
IRRI (1968). "Annual Report." Int. Rice Res. Inst., Los Baños, Philippines.
Ishihara, T. (1969). *In* "Viruses, Vectors and Vegetation" (K. Maramorosch, ed.), p. 235. Wiley (Interscience), New York.
Ishii, M., Takahashi, H., and Ono, K. (1966). *Proc. Kanto-Tosan Plant Prot. Soc.* **13,** 29.
Ishii, T., and Yasuo, S. (1967). *Proc. Kanto-Tosan Plant Prot. Soc.* **14,** 35.
Jamalainen, E. A. (1961). *Horticultura* **15,** 90.
Jensen, D. D. (1969). *In* "Viruses, Vectors and Vegetation" (K. Maramorosch, ed.), p. 505. Wiley (Interscience), New York.
Joklik, W. K. (1972). *Virology* **49,** 700.
Jordan, L. E., and Mayor, H. D. (1962). *Virology* **17,** 597.
Kapikian, A. Z., Kim, H. W., Wyatt, R. G., Rodriguez, W. J., Ross, S., Cline, W. L., Parrott, R. H. and Chanock, R. M. (1974). *Science* **185,** 1049.
Kashiwagi, Y. (1966). *Nippon Shokubutsu Byori Gakkaiho* **32,** 168.
Kempiak, G. (1972). *Tagungsber., Deut. Akad. Landwirtschaftswiss. Berlin* **121,** 99.
Kimura, I. (1973). *Rept. Tottori Mycol. Inst.* **10,** 779.
Kisimoto, R. (1967). *Virology* **32,** 144.
Kisimoto, R. (1969). *In* "The Virus Diseases of the Rice Plant," p. 243. Johns Hopkins Press, Baltimore, Maryland.
Kisimoto, R. (1973). *In* "Viruses and Invertebrates" (A. J. Gibbs, ed.), p. 137. North-Holland/Amer. Elsevier, New York.
Kisimoto, R., and Watson, M. A. (1965). *J. Invertebr. Pathol.* **7,** 297.
Kislev, N., Harpaz, I., and Klein, M. (1968). *Acta Phytopathol. Acad. Sci. Hung.* **3,** 3.
Kitagawa, Y., and Shikata, E. (1969a). *Mem. Fac. Agric., Hokkaido Univ.* **6,** 439.
Kitagawa, Y., and Shikata, E. (1969b). *Mem. Fac. Agric., Hokkaido Univ.* **6,** 446.
Kitagawa, Y., and Shikata, E. (1971). *Rev. Plant Prot. Res.* **4,** 113. (English summary of Kitagawa and Shikata 1969a,b.)
Kitagawa, Y., and Shikata, E. (1973). *Rep. Tottori Mycol. Inst.* **10,** 787.
Kitajima, E. W., and Costa, A. S. (1970). *Proc. Int. Congr. Electron Microsc., 7th, Grenoble* **3,** 323.
Klein, M. (1967). Ph.D. Thesis, Hebrew Univ., Jerusalem.
Klein, M., and Harpaz, I. (1969). *Z. Angew. Entomol.* **64,** 39.
Klein, M., and Harpaz, I. (1970). *Virology* **41,** 72.
Kodama, T., and Suzuki, N. (1973). *Nippon Shokubutsu Byori Gakkaiho* **39,** 251.
Kulkarni, H. Y. (1973). *Ann. Appl. Biol.* **75,** 205.
Kunkel, L. O. (1924). *Bull. Exp. Sta. Hawaiian Sugar Planter's Ass., Bot. Ser.* **3,** 99.
Kuribayashi, K., and Shinkai, A. (1952). *Nippon Shokubutsu Byori Gakkaiho* **16,** 41.
Lesemann, D. (1972). *J. Gen. Virol.* **16,** 273.
Lesemann, D., and Huth, W. (1975). *Phytopathol. Z.* **82,** 246.
Levin, D. H., Mendelsohn, N., Schonberg, M., Klett, H., Silverstein, S., Kapuler, A. M., and Acs, G. (1970). *Proc. Natl. Acad. Sci. U.S.A.* **66,** 890.
Lewandowski, L. J., and Traynor, B. L. (1972). *J. Virol.* **10,** 1053.
Lewandowski, L. J., Kalmakoff, J., and Tanada, Y. (1969). *J. Virol.* **4,** 857.
Lindsten, K. (1959). *Phytopathol. Z.* **35,** 420.
Lindsten, K. (1961a). *K. Lantbrukshoegsk. Ann.* **27,** 137.
Lindsten, K. (1961b). *K. Lantbrukshoegsk. Ann.* **27,** 199.
Lindsten, K. (1962). *K. Lantbrukshoegsk. Ann.* **28,** 135.
Lindsten, K. (1966). *Acta Agric. Scand., Suppl.* **16,** 150.

Lindsten, K. (1970). *Natl. Swed. Inst. Plant Prot. Contrib.* **14:134,** 407.
Lindsten, K. (1973a). *Vaxtskyddsnotiser* **37,** 55.
Lindsten, K. (1973b). *Nord. Jordbrugsforsk.* **55.**
Lindsten, K. (1974). *Mikrobiologija* **11,** 55.
Lindsten, K. (1976). In preparation.
Lindsten, K., and Gerhardson, B. (1971). *Natl. Swed. Inst. Plant Prot. Contrib.* **14:128,** 281.
Lindsten, K., Gerhardson, B., and Petterson, J. (1973). *Natl. Swed. Inst. Plant Prot. Contrib.* **15:151,** 373.
Ling, K. C. (1969). *In* "The Virus Diseases of the Rice Plant," p. 139. Johns Hopkins Press, Baltimore, Maryland.
Ling, K. C., Aguiero, V. M., and Lee, S. H. (1970). *Plant Dis. Rep.* **54,** 565.
Loening, U. E. (1969). *Biochem. J.* **113,** 131.
Loh, P. C., and Shatkin, A. J. (1968). *J. Virol.* **2,** 1353.
Loh, P. C., Hohl, H. R., and Soergel, M. (1965). *J. Bacteriol.* **89,** 1140.
Long, D. L., and Timian, R. G. (1975). *Phytopathology* **65,** 848.
Lovisolo, O. (1967). *Atti Accad. Sci. Torino, Cl. Sci. Fis., Mat. Nat.* **101,** 615.
Lovisolo, O. (1971a). *Tagungsber. Deut. Akad. Landwirtschaftswiss. Berlin* **115,** 83.
Lovisolo, O. (1971b). "CMI/AAB Descriptions of Plant Viruses," No. 72.
Lovisolo, O., and Conti, M. (1966). *Atti Accad. Sci. Torino, Cl. Sci. Fis., Mat. Nat.* **100,** 63.
Lovisolo, O., Vidano, C., and Conti, M. (1966). *Atti Accad. Sci. Torino, Cl. Sci. Fis., Mat. Nat.* **100,** 351.
Lovisolo, O., Luisoni, E., Conti, M., and Wetter, C. (1967). *Naturwissenschaften* **54,** 73.
Lovisolo, O., Luisoni, E., and Conti, M. (1974). *Mikrobiologija* **11,** 1.
Luftig, R. B., Kilham, S. S., Hay, A. J., Zweerink, H. J., and Joklik, W. K. (1972). *Virology* **48,** 170.
Luisoni, E., and Conti, M. (1970). *Phytopathol. Mediterr.* **9,** 102.
Luisoni, E., Wetter, C., Lovisolo, O., and Conti, M. (1968). *Atti. Accad. Sci. Torino, Cl. Sci. Fis., Mat. Nat.* **102,** 539.
Luisoni, E., Lovisolo, O., Kitagawa, Y., and Shikata, E. (1973). *Virology* **52,** 281.
Luisoni, E., Milne, R. G., and Boccardo, G. (1975) *Virology* **68,** 86.
Lung, M. C. Y., and Pirone, T. P. (1974). *Virology* **60,** 260.
Lyon, H. L. (1910). *Hawaii. Plant. Rec.* **3,** 200.
McDonald, J. G., and Hiebert, E. (1975). *Virology* **63,** 295.
McWhorter, F. P. (1922) *Philipp. Agric.* **11,** 103.
Maillet, P.-L., and Folliot, R. (1967). *C. R. Acad. Sci., Ser. D* **264,** 2828.
Maramorosch, K. (1953). *Plant Dis. Rep.* **37,** 612.
Maramorosch, K. (1957). *Int. Congr. Crop Prot., 4th.* (Proposal.)
Maramorosch, K. (1959). *Entomol. Exp. Appl.* **2,** 169.
Maramorosch, K. (1960). *Protoplasma* **52,** 457.
Maramorosch, K., ed. (1969a). "Viruses, Vectors and Vegetation." Wiley (Interscience), New York.
Maramorosch, K. (1969b). *In* "The Virus Diseases of the Rice Plant," p. 179. Johns Hopkins Press, Baltimore, Maryland.
Maramorosch, K., and Jensen, D. D. (1963). *Annu. Rev. Microbiol.* **17,** 495.
Maramorosch, K., Calica, C. A., Agati, J. A., and Pableo, G. (1961). *Entomol. Expl. Appl.* **4,** 86.
Martin, M. L., Palmer, E. L., and Middleton, P. J. (1975). *Virology* **68,** 146.

Martin, S. A., and Zweerink, H. J. (1972). *Virology* **50,** 495.
Martinez, G. (1976). In preparation.
Martinson, H. G., and Lewandowski, L. J. (1974). *Intervirology* **4,** 91.
Mayor, H. D. (1975). *Abstr., Int. Congr. Virol., 3rd, Madrid* p. 151.
Mediayedu, J. A. O., and Plumb, R. T. (1976). *Rothamsted Annu. Rep.* 1975, in press.
Milne, R. G., and Lovisolo, O. (1974). *New Sci.* **64,** 252.
Milne, R. G., and Luisoni, E. (1975). *Virology* **68,** 270.
Milne, R. G., and Luisoni, E. (1976). *Meth. Virol.* (in press).
Milne, R. G., Conti, M., and Lisa, V. (1973). *Virology* **53,** 130.
Milne, R. G., Kempiak, G., Lovisolo, O., and Mühle, E. (1974). *Phytopathol. Z.* **79,** 315.
Milne, R. G., Lindsten, K., and Conti, M. (1975). *Ann. Appl. Biol.* **79,** 371.
Mitsuhashi, J. (1974). *Rev. Plant Prot. Res.* **7,** 57.
Miura, K.-I., Kimura, I., and Suzuki, N. (1966). *Virology* **28,** 571.
Mochida, O., and Kisimoto, R. (1971). *Rev. Plant Prot. Res.* **4,** 1.
Moffitt, E. M., and Lister, R. M. (1975). *Phytopathology* **65,** 851.
Morinaka, T., and Sakurai, Y. (1965a). *Chugoku Agric. Res.* **33,** 17.
Morinaka, T., and Sakurai, Y. (1965b). *Nippon Shokubutsu Byori Gakkaiho* **30,** 299.
Morinaka, T., and Sakurai, Y. (1966). *Nippon Shokubutsu Byori Gakkaiho* **32,** 89.
Morinaka, T., and Sakurai, Y. (1967). *Chugoku Nogyo Shikenjo Hokoku E* **1,** 25.
Morinaka, T., Toriyama, K., and Sakurai, Y. (1969). *Jpn. J. Breed.* **19,** 74.
Mühle, E., and Kempiak, G. (1971). *Phytopathol. Z.* **72,** 269.
Muir, F. (1910). *Hawaii. Plant. Rec.* **3,** 186.
Mungomery, R. W., and Bell, A. F. (1933). *Bur. Sugar Exp. Stn., Queensl., Div. Pathol. Bull.* **4,** 1.
Nagaraj, A. N., and Black, L. M. (1962). *Virology* **16,** 152.
Nasu, S. (1963). *Kyushu Nogyo Shikenjo Iho* **8,** 153.
Nasu, S. (1965). *Nippon Shokubutsu Byori Gakkaiho* **30,** 265.
Nasu, S. (1969). *In* "The Virus Diseases of the Rice Plant," p. 93. Johns Hopkins Press, Baltimore, Maryland.
Newman, J. F. E., Brown, F., Bridger, J. C., and Woode, G. N. (1975). *Nature (London)* **258,** 631.
North, D. S., and Barber, E. G. (1935). *Proc. Congr. Int. Soc. Sugar-Cane Technol., 5th* p. 498.
Nuorteva, P. (1962) *Ann. Zool. Soc. Zool.-Bot. Fenn. Vanamo* **23,** 1.
Nuorteva, P. (1965). *Zool. Beitr.* **11,** 191.
Ocfemia, G. O. (1934a). *Univ. Philipp. Nat. Appl. Sci. Bull.* **3,** 277.
Ocfemia, G. O. (1934b). *Am. J. Bot.* **21,** 113.
Ogawa, M., Nishiuchi, T., Yamoto, I., and Kawamura, M. (1957). *Kochi Agr. Exp. Stn. Bull.* **11,** 1.
Ono, K. (1964). *Plant Prot.* **18,** 328.
Ou, S. H. (1972). "Rice Diseases." Commonw. Mycol. Inst., Kew, England.
Ou, S. H., and Rivera. C. T. (1969). *In* "The Virus Diseases of the Rice Plant," p. 23. Johns Hopkins Press, Baltimore, Maryland.
Outridge, R., and Teakle, D. S. (1976). In preparation.
Payne, C. C., and Kalmakoff, J. (1974). *Intervirology* **4,** 354.
Payne, C. C., and Tinsley, T. W. (1974). *J. Gen. Virol.* **25,** 291.
Peters, D. (1970). "CMI/AAB Descriptions of Plant Viruses," No. 36.
Pirone, T. P., and Kassanis, B. (1975). *J. Gen. Virol.* **29,** 257.
Pirone, T. P., and Shaw, J. G. (1973). *Virology* **53,** 274.

Protsenko, A. E. (1958). *Vopr. Virusol.* **4,** 481. (See *Rev. Appl. Mycol.* **39,** 22.)
Průša, V. (1958). *Phytopathol. Z.* **33,** 99.
Průša, V., Jermoljev, E., and Vacke, J. (1959). *Biol. Plant.* **1,** 223.
Raatikainen, M. (1967). *Ann. Agric. Fenn.,* **6** *Suppl.* **2,** 1.
Raatikainen, M., and Vasarainen, A. (1973). *Ann. Agric. Fenn.* **12,** 77.
Ralph, R. K. (1969). *Adv. Virus Res.* **15,** 61.
Reddy, D. V. R. (1974). *Proc. Am. Phytopathol. Soc.* **1,** 82. (Abstr.)
Reddy, D. V. R., and Black, L. M. (1972). *Virology* **50,** 412.
Reddy, D. V. R., and Black, L. M. (1973a). *Virology* **54,** 150.
Reddy, D. V. R., and Black, L. M. (1973b). *Virology* **54,** 557.
Reddy, D. V. R., and Lesnaw, J. A. (1971). *Phytopath.* **61,** 907. (Abstr.)
Reddy, D. V. R., and McLeod, R. (1976). *Virology* **70,** 274.
Reddy, D. V. R., Kimura, I., and Black, L. M. (1974). *Virology* **60,** 293.
Reddy, D. V. R., Boccardo, G., Outridge, R., Teakle, D. S., and Black, L. M. (1975a). *Virology* **63,** 287.
Reddy, D. V. R., Shikata, E., Boccardo, G., and Black, L. M. (1975b). *Virology* **67,** 279.
Redolfi, P., and Boccardo, G. (1974). *Virology* **59,** 319.
Redolfi, P., and Pennazio, S. (1972). *Acta Virol. (Engl. Ed.)* **16,** 369.
Redolfi, P., Pennazio, S., and Paul, H.-L. (1973). *Acta Virol. (Engl. Ed.)* **17,** 243.
Rivera, C. T., Ou, S. H., and Iida, T. T. (1966). *Plant Dis. Rep.* **50,** 453.
Rubio-Huertos, M., and Bos, L. (1969). *Neth. J. Plant Pathol.* **75,** 329.
Ruppel, R. F. (1965). *Mich. State Univ. Publ.* **2,** 385.
Sakurai, Y. (1969). *In* "The Virus Diseases of the Rice Plant," p. 257. Johns Hopkins Press, Balitmore, Maryland.
Sasaba, T. (1974). *Rev. Plant Prot. Res.* **7,** 81.
Schank, S. C. (1968). *Sunshine State Agric. Res. Rep.* **13,** 5.
Schank, S. C., and Edwardson, J. R. (1968). *Crop. Sci.* **8,** 118.
Schank, S. C., Edwardson, J. R., Christie, R. G., and Overman, M. A. (1972). *Euphytica* **21,** 344.
Schindler, A. J. (1942). *J. Aust. Inst. Agric. Sci.* **8,** 35.
Schumann, K. (1971). *In* "Krankheiten und Schädlinge der Futtergräser" (E. Mühle, ed.) p. 94. Hirzel Verlag, Leipzig.
Scossirolli, R. (1950). *Not. Mal. Piante* **9,** 6.
Scossirolli, R. (1951). *Ann. Sper. Agrar.* **5,** 157.
Shatkin, A. J., (1968). *In* "Molecular Basis of Virology" (H. Fraenkel-Conrat, ed.), p. 351. Reinhold, New York.
Shatkin, A. J., (1974). *Annu. Rev. Biochem.* **43,** 643.
Shatkin, A. J., and Lafiandra, A. J. (1972). *J. Virol.* **10,** 698.
Shatkin, A. J., and Rada, B. (1967). *J. Virol.* **1,** 24.
Shatkin, A. J., and Sipe, J. D. (1968). *Proc. Natl. Acad. Sci. U.S.A.* **61,** 1462.
Sheffield, F. M. L. (1968). *East Afr. Agric. For. Res. Organ., Annu. Rep.* 1967.
Sheffield, F. M. L. (1969). *Sugarcane Pathol. Newsl.* **3,** 25.
Shikata, E. (1969). *In* "The Virus Diseases of the Rice Plant," p. 223. Johns Hopkins Press, Baltimore, Maryland.
Shikata, E. (1974). "CMI/AAB Descriptions of Plant Viruses," No. 135.
Shikata, E., and Kitagawa, Y. (1976). To be published.
Shikata, E., and Maramorosch, K. (1965). *Virology* **27,** 461.
Shikata, E., and Maramorosch, K. (1967). *Virology* **32,** 363.
Shinkai, A. (1957). *Nippon Shokubutsu Byori Gakkaiho* **22,** 34.

Shinkai, A. (1962). *Nogyo Gijutsu Kenkyusho Hokoku C* **14,** 1.
Shinkai, A. (1966). *Nogyo Gijutsu Kenkyusho Hokoku* **32,** 317.
Shinkai, A. (1967). *Nogyo Gijutsu Kenkyusho Hokoku,* **33,** 318.
Sinclair, E. R., and Osborn, A. W. (1970). *J. Econ. Entomol.* **63,** 1974.
Sinha, R. C. (1967). *Virology* **31,** 746.
Sinha, R. C. (1968). *Adv. Virus Res.* **13,** 181.
Sinha, R. C. (1973). *In* "Viruses and Invertebrates" (A. J. Gibbs, ed.), p. 493. North-Holland/Amer. Elsevier, New York.
Sinha, R. C., and Reddy, D. V. R. (1964). *Virology* **24,** 626.
Slykhuis, J. T. (1967). *Rev. Appl. Mycol.* **46,** 401.
Smith, R. E., Zweerink, H. J., and Joklik, W. K. (1969). *Virology* **39,** 791.
Stollar, B. D., (1973). *In* "The Antigens" (M. Sela, ed.), Vol. 1, p. 1. Academic Press, New York.
Streissle, G., and Granados, R. R. (1968). *Arch. Gesamte Virusforsch.* **25,** 369.
Sukhov, K. S. (1940a). *Mikrobiologiya* **9,** 188.
Sukhov, K. S. (1940b). *C. R. Acad. Sci. URSS* **27,** 377.
Sukhov, K. S., and Vovk, A. M. (1938). *C. R. Acad. Sci. URSS* **19,** 207.
Suzuki, N. (1969). *In* "Viruses, Vectors and Vegetation" (K. Maramorosch, ed.), p. 557. Wiley (Interscience), New York.
Suzuki, N., and Kimura, I. (1969). *In* "The Virus Diseases of the Rice Plant," p. 207. Johns Hopkins Press, Baltimore, Maryland.
Teakle, D. S., and Steindl, D. R. L. (1969). *Virology* **37,** 139.
Thornton, A., and Zuckerman, A. J. (1975). *Nature (London)* **254,** 557.
Toriyama, K. (1969). *In* "The Virus Diseases of the Rice Plant," p. 313. Johns Hopkins Press, Baltimore, Maryland.
Toyoda, S., Kimura, I., and Suzuki, N. (1965). *Nippon Shokubutsu Byori Gakkaiho* **30,** 225.
Trujillo, G. E., Acosta, J. M., and Piñero, A. (1974). *Plant Dis. Rep.* **58,** 122.
Tryon, H. (1910). *Rep. Dep. Agric. Stock, Queensland,* 1909–1910, p. 81.
Vacke, J. (1960). *Ann. Acad. Sci. Czech. Agric.* **33,** 1049.
Vacke, J. (1964). *Proc. Conf. Czech. Plant Virol. 5th, Prague 1962* p. 335.
Vacke, J. (1966). *Biol. Plant.* **8,** 127.
Vacke, J. (1970). *Ved. Pr. Vyzk. Ustavu Rostl. Vyroby Praze-Ruzyni* **16,** 21.
Vacke, J. (1974). *Ved. Pr. Vyzk. Ustavu Rostl. Vyroby Praze-Ruzyni* **18,** 81.
Vacke, J., and Průša, V. (1962). *Rostl. Vyroba* **8,** 463.
Vacke, J., and Vostřák, J. (1975). *Proc. Natl. Conf. Plant Prot., 5th, Brno, 1974.*
Vasquez, C., and Tournier, P. (1962). *Virology* **17,** 503.
Vasquez, C., and Tournier, P. (1964). *Virology* **24,** 128.
Verwoerd, D. W. (1970). *Prog. Med. Virol.* **12,** 192.
Verwoerd, D. W., and Huismans, H. (1972). *Onderstepoort J. Vet. Res.* **39,** 185.
Verwoerd, D. W., Louw, H., and Oellermann, R. A. (1970). *J. Virol.* **5,** 1.
Verwoerd, D. W., Els, H. J., de Villiers, E.-M., and Huismans, H. (1972). *J. Virol.* **10,** 783.
Vidano, C. (1966). *Atti Accad. Sci. Torino, Cl. Sci. Fis., Mat. Nat.* **100,** 731.
Vidano, C. (1967a). *Atti Accad. Sci. Torino, Cl. Sci. Fis., Mat. Nat.* **101,** 717.
Vidano, C. (1967b). *Ann. Fac. Sci. Agrar. Torino* **4,** 33.
Vidano, C. (1968). *Atti Accad. Sci. Torino, Cl. Sci. Fis., Mat. Nat.* **102,** 641.
Vidano, C. (1969a). *Atti Accad. Sci. Torino, Cl. Sci. Fis., Mat. Nat.* **103,** 559.
Vidano, C. (1969b). *Mem. Soc. Entomol. Ital.* **48,** 94.
Vidano, C. (1970). *Virology* **41,** 218.

Vidano, C., and Bassi, M. (1966). *Atti Accad. Sci. Torino, Cl. Sci. Fis., Mat. Nat.* **100,** 73.

Vidano, C., Lovisolo, O., and Conti, M. (1966a). *Atti Accad. Sci. Torino, Cl. Sci. Fis., Mat. Nat.* **100,** 125.

Vidano, C., Lovisolo, O., and Conti M. (1966b). *Atti Accad. Sci. Torino, Cl. Sci. Fis., Mat. Nat.* **100,** 699.

Watanabe, Y., and Graham, A. F. (1967). *J. Virol.* **1,** 665.

Weber, K., and Osborn, M. (1969). *J. Biol. Chem.* **244,** 4406.

Wetter, C., Luisoni, E., Conti, M., and Lovisolo, O., (1969). *Phytopathol. Z.* **66,** 197.

Wit, F. (1956). *Euphytica* **5,** 119.

Wood, H. A. (1973). *J. Gen. Virol.* **20,** Suppl. 61.

Yasu, M., Zenbayashi, R., and Suzuki, K. (1966). *Proc. Kanto-Tosan Plant Prot. Soc.* **13,** 30.

Yasuo, S. (1969). *In* "The Virus Diseases of the Rice Plant," p. 167. Johns Hopkins Press, Baltimore, Maryland.

Zweerink, H. J., and Joklik, W. K. (1970). *Virology* **41,** 501.

Note Added in Proof

The work of Reddy *et al.* [*Virology* **73,** 36 (1976)] now shows that MWEV has 10 genome segments which resemble those of MRDV, RBSDV, and FDV. However, MWEV appears to be less closely related to the other three viruses than they are to each other.

Hatta and Francki [*Physiol. Plant Pathol.* (1977), in press] have reported in detail on the gall anatomy of sugarcane infected with FDV and Hatta and Francki [*Virology* (1977), in press] have described the structure of the FDV virion.

VIRUSES OF MYCOPLASMAS AND SPIROPLASMAS

Jack Maniloff,*† Jyotirmoy Das,* and J. R. Christensen*

University of Rochester School of Medicine and Dentistry,
Rochester, New York

I. Historical Background

The term *mycoplasma* refers to a group of microorganisms (Class Mollicutes, Order Mycoplasmatales) previously known as pleuropneumonia-like organisms, or PPLO. The many different isolates have been cataloged into the genera *Mycoplasma, Acholeplasma, Ureaplasma* or T-strains, *Spiroplasma, Thermoplasma,* and *Anaeroplasma* (Table I). Mycoplasmas are clinically important microorganisms and include the etiologic agents of a variety of plant and animal diseases (e.g., Davis and Whitcomb, 1971; Freundt, 1974). In human studies, *Mycoplasma pneumoniae* has been found to be a pathogen, and other mycoplasmas (of unknown etiology) have been isolated in clinical situations [reviewed by Thomas (1970) and McCormack *et al.* (1973)]. Mycoplasmas are also frequently found as contaminants in tissue cultures (e.g., Barile, 1973).

The biology of the mycoplasmas has been reviewed recently by Smith (1971), Maniloff and Morowitz (1972), and Razin (1973), and some of

* Department of Microbiology.
† Department of Radiation Biology and Biophysics.

TABLE I
GENERA OF THE CLASS MOLLICUTES

Genus	Genome: GC content (%)	Genome: Molecular weight	Habitat	Sterol required for growth
Mycoplasma[a]	23–41	4.6×10^8	Animals	Yes
Acholeplasma[a]	30–36	1.0×10^9	Saprophytic	No
Ureaplasma[a] (T-strains)	28	4.6×10^8	Animals	Yes
Spiroplasma[b]	26	1.0×10^9	Plants, insects	Yes
Thermoplasma[c]	46	0.9×10^9	Burning coal refuse piles	No
Anaeroplasma[d]	29–33	—	Animals	Yes

[a] Reviewed by Smith (1971), Maniloff and Morowitz (1972), and Razin (1973).
[b] Bové *et al.* (1973).
[c] Belly *et al.* (1973); Searcy and Doyle (1975).
[d] Robinson *et al.* (1975); Robinson and Allison (1975).

their properties are summarized in Table I. These prokaryotes do not have cell walls and are bounded by a single lipoprotein cell membrane. *Mycoplasma, Ureaplasma, Spiroplasma,* and *Anaeroplasma* require sterol for growth. Some *Mycoplasma* isolates are the smallest cells which have been found, and the genome sizes of the *Mycoplasma* and *Ureaplasma* are among the smallest reported (Morowitz, 1969).

The possible existence of viruses that can infect mycoplasmas was suggested by several observations of electron-dense "virus-like" particles in electron micrographs of mycoplasma cells (Edward and Fogh, 1960; Robertson *et al.*, 1972; Gourrett *et al.*, 1973). Extrachromosomal DNA in mycoplasmas has also been reported. These studies involved electron microscopic observations of small circles (Morowitz, 1969; Zouzias *et al.*, 1973), DNA satellite bands in CsCl density gradients (Haller and Lynn, 1968), and velocity sedimentation of cell DNA in neutral (Dugle and Dugle, 1971) and alkaline (Das *et al.*, 1972) sucrose gradients. However, there was no evidence in any of these studies that the "virus-like" particles in thin-section electron micrographs or the extrachromosomal DNA's were of viral origin and no infectivity was demonstrated.

The isolation of a virus that could infect a mycoplasma was first reported by Gourlay (1970). The original isolate was designated MVL1 (*Mycoplasmatales* virus *laidlawii* 1). Since then more than 50 virus isolates have been reported (Gourlay, 1971, 1972, 1973, 1974; Liss and Maniloff, 1971; Liska, 1972; Gourlay and Wyld, 1973). These have been serologically and morphologically classified into three groups: (1) Group 1, consisting of naked bullet-shaped particles; (2) Group 2, consisting of roughly spherical enveloped viruses; and (3) Group 3, polyhedral particles with tails. (These groups correspond to those previously designated Groups L1, L2, and L3.) All three groups contain DNA. The general properties of the three mycoplasmavirus groups are summarized in Table II. Most of the isolates to date are in Group 1, three are in Group 2, and one is in Group 3.

All Group 1 virus isolates are morphologically identical and serologically related. Some Group 1 isolates differ from the original MVL1 isolate and from each other in properties such as host range, ultraviolet- or antisera-inactivation-rate constants, and one-step growth parameters (Liss and Maniloff, 1971; Maniloff and Liss, 1974). Also, although most mycoplasmaviruses have been isolated from *Acholeplasma laidlawii* strains, some Group 1 viruses have been isolated from other *Mycoplasma* and *Acholeplasma* species (Liss and Maniloff, 1971; Gourlay, 1972; Clyde, 1974a,b). This creates a problem in referring to all Group 1 isolates as MVL1 as suggested by Gourlay (1974) and necessitates including more information in the virus nomenclature. For a particular isolate, Liss and Maniloff (1971) suggested following the "MV" designation by a letter or letters indicating the virus origin (e.g., L for *laidlawii* and Gs for *gallisepticum*) then followed by an isolate number.

All mycoplasma strains presently used to propagate viruses probably also carry viruses. No "cured" strain has been demonstrated. In fact,

TABLE II
PROPERTIES OF MYCOPLASMAVIRUSES[a]

Virus groups	Virion		Nucleic acid		Progeny virus release
	Morphology	Size	Type	Molecular weight	
Group 1	Naked bullet-shaped particles	14–16 nm by 70–90 nm	Single-stranded circular DNA	2×10^6	Nonlytic
Group 2	Roughly spherical enveloped particles	80 nm[b]	Double-stranded DNA	—	Nonlytic
Group 3	Polyhedral particles with short tails	57×61 nm head; 25 nm long tail	Double-stranded DNA	—	Lytic

[a] References in text.
[b] Range: 52–125 nm.

Gourlay (personal communication) was able to isolate all three groups of viruses from a single *A. laidlawii* strain, M1305/68. The nature of the mycoplasmavirus carrier state is not known. The carrier state does not appear to interfere with virological studies, however, since indicator hosts are chosen for which the frequency of spontaneous virus release is extremely low. It should be noted that some *A. laidlawii* strains have been found to produce plaques when grown on lawns without addition of exogenous virus (Liss and Maniloff, unpublished data).

In addition to the three groups of mycoplasmaviruses that have been isolated from *Mycoplasma* and *Acholeplasma* species and propagated on *A. laidlawii* strains, three morphological types of "virus-like" particles have been observed in plant *Spiroplasma* cultures (Cole *et al.,* 1974). Since it has not been possible to propagate these particles, no biochemical or virological data about them are available. The morphological data on these particles are given in Table III. The *Spiroplasma* associated with the sex-ratio condition in *Drosophila* has also been shown to carry a virus (Oishi and Poulson, 1970; Oishi, 1971). Finally, there have been several reports of "virus-like particles" in plants and insect tissues carrying "mycoplasma-like organisms" (Ploaie, 1971; Allen, 1972; Gourret *et al.,* 1973; Giannotti *et al.,* 1973). The electron micrographs show bacilliform particles, generally about 25–35 nm by 70–150 nm. Since these particles do not resemble the known mycoplasma- and spiroplasmaviruses, their relationship to the mycoplasmas is not clear, and as nothing further is known of their biology they cannot be considered further in this review.

II. Mycoplasmaviruses

A. *Biophysical and Biochemical Properties*

1. *Purification*

High titer stocks are obtained by propagating mycoplasmaviruses on lawns of *A. laidlawii* indicator hosts on nutrient agar plates. From 2 to 5×10^{10} viruses per milliliter are obtained by overlaying the plates with 10 ml of medium or phosphate buffered saline at room temperature for 16 to 18 hours. The fluid is decanted and concentrated by either ammonium sulfate (Gourlay, 1974) or polyethylene glycol precipitation (Fig. 1). The precipitate is resuspended in the desired volume of phosphate buffered saline or Tris–HCl buffer (pH 7.4).

Group 1 and 3 virus preparations are treated with either 0.4% Nonidet P40 (Gourlay *et al.,* 1971; Gourlay and Wyld, 1973) or 0.2% Triton X-100 (Liss and Maniloff, 1972). At these concentrations, the detergents have no effect on the viruses. The virus suspensions are dialyzed against

TABLE III

STRUCTURE OF *Spiroplasma* "VIRUS-LIKE" PARTICLES [a]

Particles	Morphology	Size
Type SV–C1	Rodlike particles	10–15 nm by 230–280 nm
Type SV–C2	Polyhedral particles with long tails	52–58 nm heads; 6–8 × 75–83 nm tail
Type SV–C3	Polyhedral particles with short tails	37–44 nm head; 13–18 nm long tail

[a] From Cole *et al.* (1974.)

phosphate-buffered saline or Tris-HCl buffer to remove the detergent and centrifuged to equilibrium in CsCl density gradients. Gradient fractions are assayed for infectivity, and fractions containing virus are dialyzed against buffer to remove CsCl.

Group 2 viruses are enveloped and hence sensitive to detergents and to

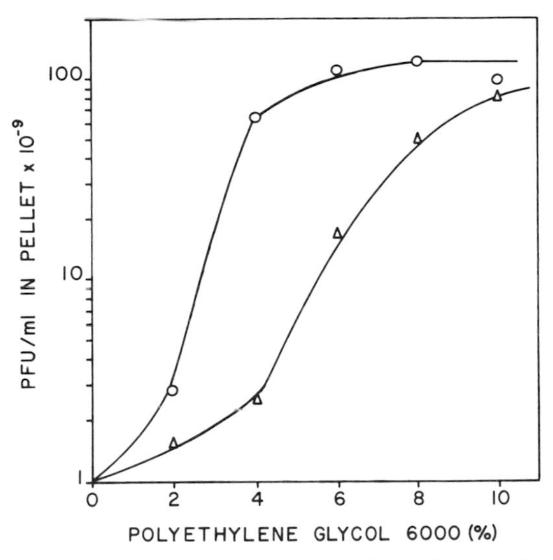

FIG. 1. Polyethylene glycol (PEG) precipitation of mycoplasmaviruses. PEG 6000 was added at varying concentration to virus suspensions in tryptose broth at 4°C and dissolved by brief stirring. After standing for 16 hours at 4°C, 10–ml samples were centrifuged for 10 minutes at 8000 *g*. The pellets were resuspended in 1 ml of Tris–EDTA–NaCl buffer (0.01 *M* Tris; 0.001 *M* EDTA; 0.1 *M* NaCl), pH 7.4. The virus recovery was measured by assaying the pellet for PFU. For a Group 1 virus MVL51 (○) and a Group 2 virus MVL2 (△) the recovery at high PEG concentrations ranged from 80 to 125%.

CsCl (Gourlay *et al.*, 1973). For purifying these viruses an ammonium sulfate–precipitated virus suspension is centrifuged at 100,000 *g* for 45 minutes at 4°C. The precipitate is dissolved in a small volume of 0.15*M* Tris–HCl buffer (pH 7.4) and dialyzed against the same buffer in the cold. The dialyzed suspension is layered over a two-step sucrose gradient containing 20% and 60% (w/w) sucrose in Tris–HCl buffer and centrifuged for 3.5 hours at 94,000 *g*. The virus band is collected by infectivity assay and centrifuged in a linear 20–60% (w/w) sucrose gradient for 17 hours at 59,000 *g*. The virus appears as a narrow band corresponding to a density of 1.2 gm/cm^3 (Gourlay *et al.*, 1973).

2. Ultrastructure

a. Group 1 Virus. Viruses in Group 1 have a density of about 1.37 gm/cm^3 in CsCl (D. J. Garwes, unpublished data*). The virions (Fig. 2) are unenveloped particles with helical symmetry, 14–16 nm wide and 70–90 nm long (Gourlay *et al.*, 1971; Bruce *et al.*, 1972; Milne *et al.*, 1972; Gourlay, 1973, 1974; Liss and Maniloff, 1973b; Maniloff and Liss, 1973). The size ranges reflect variations in measurements from different laboratories and differences in negative staining with phosphotungstate, uranyl acetate, and uranyl formate.

Two different interpretations of Group 1 virus morphology have been proposed. All workers agree that one end of the rod-shaped particle is rounded. Liss and Maniloff (1973b) report that the virions have a bullet shape and that the other end of the virion (the basal end that presumably makes contact with the cell membrane) is relatively flat with some sort of short flexible protrusion. The basal structure could not be further resolved in their studies but would add about 7 nm to the particle length. However, Bruce *et al.* (1972) state that the particle is bacilliform and that the basal end is round but may become degraded to give short protruberances or even a flattened end. It should be noted that the ends of these thin rods have such a small radius of curvature that the drying pattern of the negative stain around any end structure can artifactually give a rounded appearance.

Optical analysis of electron micrographs of negatively stained Group 1 virions (Bruce *et al.*, 1972; Liss and Maniloff, 1973b) has shown helical symmetry, and since the optical transforms have a near-meridional reflection at 4.8 nm, the distance between helix turns is 4.8 nm. Using the diffraction data, Bruce *et al.* (1972) have proposed a structure for the virion based on models of Hull *et al.* (1969) for helical structures with rounded ends. The Group 1 virus model is an icosahedron with $T = 1$ and diameter of 14–16 nm, cut across its 2-fold axis, and structure units

* Presented at the Third Intern. Congr. Virol., Madrid, 1975.

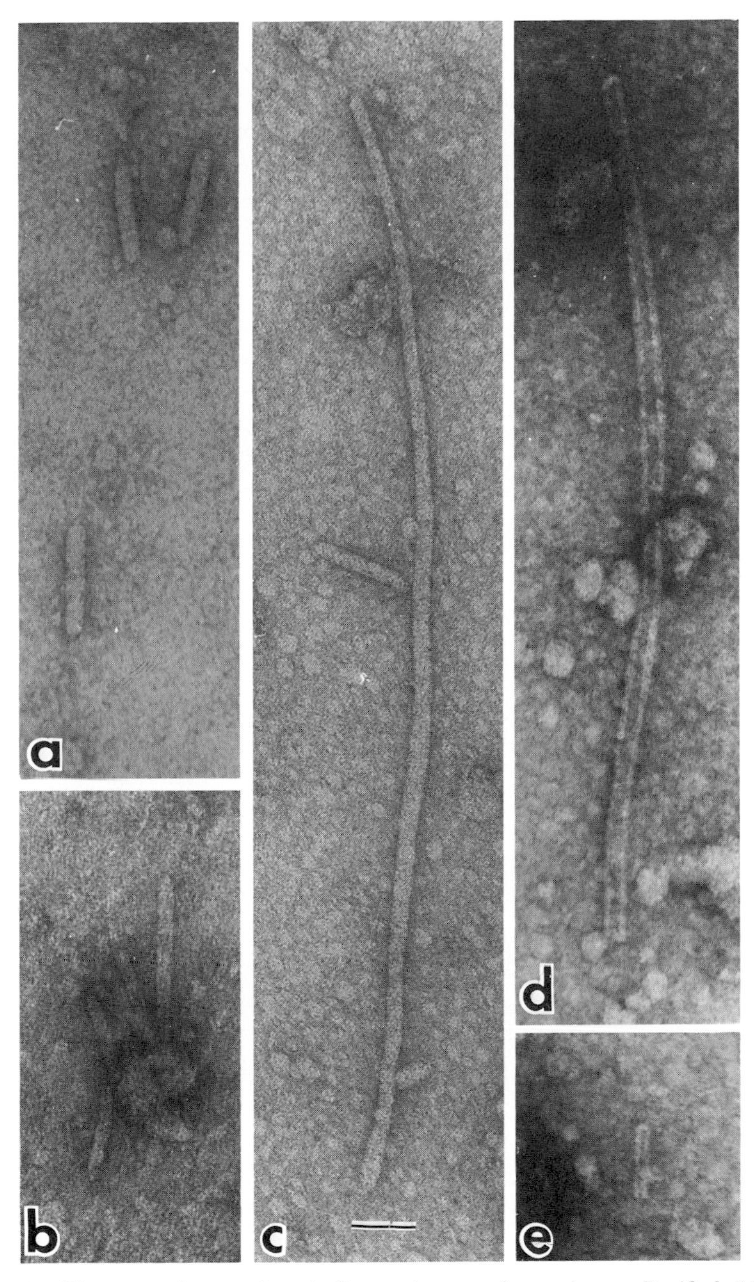

FIG. 2. Electron micrographs of Group 1 mycoplasmaviruses. Purified MVL51 preparations were negatively stained with uranyl formate. (a) Bullet-shaped virions. (b) Double-length particle, near some virions and debris. (c) Long rod-shaped structure and two virions. (d) Long tubular structure. (e) Particle penetrated by negative stain. The various morphological forms are discussed in the text. Bar denotes 50 nm. (From Maniloff *et al.*, 1977a).

then added to form a helix. This is a two-start helix with 5.6 structure units per turn and a pitch of about 20°.

In addition to infectious virus particles, preparations of Group 1 viruses or virus-infected cells contain other virus-related structures, the study of which may be useful in elucidating the mechanism of virus assembly: (1) Unpurified preparations contain 13–14 nm rings (Liss and Maniloff, 1973b; Maniloff and Liss, 1973; Gourlay, 1974), which may be intracellular viral protein subassemblies. (2) A few particles which have been penetrated by negative stain have been seen to have wall thickness of 5.2–6.2 nm (Bruce *et al.*, 1972; Maniloff, unpublished data); these may be incomplete or damaged virions. (3) Rod-shaped particles have been reported with diameters and optical transforms similar to those of virions, but with lengths varying from twice that of virus particles, i.e., about 150 nm (Liss and Maniloff, 1973b), to more than 500 nm (Bruce *et al.*, 1972; Milne *et al.*, 1972). It is interesting that no particles have been reported with a size between that of single- and double-length viruses and that the double-length particles have both ends rounded. (4) Long tubular structures have been observed (Bruce *et al.*, 1972; Milne *et al.*, 1972; Liss and Maniloff, 1973b); most widths are 13–18 nm, but some as wide as 30 nm have been seen, and lengths vary up to 1000 nm. Bruce *et al.* (1972) noted a greater frequency of these hollow structures in both older cells (late logarithmic phase) infected with virus and infected cells late (over 4 hours) in the infection; hence, these structures may be some aberrant polymerization product formed in aged infected cells. (5) Milne *et al.* (1972) found striated masses in infected cells examined 72 hours after the start of infection. Since their culture was probably no longer viable when the sample was taken, and old mycoplasma cultures are known to form many degradative products [reviewed by Maniloff and Morowitz (1972)], it is not possible to evaluate this report. Other laboratories have been unable to reproduce these results (Maniloff and Liss, 1974; Gourlay, 1974).

b. Group 2 Virus. These virions have a lipid-containing envelope (described in Section II, A, 3) and a density of 1.19 gm/cm^3 in sucrose (Gourlay *et al.*, 1973).

Negative-staining studies of MVL2, the original Group 2 isolate, have

FIG. 3. Electron micrographs of Group 2 mycoplasmaviruses. *Acholeplasma laidlawii* cells were infected with MVL2 and prepared for electron microscopy after glutaraldehyde fixation. (a) Thin-sectioned cells, stained with uranyl acetate and lead citrate. (b) Cells negatively stained with phosphotungstate. (c) Cells freeze etched and platinum shadowed. Large arrows show areas of cell surface where virus budding may be occurring. Small arrows show extracellular progeny viruses. Bar denotes 100 nm. (From Maniloff *et al.*, 1977a).

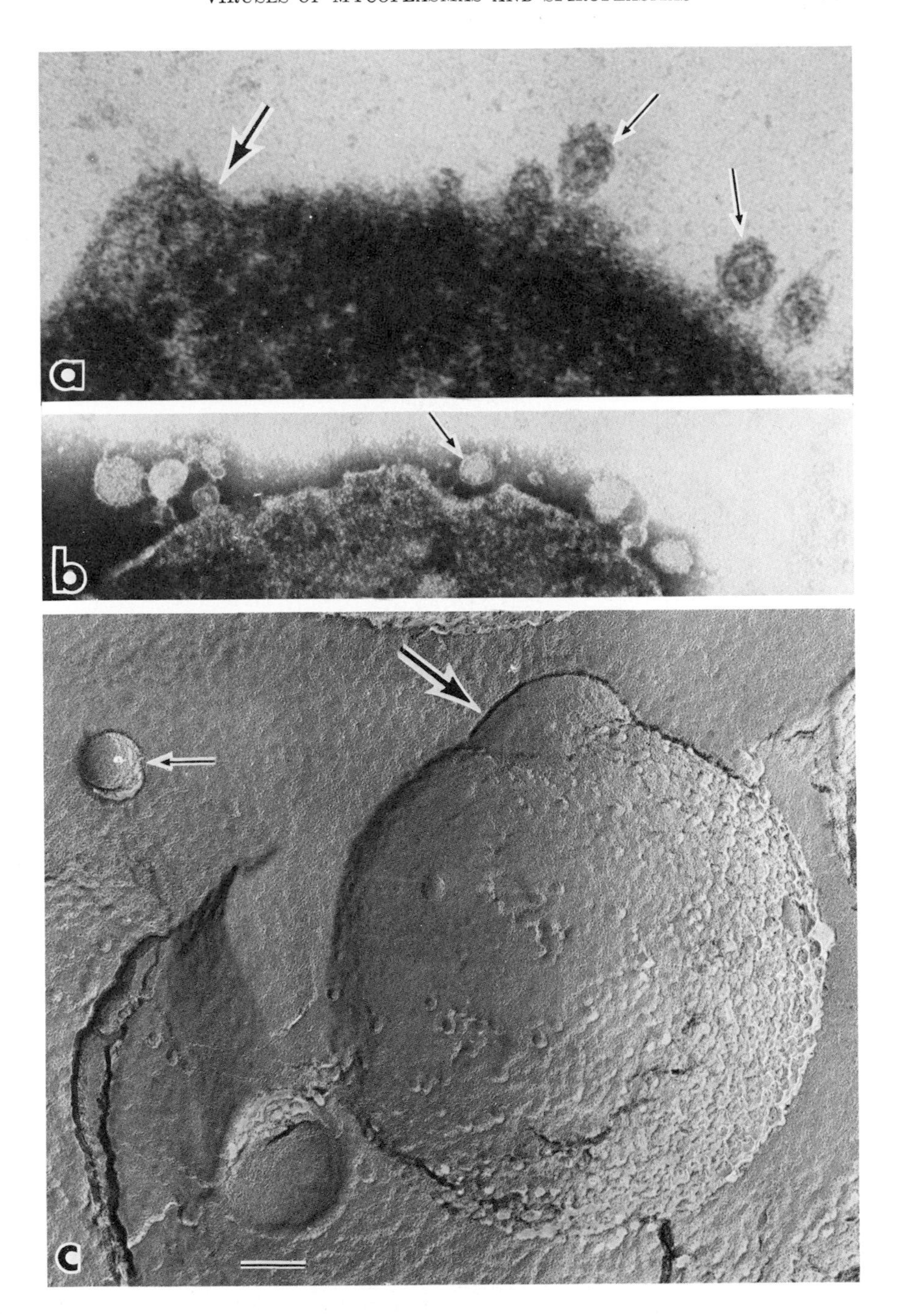
a
b
c

shown approximately spherical particles (Gourlay *et al.*, 1973), with a size range of 52–125 nm in diameter and a mean diameter of about 80 nm (Fig. 3). No internal virion structure is apparent. This may explain the observed polymorphism. Another isolate has given similar results, with a size range of 50–90 nm when sized by thin-section electron microscopy (Liska and Tkadlecek, 1975) and 62–96 nm when sized by ultrafiltration (Hutkova *et al.*, 1973).

Micrographs of thin-sectioned virus (Fig. 3) (Gourlay *et al.*, 1973) have shown the envelope to have a "unit membrane" appearance. The internal morphology appears fibrillar, with a densely staining core. These data suggest that, unlike other enveloped viruses, Group 2 particles may not have a rigid internal helical or icosahedral capsid structure. Instead the virion may be simply double-stranded DNA with some packing proteins inside a membrane. Freeze-etching studies have shown the virus to have fewer particles on the membrane fracture faces than do cell membranes (J. Maniloff and R. M. Putzrath, unpublished data).

c. Group 3 Virus. The single isolate has a density of 1.477 gm/cm^3 in CsCl and a composition of 35% DNA (double-stranded), 63% protein, and 2% fucose [(Garwes et al., 1975); described in Section II, A, 3].

Electron micrographs of negatively stained virus (Fig. 4) show uniform-sized particles with a polyhedral head, about 57 nm by 61 nm, and a short tail, about 25 nm long and 9 nm wide, attached with a collar at

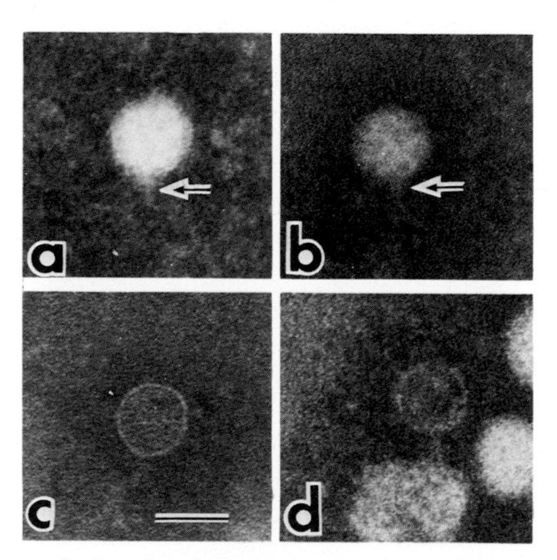

FIG. 4. Electron micrographs of Group 3 mycoplasmaviruses. A wash of MVL3-infected cell lawns was negatively stained with phosphotungstate. (a,b) Intact virus particles with tails (arrows). (c,d) Empty virion heads. Bar denotes 50 nm. (From Maniloff *et al.*, 1977a.)

one vertex of the head (Garwes *et al.*, 1975). The collar was resolved better by phosphotungstate negative staining than by uranyl acetate negative staining. Elongated (perhaps polyhead-like) particles were also sometimes found.

3. Chemical Composition

a. Nucleic Acids. All three mycoplasmavirus groups are DNA viruses. The DNA of Group 1 virus is a single-stranded, covalently closed, circular molecule of molecular weight 2×10^6 (Liss and Maniloff, 1973a). Circularity of the DNA was shown by the sensitivity of DNA infectivity to different nuclease treatments: deoxyribonuclease I (an endonuclease) destroyed infectivity, but exonuclease I, specific for single-stranded DNA, and endonuclease III, specific for double-stranded DNA, had no effect.

Electron micrographs of Group 2 viral DNA indicate that it is double stranded (J. Maniloff, R. M. Putzrath, and J. Das, unpublished data).

The DNA of MVL3, the Group 3 virus, has been studied by Garwes *et al.* (1975) and gives conflicting GC (guanine–cytosine) contents when examined by different procedures. The DNA exhibits a sharp thermal transition with a mean melting temperature of 85.5°C and hypochromicity of 42–50%. This indicates that the viral DNA is double stranded and has a GC content of about 39.5%. Chromatographic analysis of the base composition using isopropanol–HCl solvent, however, yielded a value of about 30% for the GC content, whereas buoyant density of the DNA (1.67 gm/cm^3) is consistent with a GC content of about 14%. Analogous inconsistencies have been reported for *Bacillus subtilis* phages [reviewed by Hemphill and Whiteley (1975)]. In these cases, different GC values determined by different methods are believed to reflect the presence of unusual or modified bases. For MVL3 viral DNA, Garwes *et al.* (1975) could not detect any such bases, so a question remains about the Group 3 DNA base composition.

b. Proteins. Purified preparations of a Group 1 mycoplasmavirus, MVL51, were found to contain four proteins of molecular weights 70,000, 53,000, 30,000, and 19,000 (Maniloff and Das, 1975). The approximate stoichiometric ratios of these proteins were 6.9:1.0:1.8:8.8. It is not known whether each protein is a unique gene product or whether one or more are cleavage products.

The Group 3 virus, MVL3, was found to contain five proteins with molecular weights 172,000, 81,000, 73,000, 68,000, and 43,000 (Garwes *et al.* 1975), which account for 1, 7.5, 1, 10.5, and 80%, respectively, of the total virus protein. As is true for Group 1 viral proteins, it is not known which MVL3 proteins are unique or are cleavage products.

c. Other Components. Group 2 viruses contain lipids, as shown by their sensitivity to detergents and organic solvents (Gourlay, 1971), low buoyant density of 1.19 gm/cm^3 (Gourlay *et al.*, 1973), and unit membrane of the viral envelope in electron micrographs (Gourlay *et al.*, 1973).

Garwes *et al.* (1975) found that fucose accounted for 1.8% (w/w) of the composition of the Group 3 virus. However, it was not possible to determine whether fucose is present as glycoprotein or is associated with the DNA.

4. Antiserum Inactivation

Mycoplasmaviruses have been serologically classified into three groups by use of specific antiserum that inhibits plaque formation (Gourlay, 1972). Antisera were prepared in rabbits against a purified virus of each of the three groups. The antiserum against any one group inhibited plaque formation of only that group (Gourlay, 1974). Gourlay, (1972, 1974) and Clyde (1974a) noted that long incubations with antiserum prepared against purified MVL1, the original Group 1 isolate, inhibited plaque formation by all Group 1 isolates and concluded that all Group 1 isolates are serologically identical. However, such a procedure does not differentiate between serologically related and serologically identical viruses. For this, kinetic data are needed. Maniloff and Liss (1974) observed a 3-fold range in inactivation rates among four Group 1 viruses using antiserum made aginst MVL51. This indicates some degree of serological heterogeneity among Group 1 isolates.

In studies of inactivation kinetics of various virus isolates, using antisera again several viruses isolated from *A. laidlawii*, both Maniloff and Liss (1974) and Clyde (1974a) found that an isolate from *Mycoplasma gallisepticum* A5969 had the slowest inactivation rate. This suggests that, although these are serologically similar viruses, their lack of identity may reflect the taxonomic distance between their original hosts.

5. Ultraviolet Inactivation

All three groups of mycoplasmaviruses can be inactivated by ultraviolet (UV) light [reviewed by Gourlay, (1974) and Maniloff and Liss (1974)]. The UV doses which give survival of 37% of the viruses are (in order of increasing UV sensitivity): 0.109 kJ/m^2 for Group 2, 0.048 kJ/m^2 for Group 3, and 0.028 kJ/m^2 for Group 1 viruses (Das *et al.*, 1977). Maniloff and Liss (1971) found that three Group 1 isolates had small but significantly different UV-inactivation cross sections, indicating that the Group 1 isolates were not identical.

Recently it has been shown that MVL2, a Group 2 mycoplasmavirus, can be host cell reactivated (Das *et al.*, 1977). In *Escherichia coli,* this mode of repair of UV-damaged viral DNA primarily depends on a host excision repair mechanism (Boyle and Setlow, 1970). Excision repair has also been demonstrated in *A. laidlawii* (Das *et al.*, 1972). Since Group 1 viruses have single-stranded DNA, they cannot be host cell reactivated (Das *et al.*, 1977).

Ultraviolet reactivation of Groups 1 and 2 viruses has also been shown (Das *et al.*, 1977). This is measured as the enhanced survival of UV-inactivated virus when the virus is assayed on a lightly UV-irradiated host compared to assay on an unirradiated host (Weigle, 1953). In *E. coli,* UV reactivation of viral DNA has been proposed as being due to an inducible, error-prone repair mechanism (Radman, 1974). This may indicate the existence of such an inducible enzyme system in mycoplasmas.

6. *Thermal Effects.*

Group 1 viruses are relatively heat stable. At 60°C and pH 8.0, the surviving fraction of MVL51 is 0.37 in about 3 minutes and 0.03 in 30 minutes (Das and Maniloff, unpublished data). The inactivation curve is exponential only to about 15% survival. Gourlay and Wyld (1972) showed that inactivation is exponential at temperatures above 70°C. Clyde (1974b) reported Group 1 viruses to be resistant to 20 cycles of freezing and thawing.

Group 2 viruses are extremely heat labile (Gourlay, 1971). At 60°C and pH 8.0, the survival of MVL2 is about 4×10^{-4} in 5 minutes and the inactivation is exponential (Das and Maniloff, unpublished data). The Group 3 virus is relatively heat stable (Gourlay and Wyld, 1973).

7. *Detergent and Organic Solvent Inactivation*

Group 1 viruses are resistant to relatively high concentrations of detergents. No loss of virus titer was reported after treatment with 0.4% Nonidet P40 (Gourlay, 1970), 0.4% Sarkosyl NL97 (Das and Maniloff, 1975), or 0.4% Triton X-100 (Liss and Maniloff, 1971). These viruses are also resistant to ether but are sensitive to chloroform (Gourlay, 1970; Gourlay and Wyld, 1972).

Group 2 viruses are sensitive to detergents, ether, and chloroform (Gourlay, 1971). This sensitivity is consistent with the fact that these viruses have a membrane envelope.

Group 3 viruses are resistant to Nonidet P40 and ether (Gourlay and Wyld, 1973).

B. *Biological Properties*

1. *PFU Assay*

Mycoplasmaviruses are assayed as plaque-forming units (PFU) on lawns of indicator host cells. Thus far, only some *A. laidlawii* strains have been shown to produce plaques. Group 1 and 2 viruses are nonlytic (discussed in Section II, C and E), and the infected cells continue to grow and divide while producing viruses. However, the growth of infected cells is slower than uninfected ones, and colonies arising from infected cells are smaller than from uninfected cells. It is this differential growth that gives rise to plaques (Liss and Maniloff, 1973b). Group 1 virus plaques are turbid, sometimes having a clear center, with sizes varying from 0.5 to 6 mm in diameter (Liss and Maniloff, 1971; Gourlay and Wyld, 1972). Group 2 virus plaques are turbid, but less so than those produced by Group 1 viruses, and are about 1 mm in diameter. Group 3 virus plaques are clear and smaller than Group 2 plaques.

Plaque assays are carried out at 37°C, although all three groups of mycoplasmaviruses can produce plaques at room temperature, 20–22°C (Maniloff and Liss, 1974). It has been shown that one Group 1 or Group 2 virus is sufficient to produce a plaque, so PFU can be used to quantitate these viruses (Maniloff and Liss, 1974).

The size and morphology of mycoplasmavirus plaques are affected by the host strain, age of the cell culture used to make lawns, water content of the agar plates, whether or not incubation is under anerobic or high-CO_2 incubation conditions, and the growth medium used (Gourlay and Wyld, 1972; Liss and Maniloff, 1972; Maniloff and Liss, 1973; Fraser and Fleischmann, 1974). Group 1 viruses will form plaques on lawns that are up to 24 hours old (Liss and Maniloff, 1972), whereas Group 2 viruses only plaque on lawns less than 4 hours old (Maniloff, unpublished data). It is necessary to incubate plates with Group 3 viruses for 48 hours, in order to improve plaque visibility. Gourlay and Wyld (1973) observed that 100- to 1000-fold dilution of the culture used to make lawns improved the plaque visibility of these viruses.

Maniloff and Liss (1974) discussed the fact that, since plaque formation is so dependent on the handling of the indicator culture, it may be difficult to compare results from laboratories using different culture methods.

Although mycoplasmavirus plaques can be seen without staining, staining increases visibility and facilitates counting. Two staining methods have been reported. Quinlan *et al.* (1972) have used 1:20 dilution of Dienes stain in ethanol to stain the lawn and increase the contrast of the plaques. Fraser and Crum (1975) incorporated tetrazolium in their medium, permitting clear plaques to be seen against a red background of reduced tetrazolium.

2. Host Range

Several *Acholeplasma* and *Mycoplasma* species have been tested as indicators for mycoplasmaviruses. So far, only some *A. laidlawii* strains have been shown to produce plaques with any of the three groups of viruses (Maniloff and Liss, 1974). Gourlay and Wyld (1972) found 62.5% (10 out of 16) *A. laidlawii* strains tested produced plaques when tested with the Group 1 virus MVL1. In a similar study, Maniloff and Liss (1974) reported that 29% (23 out of 78) of the *A. laidlawii* strains they examined made plaques with the Group 1 virus MVL51.

3. Host Modification and Restriction

Most viruses have a restricted host range. Many temperate and virulent bacteriophages have been shown to exhibit an altered host range as a result of a single cycle of phage growth in a particular host [reviewed by Arber (1974)]. This phenomenon, called host-induced modification, differs from mutational modification of host range, since the property is not heritable and the modification is lost when the phage is grown in some other host.

Host modification and restriction of mycoplasmaviruses have been examined using MVL51 (a Group 1 virus), MVL2 (a Group 2 virus), and *A. laidlawii* strains JA1 and M1305/68 (Maniloff and Das, 1975); see Table IV. JA1 neither modifies nor restricts MVL51. M1305/68 slightly restricts MVL51 propagated on JA1 and modifies the virus so that the plating on M1305/68 is improved. Even the modified virus plates less well on M1305/68 than on JA1. MVL2 showed host-controlled modification and restriction in both hosts, giving 1000-fold more plaques on the host in which it had been grown.

The number of plaques produced on the restricting host depends upon the physiological state of the cells. For example, if the host is irradiated with UV light before MVL2 infection, there is an increase in the plating efficiency on the restricting host (Maniloff *et al.*, 1977b).

This host modification and restriction property of mycoplasmaviruses

TABLE IV
RELATIVE NUMBER OF VIRUS PLAQUES ON *Acholeplasma laidlawii* LAWNS[a]

Virus[b]	Relative PFU on *A. laidlawii*	
	Strain JA1	Strain 1305
MVL51 · JA1	1.0	0.2
MVL51 · 1305	1.0	0.6
MVL2 · JA1	1.0	0.015
MVL2 · 1305	0.011	1.0

[a] From Maniloff and Das (1975) and Nowak *et al.* (1976).

[b] The symbols after the dot indicate the last host on which the virus was grown.

is important as a marker for genetic studies in mycoplasmas. Moreover, it suggests that *A. laidlawii* cells possess enzyme systems responsible for DNA restriction and modification, though the biochemical basis for the phenomenon remains to be elucidated in mycoplasmas. Whatever the basis may be, it should be possible to isolate mycoplasma mutants deficient in modification and/or restriction.

C. Replication of Group 1 Viruses

1. "One-Step Growth"

Since progeny virus release is not lytic, a classical one-step growth experiment is not possible; but the experiments using the one-step virus growth protocol do reveal some features of viral growth. For such experiments cells are mixed with viruses at a multiplicity of infection (MOI) less than 1, so that each infected cell receives only one virus, and incubated at 37°C for 5 minutes to allow adsorption. The infected cell suspension is then diluted to eliminate reinfection and plated at various times for assay of the number of PFU in the culture. Group 1 virus growth curves have a short latent period of about 10 minutes followed by the gradual increase in virus titer over several hours (Liss and Maniloff, 1971, 1973b). This suggests continual viral production rather than release of virus by a cell lysis mechanism. After the rise period the virus titer reaches a plateau. It is not known whether this plateau is inherent in the virus life cycle or whether it reflects altered culture conditions. For

MVL51 infection in tryptose broth, the plateau is reached about 2 hours after infection, and at this time about 150–200 virus particles have been released per infected cell. The one-step growth curves for other Group 1 isolates exhibit the same pattern, but the growth parameters (such as latent period or amount of progeny released per infected cell) are different for different isolates (Liss and Maniloff, 1971, 1973b).

2. *Artificial Lysis*

The possible formation of intracellular viruses during the viral latent period has been examined by artificially lysing infected cells (Liss and Maniloff, 1973b). These experiments were carried out like one-step growth experiments, except that each time a sample was removed for PFU assay a parallel sample was removed, lysed with 0.2% Triton X-100, and assayed for PFU. Hence, the unlysed sample measures the number of infected cells plus the number of extracellular viruses, and the lysed sample measures the number of intracellular viruses plus the number of extracellular viruses. The results during the latent period showed that the PFU of the lysed sample was equal to the background number of free unadsorbed viruses, indicating no infectious intracellular viruses. During the rise period the number of PFU in the lysed sample increased to equal, but never exceed, the number of PFU in the unlysed sample, reflecting the release of the progeny viruses. These experiments are consistent with the interpretation that there is no intracellular pool of completed viruses and that virus maturation and release are coincident.

3. *"Single Burst" Experiments*

Although infection is nonlytic and viruses are released as they are made, rather than in a burst at cell lysis, the term "burst" is used in discussing virus yield. The number of progeny viruses released from individual infected cells was determined by the protocol of Ellis and Delbrück (1939). The average virus yield is 88 per cell at 60 minutes after infection, and 154 per cell at 120 minutes (Liss and Maniloff, 1973b). These data confirm that each infected cell releases virus continuously at a constant rate, rather than in a short burst.

The "burst size" measured in one-step growth experiments is an average over all infected cells and for Group 1 viruses ranges from 3 for the MVL1 isolate to 150–200 for the MVL51 isolate (Liss and Maniloff, 1971, 1973b). Both the progeny virus yield and the growth rate of infected cells vary as a function of MOI (Maniloff and Liss, 1973; Liss and

Maniloff, 1973b). At an MOI above 10, virus yield is variable and usually less than that measured at low MOI.

4. Growth of Infected Cells

MVL51-infected cells (at an MOI less than 10) produce virus without a loss of cell titer (Liss and Maniloff, 1973b). Above MOI 10, there is a decrease in the number of viable colony-forming units (Maniloff and Liss, 1973). Cells infected at an MOI less than 10 grow slower than uninfected cells: the infected cell doubling time is 160 minutes, compared with 110 minutes for uninfected cells (Liss and Maniloff, 1973b). Infected cells also produce smaller colonies than do uninfected cells: infected cell colony size is about 0.3 mm, and uninfected cell colony size about 0.5 mm (Maniloff and Liss, 1973).

In addition to viable cell measurements, growth of MVL51-infected cells was followed by electron microscopy (Maniloff and Liss, unpublished data). Cells infected at an MOI sufficient to assure that each cell received a virus were grown on electron microscope grids and examined at various times. Increasing sizes of microcolonies were observed, and each cell in a microcolony could be seen releasing virus. Hence, each infected cell had given rise to a microcolony, each cell of which was infected and producing virus.

5. Plaque Formation

Some early reports on these viruses considered the infection to be lytic because plaque formation was assumed to imply lysis of infected cells. Studies of the filamentous bacterial viruses, which are also nonlytic, have shown that plaques can be due to infected cells growing slower and making smaller colonies than uninfected ones (Hsu, 1968). Similarly, mycoplasmas infected with Group 1 virus grow slower and make smaller colonies (Maniloff and Liss, 1973) than do uninfected mycoplasmas. Thus, Group 1 virus plaques are produced by the differential growth rate and size of infected cell colonies, so that the uninfected cell lawn is more dense than the area of infected cells.

D. Details of MVL51 Replication

1. Introduction

In order to allow a fuller understanding of the replication of Group 1 viruses, a single isolate MVL51 has been chosen for detailed studies. MVL51 was chosen because it gives the greatest yield of progeny viruses.

2. Adsorption and Penetration

The process of infection by MVL51 involves two steps: (1) attachment or adsorption of the virus to the cell; (2) penetration of viral DNA into the cell. Adsorption of viruses to *A. laidlawii* cells is ionic and requires mono- or divalent cations (Fraser and Fleischmann, 1974). Adsorption takes place whether the cation is Na^+, K^+, NH_4^+, Ca^{2+}, or Mg^{2+}. The optimum concentrations of mono- and divalent cations are about 0.08 and 0.02*M*, respectively. The adsorption follows first-order kinetics, with a rate constant per cell of about 3×10^{-9} cm^3/minute. The theoretical value of the rate constant ($2\text{–}5 \times 10^{-9}$ cm^3/minute), calculated assuming single-hit collision kinetics (Fraser and Fleischmann, 1974; Maniloff and Liss, 1974), is in agreement with the experimental value indicating that nearly every collision between cell and virus results in adsorption.

It has been found (Fraser and Fleischmann, 1974) that the adsorption rate constant is not very temperature dependent over the range of 0–42°C, indicating that cell metabolism is probably not required for this process. The adsorption rate increases 2-fold between 0 and 42°C, with 30–42°C being the optimum temperature range. Garen and Puck (1951) pointed out that such adsorption rate changes can be explained by the temperature dependence of the solvent vicosity. Hence, the 2-fold change in the MVL51 adsorption rate is probably due to the approximate 2-fold change in water viscosity over the temperature range studied.

The pH optimum for virus adsorption is about 6. It should be noted that for virus growth studies adsorption is normally carried out at a pH between 7.5 and 8.0, where the adsorption rate is two to three times less than at pH 6.0 (Fraser and Fleischmann, 1974).

Fraser and Fleischmann (1974) estimated that each cell has about 10 MVL51 adsorption sites. Das and Maniloff (1976a) found that, although infected cells can take up several viral DNA molecules, only two to three parental viral DNA molecules per cell can bind to the cell membrane and participate in DNA replication.

Treatment of cells before infection with chloramphenicol or rifampicin has little effect on penetration and binding of viral DNA to the cell membrane sites (Das and Maniloff, 1976c). These data indicate that neither viral transcription nor the synthesis of new proteins is required for DNA penetration. This is in contrast to filamentous bacteriophage, which, like Group 1 mycoplasmaviruses, have single-stranded circular DNA of molecular weight about 2×10^6. Burtlag, Schekman, and Kornberg (1971) showed that penetration of filamentous phage M13 DNA is inhibited by pretreatment of cells with rifampicin. They suggest that, since M13 DNA is replicated to the double-stranded form during penetration, this step requires an RNA primer. However, in the case of ϕX174, another small single-stranded circular DNA phage, DNA penetration is resistant to rifampicin; thus, it resembles MVL51.

3. Intracellular DNA Replication

a. Replicative Intermediates. Intracellular replication of MVL51 can be divided into three distinct phases (Das and Maniloff, 1975), similar to those reported for the replication of single-stranded DNA bacteriophages (Marvin and Hohn, 1969; Kornberg, 1974). First, the parental viral DNA is converted to double-stranded replicative form (Das and Maniloff, 1976a). This step presumably is accomplished by preexisting host cell enzymes, since this conversion is unaffected by pretreatment of cells with chloramphenicol (Das and Maniloff, 1976c).

The double-stranded replicative form (RF) exists in two configurations, RFI and RFII, which can be separated by velocity sedimentation in sucrose gradients (Das and Maniloff, 1975). At neutral pH and 1 *M* NaCl the sedimentation rates for RFI and RFII relative to SSI (the single-stranded circular viral DNA) are 0.71 and 0.52, respectively (Fig. 5). This is consistent with RFI being a covalently closed, double-stranded, circular derivative of the 2×10^6 daltons single-stranded, circular viral DNA and RFII being a nicked form of RFI. These identifications are confirmed by ethidium bromide–CsCl equilibrium gradient analysis (Das and Maniloff, 1975) and electron microscopy of RFI and RFII (Fig. 6). RFI can be converted to RFII by phosphodiester cleavage by pancreatic deoxyribonuclease (Das and Maniloff, 1975). Electron micrographs (Fig. 6) show that RFI has a superhelical configuration.

During the second phase of replication, the parental RF replicates to produce a pool of progeny RF molecules. In tryptose broth, this occurs during the first 30 to 40 minutes after infection, when most of the virus-specific DNA synthesis consists of RF replication (Fig. 5). This step requires some cellular function, since a host cell variant (REP^- phenotype; see Section II, D, 5) is capable of converting the parental single-stranded DNA to parental RF, but is blocked in further replication (Nowak *et al.*, 1976). The possible requirement for viral gene products in RF replication is not known.

In the third stage of replication, the RFII molecules serve as precursors for the synthesis of progeny single-stranded viral DNA (SSI). About 30 to 40 minutes after infection, the amount of progeny viral chromosomes (SSI) begins to increase (Fig. 5).

Rifampicin treatment of infected cells, before or during infection, has no effect on any of the stages of viral DNA replication (Das and

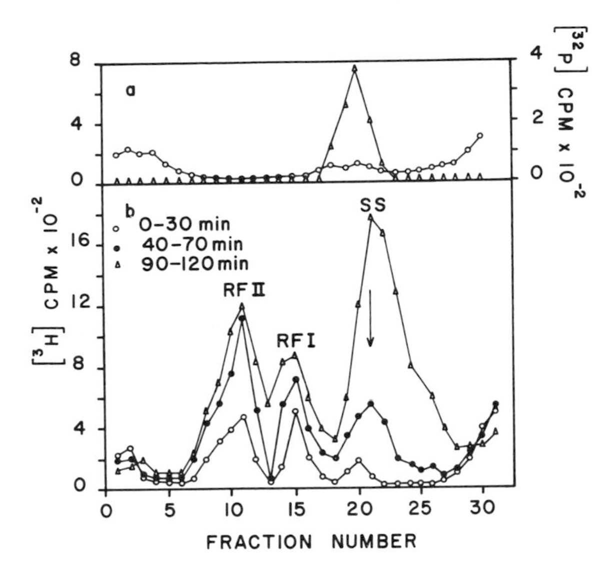

Fig. 5. Velocity sedimentation of *A. laidlawii* cells infected with MVL51, a Group 1 mycoplasmavirus. Cells were infected with virus and nascent DNA was labeled by the addition of [^{3}H]-thymidine at 0–30, 40–70, or 90–120 minutes after infection; ^{32}P-labeled marker viral DNA, shown by △ in (a) and arrow in (b), was added, and sedimentation was carried out through high salt (1 *M* NaCl) sucrose gradients. (a) Gradient from uninfected control cells. (b) Cells infected at an MOI of 20. Sedimentation is from left to right. (From Das and Maniloff, 1975.)

Maniloff, 1976c). Therefore, virus-specific transcription is not required for the replication of virus DNA.

Growth in poor nutrient media like Eagle's basal medium, rather than richer tryptose broth, allows the identification of an intermediate between nascent progeny chromosomes (SSI) and mature viruses (Das and Maniloff, 1976b). This intermediate is a protein-associated form of SSI and is designated as SSII. The SSII DNA-protein complex is stable in 1 *M* NaCl. The conversion of SSI to SSII is inhibited by chloramphenicol. SSII contains two proteins, which are electrophoretically identical to two of the four virion proteins (Maniloff and Das, 1975). The two proteins found in SSII are 70,000- and 53,000-dalton viral proteins and have an approximate stochiometric ratio of 1.4:1.0.

b. Membrane-Associated Parental RF Replication. Viral DNA becomes associated with host cell membrane after infection, and this association involves protein (Das and Maniloff, 1976a). The association was shown by infecting cells with ^{32}P-labeled viruses, lysing the cells by freezing and thawing, and sedimenting the lysate through a sucrose

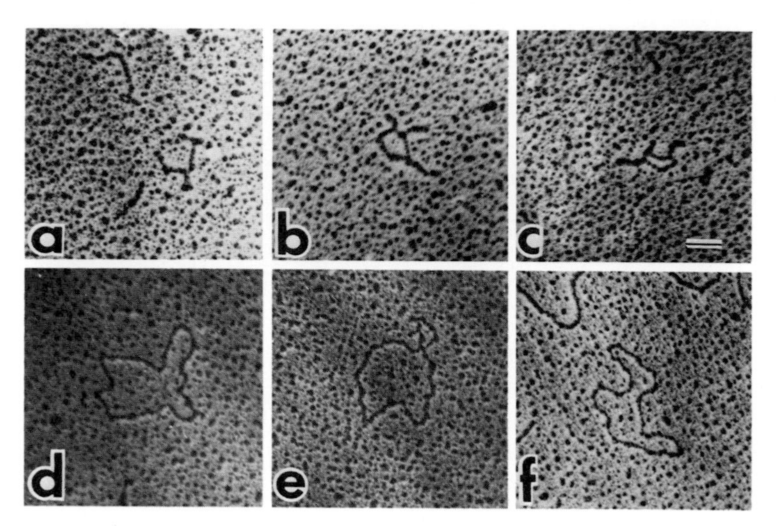

FIG. 6. Electron micrographs of RFI and RFII molecules. RFI and RFII were isolated from gradients as in Fig. 5 and prepared for electron microscopy by the method of Kleinschmidt (1968): (a,b,c) RFI; (d,e,f) RFII. Bar denotes 500 nm.

gradient over a CsCl shelf. Most of the parental label sediments with the cell membrane as a fast-sedimenting complex. The remaining parental DNA sediments slowly as free-sedimenting material.

The amount of fast- and free-sedimenting parental viral DNA was measured as a function of MOI (Fig. 7). It was found that the number of membrane-associated viral DNA molecules reached a saturation value. From this value, it was calculated that there are only two to three sites per cell (Das and Maniloff, 1976a).

c. Fate of Parental DNA. Parental viral DNA is not transferred to progeny viruses (Maniloff and Liss, 1973; Liss and Maniloff, 1974a; Das and Maniloff, 1976a). This conclusion is based on experiments in which cells were infected with ^{32}P-labeled viruses. The amount of ^{32}P-labeled viral DNA in the cell remained constant, and no extracellular acid-insoluble radioactivity could be detected. Most (70–80%) of the intracellular parental viral DNA remained in RF forms; the remainder was found in a faster-sedimenting peak in high-salt sucrose gradients and may represent defective viruses or abortive replication forms (Das and Maniloff, 1976a).

d. Membrane-Associated Replication of Progeny RF. Several types of experiments have shown that the synthesis of progeny RF molecules occurs by symmetric (semiconservative) replication at cell membrane

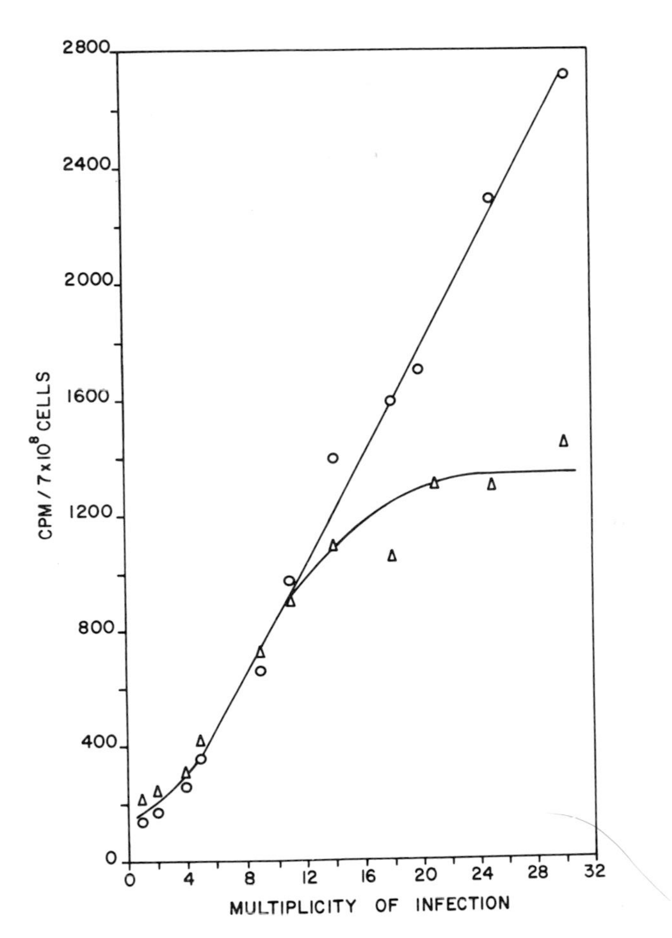

FIG. 7. Distribution of ^{32}P-labeled parental MVL51 viral DNA in fast-sedimenting (△) and free-sedimenting (O) material as a function of MOI. The specific activity was 1.25×10^6 PFU/CPM. (From Das and Maniloff, 1976a.)

sites. First, during early infection, when most synthesis of viral DNA is in RF forms [see Fig. 5; also Das and Maniloff (1975)], 80% of nascent viral DNA was in membrane-associated material. Second, equilibrium centrifugation of membrane-associated progeny RFII molecules in alkaline CsCl gradients showed nascent DNA in both viral and complementary strands [Fig. 8 (Das and Maniloff, 1976a)], indicating symmetrical replication of both strands of the double-stranded RF. Third, when infected cells were pulse labeled for short periods during early infection

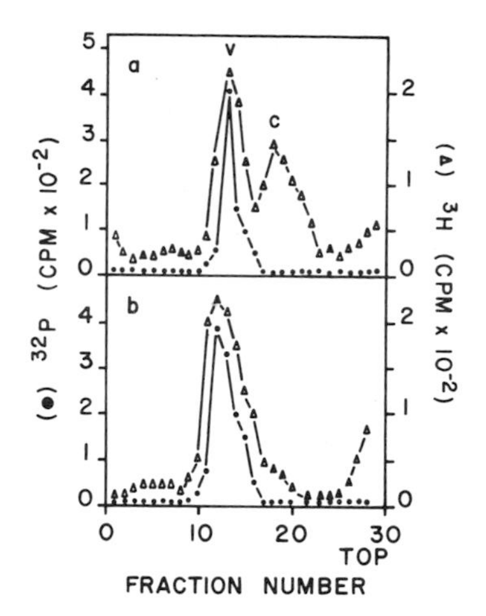

FIG. 8. Equilibrium centrifugation of Group 1 viral RFII DNA in alkaline CsCl gradients. (a) Membrane-associated RFII isolated from fast-sedimenting material. (b) Cytoplasmic RFII from free-sedimenting material. The parental virus was labeled with ^{32}P, which marked the position of the viral strand. Nascent DNA was labeled with [^{3}H]-thymidine. The complementary strand (*c*) is less dense than the viral strand (*v*). (From Das and Maniloff, 1976a.)

(Table V), the amount of nascent DNA in free-sedimenting material, relative to fast-sedimenting material, increased with increasing pulse length. This suggests that RF molecules are synthesized at the membrane and then released into the cytoplasm.

e. Cytoplasmic Replication of Progeny Viral DNA. At 30 to 40 min-

TABLE V
PULSE-LABELED DNA IN MEMBRANE-BOUND AND FREELY SEDIMENTING FRACTIONS[a]

	Percentage of pulse label	
Pulse length (seconds)	In fast-sedimenting material	In free-sedimenting material
20	47	53
60	18	82

[a] The pulse label was added 10–15 minutes after the start of infection (Das and Maniloff, unpublished data).

utes after infection, when progeny RF replication is continuing and progeny SSI synthesis increases [see Fig. 5; also Das and Maniloff (1975)], most nascent DNA is found as free-sedimenting material (Das and Maniloff, 1976a). Analysis of progeny RFII from free-sedimenting material by equilibrium centrifugation in alkaline CsCl gradients showed nascent DNA only in the viral strand [see Fig. 8; also Das and Maniloff (1976a)], indicating that an asymmetric mechanism of RF replication is involved in the synthesis of progeny SSI molecules. These results do not rule out the occurrence of some asymmetric replication on the membrane.

4. Assembly and Release

Since SSII, the protein-associated form of SSI, cannot be found in cells in richer medium, it is probably normally a short-lived intermediate. SSII contains two of the four virion proteins, but in a different stoichiometric ratio from that found in the virion. Conversion of SSI to SSII is blocked by chloramphenicol. Rifampicin blocks the conversion of SSI to mature virus particles (Das and Maniloff, 1976c), but it is not known whether this block is before or after SSII formation.

Electron micrographs show that virus assembly and release occur at a limited number of membrane sites per cell and that clusters of extracellular progeny viruses are seen at these sites (Liss and Maniloff, 1973b). Virus maturation and release involve interactions with the cell membrane that do not affect cell viability.

Gourlay and Wyld (1972) observed that, as viral infection progressed, infected cells became increasingly sensitive to inactivation by antiviral serum. This is consistent with the appearance of viral antigens in the cell membrane during virus assembly and release.

5. Abortive Infection in REP⁻ cells

A variant of *A. laidlawii* JA1 with a REP⁻ phenotype has recently been isolated and characterized (Nowak *et al.,* 1976). This variant no longer allows growth of the single-stranded DNA Group 1 mycoplasmaviruses, but retains the ability to propagate double-stranded DNA viruses of Groups 2 and 3. REP⁻ cells also have a greater sensitivity to UV light than parental cells.

Replication of MVL51 in REP⁻ cells has been examined (Nowak *et al.,* 1976). The initial steps of infection are not impaired. Infecting viral DNA can associate with the cell membrane, and the parental single-stranded DNA converted to double-stranded replicative forms (RFI and RFII). However, this parental RF does not replicate, and no progeny

RF molecules are formed. This implicates a cell function in the replication of parental viral RF to progeny RF. Interestingly *Escherichia coli rep* [3] mutants are blocked in an analogous step in their replication of single-stranded DNA phages (Denhardt *et al.*, 1972).

6. Summary

Figure 9 shows a schematic of the replication of Group 1 mycoplasmaviruses, based on the studies of MVL51 just reviewed. Upon infection the single-stranded circular parental DNA is converted to membrane-associated double-stranded replicative forms RFI and RFII. These undergo semiconservative replication at the membrane to produce progeny RF. Since this step is blocked in a REP⁻ host, it must require some cellular function. Further replication is asymmetric and produces single-stranded circular progeny viral chromosomes. This replication takes place in the cytoplasm, but no data are yet available on whether it can also occur at membrane sites. The SSI viral DNA associates with viral proteins and undergoes assembly and extrusion through the cell membrane without lysing the cell. An intermediate in this pathway, SSII, has been shown to be a complex of SSI with two of the four virion proteins. Chloram-

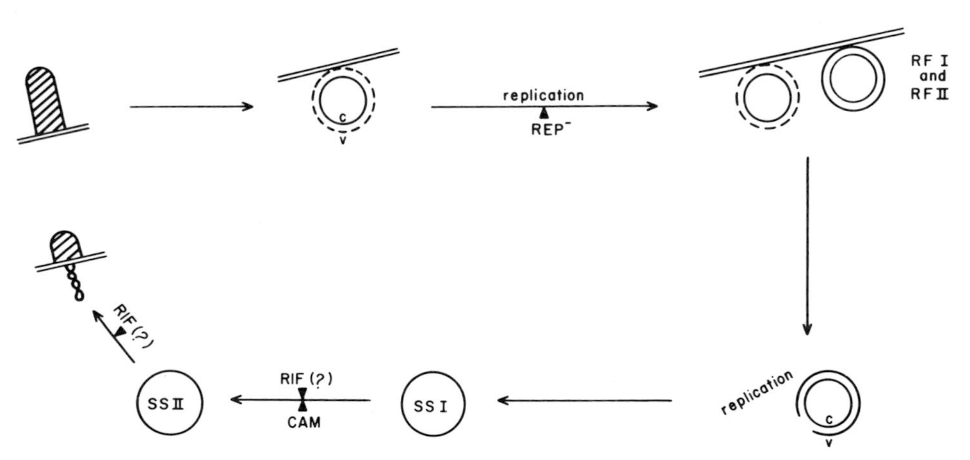

FIG. 9. Schematic of the replication of Group 1 mycoplasmaviruses, based on studies of MVL51 described in text. Dotted lines show parental viral DNA strands, and continuous lines show progeny DNA strands. Steps blocked by rifampicin (RIF), chloramphenicol (CAM), and REP⁻ cell variant are shown. Viral and complementary DNA strands are marked *v* and *c,* respectively. The parallel double lines denote the cell membrane.

phenicol inhibits conversion of SSI to SSII, and rifampicin blocks assembly somewhere between SSI and the completed virion.

Data on the role of cellular and viral functions in viral replication are incomplete. The REP^- cells showed a cellular role in the RF replication. In addition, rifampicin has no effect on viral DNA synthesis but stops virus production and blocks transcription of viral DNA in rifampicin-resistant cells (Das and Maniloff, 1976c). This indicates the involvement of some viral function in the transcription of viral DNA.

E. Replication of Group 2 and 3 Viruses

A preliminary report of one-step growth and artificial lysis experiments on a Group 2 virus concluded that the infection is not lytic (Putzrath and Maniloff, 1976). It was also observed that infected cells could give rise to colonies (R. M. Putzrath and J. Maniloff, unpublished data). Liska and Tkadlecek (1975) report that thin-section electron micrographs of infected cell cultures show what appear to be disintegrated cells and interpret this as evidence for lytic infection. However, Maniloff and Morowitz (1972) have pointed out the difficulties in obtaining good cytological preservation of mycoplasmas in general, and Maniloff (1970) has shown that as *A. laidlawii* cells age they produce intracytoplasmic aggregates, swollen cells, and cellular debris.

No data are available on the intracellular replication of Group 2 viruses. Thin-section electron micrographs of Group 2 virus infected cells show virus release by budding. Micrographs of freeze-etched preparations show that the membrane fracture face at the budding area has fewer particles than are found on the rest of the cell membrane (Putzrath and Maniloff, unpublished data). This indicates a virus-induced modification of specific membrane sites. Gourlay *et al.* (1973) reported that antiserum to *A. laidlawii* failed to inactivate a Group 2 virus, and antiserum to the virus failed to inhibit *A. laidlawii* growth. They suggested that this demonstrated a serological difference between viral and cell membranes. However, these workers used different *A. laidlawii* strains to prepare their antiserum and to grow the virus, and in view of the known serological heterogeneity of *Acholeplasma species* (e.g., see Tully, 1973), no conclusions can be drawn from the available data about the serological relatedness of Group 2 virus and cell membranes.

The clear nature of Group 3 virus plaques (Gourlay, 1974) suggests that these may be lytic viruses, but no data are available yet on Group 3 infection. Liss (1976a) has stated that the Group 3 virus infection is lytic.

F. Transfection

1. Properties

Although DNA-mediated transformation in mycoplasmas has not been achieved (Folsome, 1968; Smith, 1971), DNA isolated from two Group 1 viruses (MVL51 and MVGs51) is able to transfect mycoplasmas (Liss and Maniloff, 1972). Host cells are competent for the transfection only at the late logarithmic phase of growth. However, it has been shown that cells treated with 0.01 *M* ethylenediaminetetraacetate (EDTA) are competent for a longer period (Liss and Maniloff, 1974b). Fifty percent of the maximum transfection frequency can be obtained in mid-logarithmic-phase cells, and the number of transfectants begin to decrease only at 24 hours, paralleling cell death. Properties of the mycoplasma transfection system are summarized in Table VI. Compared to infection, transfection has a longer latent period and a smaller virus yield. The transfection system is sensitive to deoxyribonuclease treatment, but not to ribonuclease, proteolytic enzymes, or virus-specific antiserum. The time course of the DNA–cell interaction indicates that 10–15 minutes is required for the uptake of viral DNA into a deoxyribonuclease-resistant form (Liss and Maniloff, 1972). The dose–response curve (plot of the logarithm of the number of transfectants against DNA concentration) for transfection shows a slope of 2, indicating that an average of two molecules of DNA per cell may be required in order to produce an infection. The maximum transfection efficiency is 3–4 $\times$ 10^5 viral equivalents of DNA per transfectant. If EDTA is omitted from the transfection reaction mixture, the slope of the dose response curve increases, but the maximum transfection efficiency does not change (Liss and Maniloff, 1974b). A similar EDTA

TABLE VI

COMPARISON OF GROUP 1 MYCOPLASMAVIRUS INFECTION AND TRANSFECTION[a]

	Infection (by intact virus)	Transfection (by viral DNA)
Latent period	10 minutes	30–40 minutes
Progeny titer at 2 hours	150–200	85–100
Effect of DNase	None	Inhibition
Effect of RNase	None	None
Effect of pronase	None	None
Effect of anti-MVL51 serum	Inhibition	None

[a] From Maniloff and Liss (1973).

effect has been observed for *Bacillus subtilis* transfection by double-stranded DNA phages (Epstein, 1971), and in transformation of *B. subtilis* with single-stranded DNA (Tavethia and Mandel, 1970; Chilton and Hall, 1968). The EDTA effect in all these cases has been attributed to the inhibition of nuclease activity. *Acholeplasma laidlawii* has no specificity for the DNA strandedness; single- and double-stranded DNA compete with transfecting DNA with equivalent efficiencies (Liss and Maniloff, 1974b).

Fragmented bacteriophage genomes can give rise to productive infection by recombination (Spizizen *et al.*, 1966). The existence of transfection in mycoplasmas, which requires several viral DNA molecules per transfectant, suggests that this would be a useful system for detecting recombinant events in mycoplasmas in the absence of mutants.

2. *Comparison with Infection*

The indicator host *A. laidlawii* JA1 can support the growth of Group 1 viruses throughout the logarithmic phase of growth, but the cells are competent for transfection only for a relatively short time. Such a transitory physiological state of competence of recipient cells is a general phenomenon, and there are several hypotheses concerning the physiological basis and nature of competence (e.g., Hayes, 1974). Possible explanations include synthesis of specific receptor sites at the cell surface and production of an extracellular factor known as competence factor. However, the exact nature of competence is unresolved.

It has been found that *M. gallisepticum* A5969, which cannot be infected by Group 1 virus, can be transfected by the viral DNA and produce progeny viruses (Maniloff and Liss, 1973). This is most interesting in view of the requirement for a number of cell functions in viral replication (described in Section II, D).

G. Genetics

The only studies done thus far have been on MVL51, a Group 1 mycoplasmavirus. Nitrous acid inactivation of MVL51 followed first-order kinetics (Fig. 10).

Following nitrous acid mutagenesis, MVL51 samples with 10^{-4} to 10^{-5} fraction survivors were plated at 30°C, picked and replicated at permissive (30°C) and nonpermissive (37°C) temperatures. Of 2124 plaques picked, 14 (0.7%) were temperature sensitive *(ts)* mutants (Maniloff, unpublished data), similar to the frequency found for fila-

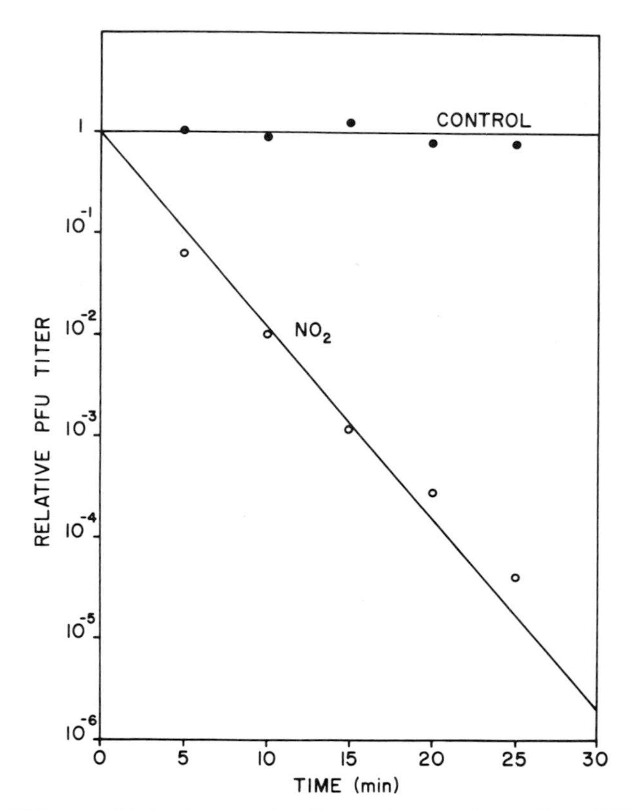

FIG. 10. Nitrous acid treatment of a Group 1 mycoplasmavirus, MVL51 (Maniloff, unpublished data). The initial virus titer was 1.2×10^9 PFU/ml. Treatment with 0.5 *M* sodium nitrite was by the protocol of Pratt *et al.* (1966). The control was maintained at pH 4.2.

mentous phages (Pratt *et al.*, 1966). These mutants should be useful for the development of mycoplasmavirus genetics.

III. VIRUSES AND "VIRUS-LIKE" PARTICLES OF SPIROPLASMAS

A. Occurrence

The spiroplasmas are a group of helical mycoplasmas which have been found in plants and insects. They are associated with a variety of plant diseases (e.g., Davis *et al.*, 1972; Bové *et al.*, 1973; Daniels *et al.*, 1973). The sex-ratio agent of *Drosophilia* (causing a maternally inherited sex-ratio trait in which affected females produce only female progeny) was shown to be a microorganism, originally identified as a spirochete (Poulson and Sakaguchi, 1961) and more recently shown to be a spiroplasma (Williamson and Whitcomb, 1974). The sex-ratio organism (SRO) has not been cultivated *in vitro*.

Cole *et al.* (1974) have found that 16 out of 23 isolates of *Spiroplasma citri* examined contained "virus-like" particles. These studies were done by electron microscopic examination of cell cultures, and three different morphological types of particles were described (see Table III); four of the 16 infected cultures contained all three types.

The eight *Drosophilia* strains carrying the SRO trait that have been studied thus far have been found to be infected by viruses (Oishi and Poulson, 1970; Oishi, 1971).

B. Ultrastructure

The three types of *S. citri* "virus-like" particles have been designated SV-C1, SV-C2, and SV-C3 (Cole *et al.*, 1974), for *Spiroplasma* virus *citri*. Since these viruses have not been propagated, ultrastructural studies have involved the examination of spiroplasma cultures by thin-section and negative-staining electron microscopic procedures.

Type SV-C1 particles are rods, 10–15 nm in diameter and usually 230–280 nm long, although a few are longer than 1000 nm (Cole *et al.*, 1974). Hollow rods are also found. No intracellular SV-C1 particles were seen.

SV-C2 particles have a morphology like that of type B bacteriophages (Cole *et al.*, 1973a,b, 1974). The particle has a polyhedral head with a hexagonal shape, 52–58 nm from vertex to vertex and 48–51 nm from flat side to flat side. There is a long tail 6–8 nm wide and 75–83 nm long, apparently unsheathed and noncontractile, attached to one vertex. The end of the tail appears wider, suggesting that there may be a base with spikes, but better resolution is needed and will require purified material.

The Type SV-C3 particles appear to be polyhedral with a short tail (Cole *et al.*, 1974). The hexagonal head measures 37–44 nm from vertex to vertex and 35–37 from flat side to flat side. The tail projecting from one vertex is 13–18 nm long and seems wider at the tip. Some micrographs show particles enclosed in a membrane, apparently budding off from the cell. In other parts of the field naked particles are seen. It is not known whether the enveloped and/or naked particle is infectious.

The two SRO viruses that have been described (designated spv-1 and spv-2) are DNA viruses with similar ultrastructural properties [Table VII; see Oishi and Poulson (1970); Oishi (1971)]: spv-1 and spv-2 cosediment in sucrose gradients and appear indistinguishable by electron microscopy, but do have different biological properties. The particles are polyhedral and are sometimes seen at the cell surface enclosed by cell

TABLE VII
PROPERTIES OF SRO VIRUSES[a]

Virus	SRO carrier strain	SRO strain lysed	Nucleic acid	Particle density in CsCl (gm/cm^3)	Particle diameter (nm)
spv-1	NSR	WSR	DNA	1.480	50–60
spv-2	WSR	NSR	DNA	—	50–60

[a] Adapted from Oishi (1971).

membrane (Oishi and Poulson, 1970), similar to observations on SV-C3 particles (Cole *et al.*, 1974). The size difference reported for spv-1 (50–60 nm) and SV-C3 (35–44 nm) may reflect the fact that the spv-1 measurements were on intracellular particles whereas the SV-C3 measurements were on extracellular particles.

C. Biology

Until recently it has not been possible to propagate "virus-like" particles associated with *S. citri* (Cole *et al.*, 1974). Varying numbers of particles can be observed in electron microscopic preparations of spiroplasma cell cultures. However, particles cannot be found in every sample, nor can they be found in every passage of a culture containing particles in prior and subsequent passages. In general, the relationship of the particles to viruses is circumstantial and depends on their striking morphological similarity to known viruses. Recently, Cole (unpublished data presented at the 76th Annual Meeting of the American Society of Microbiologists, Atlantic City, 1976) reported that he has been able to plaque SV-C3, but not SV-C1 or SV-C2, on an *S. citri* strain.

The biology of the SRO viruses has been studied in *Drosophila* (Oishi and Poulson, 1970; Oishi, 1971). The SRO strain NSR carries the spv-1 virus, such that infectious virus is always present. Virus spv-1 does not lyse the NSR strain but does lyse SRO strain WSR. Infection of *Drosophila* carrying WSR by spv-1 results in the elimination of WSR and "curing" of the sex-ratio condition. This infection by spv-1 leads to the induction of a latent virus, spv-2, which is not lytic for WSR but will lyse NSR. However, the lysis of the NSR strain by spv-2 does not lead to the immediate elimination of the sex-ratio condition. In view of the morphological similarity of spv-1 and spv-2 and their biological properties, it is possible that they are the same virion, differing only by host modification and restriction. If so, it would be an example of the unadaptive type of modification, originally described by Luria and Human (1952) for bacteriophage T2.

IV. Biological Considerations

A. Mycoplasma Plasmids

Since the observation of Haller and Lynn (1968) of a satellite DNA band in CsCl density gradients of *M. arthritidis*, there have been several reports of extrachromosomal DNA in mycoplasmas, summarizezd in Table VIII. Although the biological properties of these elements have not been described, it has generally been assumed that they are plasmids. These data indicate that double-stranded DNA plasmids might be widespread throughout the mycoplasmas and suggest that the plasmids may be about 20×10^6 daltons. Based on the amount of radioisotopically labeled thymidine recovered as cell chromosomal and plasmid DNA, Dugle and Dugle (1971), Das *et al.* (1972), and Liss (1976b) all report that about 35% of the cellular DNA is in plasmids. This would indicate 50 to 100 plasmids per cell. Clowes (1972) has pointed out the relationship between the size of a plasmid and the type of replication control to which it is subject. Large plasmids generally have "stringent" replication (few plasmid copies per cell), whereas small plasmids have "relaxed" replica-

TABLE VIII
Mycoplasma Plasmids

Mycoplasma	Procedure	Plasmid structure	Plasmid molecular weight[a]	Reference
M. arthritidis	Electron microscopy	Circles	20×10^6	Morowitz (1969)
M. hominis	Electron microscopy	Circles	18×10^6	Zouzias *et al.* (1973)
A. laidlawii	Sedimentation velocity	—	38×10^6 (16×10^6)	Dugle and Dugle (1971)
A. laidlawii	Sedimentation velocity	—	26×10^6 (19×10^6)	Das *et al.* (1972)

[a] Values in parentheses indicate molecular weights recalculated [using sedimentation coefficient ratios given by Clowes (1972)] assuming plasmids are circular molecules—instead of linear molecules, as in the references cited.

tion (many copies per cell). The molecular size of mycoplasma plasmids is near the borderline between the two classes.

Hutkova *et al.* (1973) measured a DNA peak at $25–40 \times 10^6$ daltons molecular weight in *A. laidlawii* which had been passaged in the presence of a Group 2 mycoplasmavirus. This peak was not observed in uninfected cells. If the molecular weight is recalculated using the sedimentation coefficient ratios given by Clowes (1972) instead of the molecular weight of linear molecules given by Hutkova *et al.* (1973), the DNA could be circles of molecular weight $18–28 \times 10^6$ daltons. Further studies are needed on the possible relationship between mycoplasma plasmids and the carrier states of mycoplasmavirus DNAs.

A "bacteriocin-like" agent in *M. pulmonis* has recently been reported (Gourlay *et al.*, 1976). It is possible that the genetic element for this property, as well as for other biological properties of mycoplasmas, may be plasmids.

B. Ubiquity of Viruses

Although the first mycoplasmavirus was not isolated until 1970 (Gourlay, 1970), since then they have been reported in species of *Mycoplasma, Acholeplasma,* and *Sprioplasma* (Liss and Maniloff, 1971; Oishi, 1971; Gourlay, 1972; Cole *et al.*, 1974; Maniloff and Liss, 1974). Maniloff and Liss (1974) have noted that the viral growth properties probably work against the establishment of an infection that is lethal for the culture and allow the persistence of some kind of stable carrier state. The apparent ubiquity of the mycoplasmaviruses may reflect the spread of viruses and/or infectious DNA [since mycoplasmas resistant to virus infection can be infected by viral DNA (Liss and Maniloff, 1972)] in the normal ecological situation.

It is interesting to consider the question of the origin of mycoplasma- and spiroplasmaviruses in terms of the ideas proposed for other viruses. Bradley (1971) has reviewed the proposal that the DNA bacteriophages originated as episomes and evolved in a series of steps to complete viruses. From studies of the nearest neighbor analysis of viral and cell nucleic acids, it has been suggested that small viruses had their origins within the polypeptide-specifying regions of their ancestral host DNA and could have originated at any time during the host's evolutionary history [reviewed by Subak-Sharpe *et al.* (1974)]. However, mycoplasmas are *genetic code limited organisms,* as defined by Woese and Bleyman (1972): a genetic code limited organism is one whose structural genes have as extreme a GC content as is possible, consistent with the codon frequencies necessary for the amino acid composition of the cell. Hence, nearest neighbor analysis of Group 1 mycoplasmaviruses (and any others that turn out to be small viruses) and of mycoplasma cells could give information on the origin of the viruses and the evolution of both viruses and cells.

C. Pathogenic Considerations

As discussed in Section I, the mycoplasmas are known to be the cause of a variety of animal and plant diseases. Although only *M. pneumoniae* has been proven to be a human pathogen, Thomas (1970) has pointed out that comparative pathology suggests that mycoplasmas may be involved in other human disease states. This, plus the ubiquity of mycoplasma- and spiroplasmaviruses, indicates that the role of the viruses should also be considered with respect to mycoplasma pathogenicity.

The further development of mycoplasma virology should allow us to determine whether mycoplasma pathogenicity is affected by lysogenic conversions or by virus-induced membrane antigen changes. The viruses carried by mycoplasmas may be involved in causing "virus-like" effects in mycoplasma-contaminated tissue culture cells [reviewed by Stanbridge (1971), Barile (1973), and Fogh and Fogh (1973)] and in the induction of interferon by a mycoplasma, recently reported by Rinaldo *et al.* (1974). In fact, Atanasoff (1972) has suggested that the role of mycoplasmas in disease may be as vectors for viruses. Experimental data on these questions are needed and could prove to be of medical significance.

Acknowledgments

We thank the other members of our research group, Dr. A. Liss, Ms. R. M. Putzrath, Mr. J. Nowak, and Mr. D. Gerling, who have contributed to our information and ideas concerning mycoplasmaviruses. Support for the studies in this laboratory has come from the United States Public Health Service, National Institute of Allergy and Infectious Diseases, Grant AI 10605

One of us (J.D.) is on leave of absence from the Department of Physics, Calcutta University, Calcutta, India.

References

Allen, T. C. (1972). *Virology* **47,** 491–493.

Arber, W. (1974). *Prog. Nucl. Acid Res. Mol. Biol.* **14,** 1–37.

Atanasoff, D. (1972). *Phytopathol. Z.* **74,** 342–348.

Barile, M. F. (1973). *In* "Contamination of Tissue Cultures " (J. Fogh, ed.), pp. 131–172. Academic Press, New York.

Belly, R. T., Bohlool, B. B., and Brock, T. D. (1973). *Ann. N.Y. Acad. Sci.* **225,** 94–107.

Bové, J. M., Saglio, P., Tully, J. G., Freundt, E. A., Lund, Z., Pillot, J., and Taylor-Robinson, D. (1973). *Ann. N.Y. Acad. Sci.* **225,** 462–470.

Boyle, J. M., and Setlow, R. B. (1970). *J. Mol. Biol.* **51,** 131–144.
Bradley, D. E. (1971). *In* "Comparative Virology" (K. Maramorosch and E. Kurstak, eds.), pp. 207–253. Academic Press, New York.
Bruce, J., Gourlay, R. N., Hull, R., and Garwes, D. J. (1972). *J. Gen. Virol.* **16,** 215–221.
Burtlag, D., Schekman, R., and Kornberg, A. (1971). *Proc. Natl. Acad. Sci. U.S.A.* **68,** 2826–2829.
Chilton, M. D., and Hall, B. (1968). *J. Mol. Biol.* **32,** 369–378.
Clowes, R. C. (1972). *Bacteriol. Rev.* **36,** 361–405.
Clyde, W. A. (1974a). *Colloq. Inst. Natl. Sante Rech. Med., Paris* **33,** 109–117.
Clyde, W. A. (1974b). *Ann. N.Y. Acad. Sci.* **225,** 159–160.
Cole, R. M., Tully, J. G., Popkin, T. J., and Bové, J. M. (1973a). *J. Bacteriol.* **115,** 367–386.
Cole, R. M., Tully, J. G., Popkin, T. J., and Bové, J. M. (1973b). *Ann. N.Y. Acad. Sci.* **225,** 471–493.
Cole, R. M., Tully, J. G., and Popkin, T. J. (1974). *Colloq. Inst. Natl. Sante Rech. Med., Paris* **33,** 125–132.
Daniels, M. J., Markham, P. G., Meddins, B. M., Plaskitt, A. K., Townsend, R., and Bar-Joseph, M. (1973). *Nature (London)* **244,** 523–524.
Das, J., and Maniloff, J. (1975). *Biochem. Biophys. Res. Commun.* **66,** 599–605.
Das, J., and Maniloff, J. (1976a). *Proc. Natl. Acad. Sci. U.S.A.* **73,** 1489–1493.
Das, J., and Maniloff, J. (1976b). *Microbios* **15,** 127–134.
Das, J., and Maniloff, J. (1976c). *J. Virol.* **18,** 969–976.
Das, J., Maniloff, J., and Bhattacharjee, S. B. (1972). *Biochim. Biophys. Acta* **259,** 189–197.
Das, J., Nowak, J., and Maniloff, J. (1977). *J. Bacteriol.* In press.
Davis, R. E., and Whitcomb, R. F. (1971). *Annu. Rev. Phytopathol.* **9,** 119–154.
Davis, R. E., Worley, J. F., Whitcomb, R. F., Ishijima, T., and Steere, R. L. (1972). *Science* **176,** 521–523.
Denhardt, D. T., Iwaya, M., and Larison, L. L. (1972). *Virology* **49,** 486–496.
Dugle, D. L., and Dugle, J. R. (1971). *Can. J. Microbiol.* **17,** 433–434.
Edward, G. A., and Fogh, J. (1960). *J. Bacteriol.* **79,** 267–276.
Ellis, E. L., and Delbrück, M. (1939). *J. Gen. Physiol.* **22,** 365–384.
Epstein, H. T. (1971). *J. Virol.* **7,** 749–752.
Fogh, J., and Fogh, H. (1973). *Ann. N.Y. Acad. Sci.* **225,** 311–329.
Folsome, C. E. (1968). *J. Gen. Microbiol.* **50,** 43–53.
Fraser, D., and Crum, J. (1975). *Appl. Microbiol.* **29,** 305–306.
Fraser, D., and Fleischmann, C. (1974). *J. Virol.* **13,** 1067–1074.
Freundt, E. A. (1974). *Pathol. Microbiol.* **40,** 155–187.
Garen, A., and Puck, T. T. (1951). *J. Exp. Med.* **94,** 177–189.
Garwes, D. J., Pike, B. V., Wyld, S. G., Pocock, D. H., and Gourlay, R. N. (1975). *J. Gen. Virol.* **29,** 11–24.
Giannotti, J., Devauchelle, G., Vago, C., and Marchoux, G. (1973). *Ann. Phytopathol.* **5,** 461–465.
Gourlay, R. N. (1970). *Nature (London)* **225,** 1165.
Gourlay, R. N. (1971). *J. Gen. Virol.* **12,** 65–67.
Gourlay, R. N. (1972). *Pathogenic Mycoplasmas, Ciba Found. Symp.* pp. 145–156.
Gourlay, R. N. (1973). *Ann. N.Y. Acad. Sci.* **225,** 144–148.
Gourlay, R. N. (1974). *CRC Crit. Rev. Microbiol.* **3,** 315–331.
Gourlay, R. N., and Wyld, S. G. (1972). *J. Gen. Virol.* **14,** 15–23.

Gourlay, R. N., and Wyld, S. G. (1973). *J. Gen. Virol.* **19,** 279–283.
Gourlay, R. N., Bruce, J., and Garwes, D. J. (1971). *Nature (London), New Biol.* **229,** 118.
Gourlay, R. N., Garwes, D. J., Bruce, J., and Wyld, S. G. (1973). *J. Gen. Virol.* **18,** 127–133.
Gourlay, R. N., Wyld, S. G., and Taylor-Robinson, D. (1976). *Nature (London)* **259,** 120–122.
Gourrett, J. P., Maillet, P. L., and Gouranton, J. (1973). *J. Gen. Microbiol.* **74,** 241–249.
Haller, G. J., and Lynn, R. J. (1968). *Bacteriol. Proc.* p. 68.
Hayes, W. (1974). "The Genetics of Bacteria and Their Viruses," Wiley, New York.
Hemphill, H. E., and Whiteley, H. R. (1975). *Bacteriol. Rev.* **39,** 275–315.
Hsu, Y. C. (1968). *Bacteriol. Rev.* **32,** 387–399.
Hull, R., Hills, G. J., and Markham, R. (1969). *Virology* **37,** 416–428.
Hutkova, J., Drasil, V., Melichar, A. and Smarda J. (1973). *Stud. Biophys.* **39,** 217–222.
Kleinschmidt, A. K. (1968). *In* "Nucleic Acids" (L. Grossman and K. Moldave, eds.), Methods in Enzymology, Vol. 12B, pp. 361–377. Academic Press, New York.
Kornberg, A. (1974). "DNA Synthesis." Freeman, San Francisco, California.
Liska, B. (1972). *Stud. Biophys.* **34,** 151–155.
Liska, B., and Tkadlecek, L. (1975). *Folia Microbiol.* **20,** 1–7.
Liss, A. (1976a). *Biochem. Biophys. Res. Commun.* **71,** 235–240.
Liss, A. (1976b). *Bacteriol. Proc.* p. 62.
Liss, A., and Maniloff, J. (1971). *Science* **173,** 725–727.
Liss, A., and Maniloff, J. (1972). *Proc. Natl. Acad. Sci. U.S.A.* **69,** 3423–3427.
Liss, A., and Maniloff, J. (1973a). *Biochem. Biophys. Res. Commun.* **51,** 214–218.
Liss, A., and Maniloff, J. (1973b). *Virology* **55,** 118–126.
Liss, A., and Maniloff, J. (1974a). *J. Virol.* **13,** 769–774.
Liss, A., and Maniloff, J. (1974b). *Microbios* **11,** 107–113.
Luria, S. E., and Human, M. L. (1952). *J. Bacteriol.* **64,** 569–577.
McCormack, W. M., Braun, P., Lee, Y. H., Klien, J. O., and Kass, E. H. (1973). *New Engl. J. Med.* **288,** 78–89.
Maniloff, J. (1970). *J. Bacteriol.* **102,** 561–572.
Maniloff, J., and Das, J. (1975). *In* "DNA Synthesis and Its Regulation" (M. Goulian, P. Hanawalt, and C. F. Fox, eds.), pp. 445–450. Benjamin, Reading, Massachusetts.
Maniloff, J., and Liss, A. (1973). *Ann. N.Y. Acad. Sci.* **225,** 149–158.
Maniloff, J., and Liss, A. (1974). *In* "Virus, Evolution and Cancer" (E. Kurstak and K. Maramorosch, eds.), pp. 584–604. Academic Press, New York.
Maniloff, J., and Morowitz, H. J. (1972). *Bacteriol. Rev.* **36,** 263–290.
Maniloff, J., Das, J., and Putzrath, R. M. (1977a). *In* "Insect and Plant Viruses; An Atlas." Academic Press, New York. In press.
Maniloff, J., Das, J., and Nowak, J. A. (1977b). *In* "Beltsville Symposium on Virology in Agriculture." U.S. Dept. Agr., in press.
Marvin, D. A., and Hohn, B. (1969). *Bacteriol. Rev.* **33,** 172–209.
Milne, R. G., Thompson, G. W., and Taylor-Robinson, D. (1972). *Arch. Gesamte Virusforsch.* **37,** 378–385.
Morowitz, H. J. (1969). *In* "The Mycoplasmatales and the L-Phase of Bacteria" (L. Hayflick, ed.), pp. 405–412. Appleton, New York.
Nowak, J., Das, J., and Maniloff, J. (1976). *J. Bacteriol.* **127,** 832–836.
Oishi, K. (1971). *Genet. Res.* **18,** 45–56.

Oishi, K., and Poulson, D. F. (1970). *Proc. Natl. Acad. Sci. U.S.A.* **67,** 1565–1572.
Ploaie, P. G. (1971). *Rev. Roum. Biol.—Botanique* **16,** 3–6.
Poulson, D. F., and Sakaguchi, B. (1961). *Science* **133,** 1489–1490.
Pratt, D., Tzagoloff, H., and Erdahl, W. S. (1966). *Virology* **30,** 397–410.
Putzrath, R. M., and Maniloff, J. (1976). *Biophys. J.* **16,** 48a.
Quinlan, D. C., Liss, A., and Maniloff, J. (1972). *Microbios* **6,** 179–185.
Radman, M. (1974). *In* "Molecular and Environmental Aspects of Mutagenesis" (L. Prakash, F. Sherman, M. W. Miller, C. W. Lawrence, and H. W. Taber, eds.), pp. 128–142. Thomas, Springfield, Illinois.
Razin, S. (1973). *Adv. Microbiol. Physiol.* **10,** 1–80.
Rinaldo, C. R., Cole, B. C., Overall, J. C., Ward, J. R., and Glasgow, L. A. (1974). *Proc. Soc. Exp. Biol. Med.* **146,** 613–618.
Robertson, J., Gomersall, M., and Gill, P. (1972). *Can. J. Microbiol.* **18,** 1971–1972.
Robinson, I. M., and Allison, M. J. (1975). *Int. J. Syst. Bacteriol.* **25,** 182–186.
Robinson, I. M., Allison, M. J., and Hartman, P. A. (1975). *Int. J. Syst. Bacteriol.* **25,** 173–181.
Searcy, D. G., and Doyle, E. K. (1975). *Int. J. Syst. Bacteriol.* **25,** 286–289.
Smith, P. F. (1971). "The Biology of Mycoplasmas." Academic Press, New York.
Spizizen, J., Reilly, B. E., and Evans, A. H. (1966). *Annu. Rev. Microbiol.* **20,** 371–400.
Stanbridge, E. (1971). *Bacteriol. Rev.* **35,** 206–227.
Subak-Sharpe, J. H., Elton, R. A., and Russell, G. J. (1974). *Symp. Soc. Gen. Microbiol.* **24,** 131–150.
Tavethia, M. J., and Mandel, M. (1970). *J. Bacteriol.* **101,** 844–850.
Thomas, L. (1970). *Annu. Rev. Med.* **21,** 179–186.
Tully, J. G. (1973). *Ann. N.Y. Acad. Sci.* **225,** 74–93.
Weigle, J. J. (1953). *Proc. Natl. Acad. Sci. U.S.A.* **39,** 628–636.
Williamson, D. L., and Whitcomb, R. F. (1974). *Colloq. Inst. Natl. Sante Rech. Med., Paris* **33,** 283–290.
Woese, C. R., and Bleyman, M. A. (1972). *J. Mol. Evol.* **1,** 223–229.
Zouzias, D., Mazaitis, A. J., Simberkoff, M., and Rush, M. (1973). *Biochim. Biophys. Acta* **312,** 484–491.

AUTHOR INDEX

Numbers in italics refer to the pages on which the complete references are listed.

A

B

C

D

E

F

G

H

I

J

K

N

O

P

Q

R

S

T

U

V

W

Y

Z

SUBJECT INDEX

D

E

F

R

S

T

U

V

W

Z

A 7
B 8
C 9
D 0
E 1
F 2
G 3
H 4
I 5
J 6